SM358
Science: Level 3

The Quantum World

Book 1

Wave mechanics

Edited by John Bolton and Robert Lambourne

SM358 Course Team

Course Team Chair
John Bolton

Academic Editors
John Bolton, Stuart Freake, Robert Lambourne, Raymond Mackintosh

Authors
Silvia Bergamini, John Bolton, Mark Bowden, David Broadhurst, Jimena Gorfinkiel, Robert Lambourne, Raymond Mackintosh, Nigel Mason, Elaine Moore, Jonathan Underwood

Other Course Team Members
Robert Hasson, Mike Thorpe

Consultant
Derek Capper

Course Manager
Gillian Knight

Course Team Assistant
Yvonne McKay

LTS, Project Manager
Rafael Hidalgo

Editors
Peter Twomey, Alison Cadle

TeX Specialist
Jonathan Fine

Graphic Design Advisors
Mandy Anton, Chris Hough

Graphic Artist
Roger Courthold

Picture Researcher/Copyrights
Lydia Eaton

Software Designers
Fiona Thomson, Will Rawes

Video Producers
Owen Horn, Martin Chiverton

External Assessor
Charles Adams (Durham University)

This publication forms part of an Open University course SM358 The Quantum World. The complete list of texts which make up this course can be found at the back. Details of this and other Open University courses can be obtained from the Student Registration and Enquiry Service, The Open University, PO Box 197, Milton Keynes, MK7 6BJ, United Kingdom: tel. +44 (0)870 333 4340, email general-enquiries@open.ac.uk

Alternatively, you may visit the Open University website at http://www.open.ac.uk where you can learn more about the wide range of courses and packs offered at all levels by The Open University.

To purchase a selection of Open University course materials visit http://www.ouw.co.uk or contact Open University Worldwide, Michael Young Building, Walton Hall, Milton Keynes MK7 6AA, United Kingdom for a brochure. tel. +44 (0)1908 858785; fax +44 (0)1908 858787; email ouwenq@open.ac.uk

The Open University
Walton Hall, Milton Keynes
MK7 6AA

Edited and designed by The Open University.

Typeset at The Open University.

Printed and bound in the United Kingdom at the University Press, Cambridge.

ISBN 978 07492 2515 5

2.1

THE QUANTUM WORLD

Introduction 7

Chapter 1 The quantum revolution 9

Introduction 9

1.1 The quantization of energy 9
- 1.1.1 Energy levels 10
- 1.1.2 Democritus had true insight! 12

1.2 A farewell to determinism 13
- 1.2.1 The exponential law of radioactive decay 13
- 1.2.2 An application of quantum indeterminism 15

1.3 Light as waves — light as particles 16
- 1.3.1 Light behaving as waves 18
- 1.3.2 Light behaving as particles 18
- 1.3.3 Light behaving like waves *and* particles 20

1.4 Electrons as particles — electrons as waves 20
- 1.4.1 The de Broglie relationship 21
- 1.4.2 Diffraction and the uncertainty principle 22
- 1.4.3 Electrons behaving like waves *and* particles 22
- 1.4.4 Describing electron waves: the wave function 24
- 1.4.5 Particles and the collapse of the wave function 26
- 1.4.6 The de Broglie wave function 27

1.5 The origin of quantum interference 28
- 1.5.1 Electron interference caused by a double slit 28
- 1.5.2 Probability amplitudes 31
- 1.5.3 Photon interference in the laboratory 32

1.6 Quantum physics: its scope and mystery 34

Chapter 2 Schrödinger’s equation and wave functions 37

Introduction 37

2.1 Schrödinger’s equation for free particles 38

2.2 Mathematical preliminaries 40
- 2.2.1 Operators 40
- 2.2.2 Linear operators 41
- 2.2.3 Eigenvalue equations 42

2.3 The path to Schrödinger’s equation 43
- 2.3.1 Observables and linear operators 44
- 2.3.2 Guessing the form of Schrödinger’s equation 45
- 2.3.3 A systematic recipe for Schrödinger’s equation 46

2.4 Wave functions and their interpretation 49
- 2.4.1 Born’s rule and normalization 49
- 2.4.2 Wave functions and states 51
- 2.4.3 The superposition principle 52

2.5 The time-independent Schrödinger equation 53
- 2.5.1 The separation of variables 53
- 2.5.2 An eigenvalue equation for energy 55
- 2.5.3 Energy eigenvalues and eigenfunctions 56
- 2.5.4 Stationary states and wave packets 57

2.6 Schrödinger's equation, an overview 59

Chapter 3 Particles in boxes 63

Introduction 63

3.1 The one-dimensional infinite square well 65
- 3.1.1 The classical system 65
- 3.1.2 Setting up Schrödinger's equation 66
- 3.1.3 Solving the time-independent Schrödinger equation 67
- 3.1.4 Boundary conditions for the eigenfunctions 68
- 3.1.5 Applying the boundary conditions: energy quantization 69
- 3.1.6 The normalization condition 70
- 3.1.7 The stationary-state wave functions 72
- 3.1.8 Alternative descriptions of the well 74

3.2 Two- and three-dimensional infinite square wells 78
- 3.2.1 The two-dimensional infinite square well 78
- 3.2.2 Degeneracy 81
- 3.2.3 The three-dimensional infinite square well 82
- 3.2.4 F-centres and quantum dots 82

3.3 The finite square well 84
- 3.3.1 The system and its Schrödinger equation 85
- 3.3.2 Solving the time-independent Schrödinger equation 86
- 3.3.3 Computations and generalizations 92

Chapter 4 The Heisenberg uncertainty principle 95

Introduction 95

4.1 Indeterminacy in quantum mechanics 96
- 4.1.1 The wave function and Schrödinger's equation 96
- 4.1.2 Relating the wave function to measurements 97
- 4.1.3 Stationary states and energy values 98
- 4.1.4 Linear combinations of stationary states 99

4.2 Probability distributions 100
- 4.2.1 The overlap rule 101
- 4.2.2 Extension to other observables 108

4.3 Expectation values in quantum mechanics 110
- 4.3.1 Mean values and expectation values 111
- 4.3.2 The sandwich integral rule 112

4.4 Uncertainties in quantum mechanics 115

4.5 The Heisenberg uncertainty principle 117
- 4.5.1 The uncertainty principle in action 119

4.5.2 Making estimates with the uncertainty principle 120

Chapter 5 Simple harmonic oscillators 124

Introduction 124

5.1 Classical harmonic oscillators 125
5.1.1 Diatomic molecules in classical physics 127

5.2 Quantum harmonic oscillators 128
5.2.1 Schrödinger's equation 128
5.2.2 Eigenvalues and eigenfunctions 130

5.3 Solving the time-independent Schrödinger equation 136
5.3.1 Choosing appropriate variables 136
5.3.2 Factorizing the Hamiltonian operator 136
5.3.3 Obtaining the eigenvalues and eigenfunctions 138

5.4 Quantum properties of oscillators 142
5.4.1 Expectation values and uncertainties 142
5.4.2 A selection rule 145

Chapter 6 Wave packets and motion 148

Introduction 148

6.1 Time-dependence in the quantum world 149
6.1.1 The frozen world of stationary states 149
6.1.2 The dynamic world of wave packets 150
6.1.3 A wave packet in a harmonic well 151
6.1.4 More general harmonic-oscillator wave packets 155

6.2 Ehrenfest's theorem and the classical limit 158
6.2.1 Ehrenfest's theorem 158
6.2.2 The classical limit of quantum mechanics 160

6.3 Predicting the motion of a wave packet 161

6.4 Free-particle wave packets 166
6.4.1 Free-particle stationary states 166
6.4.2 Constructing a free-particle wave packet 168
6.4.3 Interpreting $A(k)$ as a momentum amplitude 169
6.4.4 Predicting free-particle motion 170
6.4.5 Wave-packet spreading 172

6.5 Beyond free particles 175

Chapter 7 Scattering and tunnelling 178

Introduction 178

7.1 Scattering: a wave-packet approach 180
7.1.1 Wave packets and scattering in one dimension 181
7.1.2 Simulations using wave packets 182

7.2 Scattering: a stationary-state approach 183
7.2.1 Stationary states and scattering in one dimension 183
7.2.2 Scattering from a finite square step 184

7.2.3 Probability currents 189
7.2.4 Scattering from finite square wells and barriers 191
7.2.5 Scattering in three dimensions 195
7.3 Tunnelling: wave packets and stationary states 197
7.3.1 Wave packets and tunnelling in one dimension 197
7.3.2 Stationary states and barrier penetration 198
7.3.3 Stationary states and tunnelling in one dimension 200
7.3.4 Simulations using stationary states 202
7.4 Applications of tunnelling 202
7.4.1 Alpha decay 202
7.4.2 Stellar astrophysics 205
7.4.3 The scanning tunnelling microscope 206

Chapter 8 Mathematical toolkit 210

Introduction 210
8.1 Complex numbers 210
8.1.1 The arithmetic of complex numbers 211
8.1.2 Complex conjugation 212
8.1.3 The geometry of complex numbers 212
8.1.4 Euler's formula and exponential form 214
8.1.5 Multiplying and dividing complex numbers 215
8.1.6 Powers of complex numbers 216
8.2 Ordinary differential equations 217
8.2.1 Types of differential equation 217
8.2.2 General properties of solutions 218
8.2.3 Solutions of some basic differential equations 219
8.2.4 Boundary conditions and eigenvalue equations 220
8.3 Partial differential equations 222
8.3.1 Partial differentiation 222
8.3.2 Partial differential equations 224
8.3.3 The method of separation of variables 225
8.4 Probability 226
8.4.1 The concept of probability 226
8.4.2 Adding probabilities 227
8.4.3 Average values and expectation values 228
8.4.4 Standard deviations and uncertainties 230
8.4.5 Continuous probability distributions 231

Acknowledgements 233

Solutions 234

Index 251

Introduction to the course

Welcome to *The Quantum World*! The books in this series, of which this is the first, provide a wide-ranging survey of the most fundamental physical principles that govern the way the world works. It has long been recognized that those principles are rooted in *quantum physics*, and it is therefore essentially a quantum world in which we live and a quantum world that these books describe.

All emboldened terms are defined in the Course Glossary.

Quantum physics was born in 1900, when the German physicist Max Planck used the term 'quantum' to describe the minimum amount of energy that could be exchanged between matter and radiation at a specified frequency. Over the following 25 years or so, Planck's original idea was developed and elaborated by Albert Einstein, Niels Bohr, Arnold Sommerfeld and many other physicists. These pioneers tackled a variety of problems involving the interaction of matter and radiation including, most notably, but not entirely successfully, the absorption and emission of light by atoms. The subject they created, based partly on ad-hoc rules and involving an uncomfortable fusion of classical and quantum concepts, is now referred to as the **old quantum theory** and is generally regarded as representing humanity's first faltering steps towards a new kind of physics.

The need for something more systematic was recognized by Max Born, who coined the term **quantum mechanics** in 1924. Born saw, perhaps more clearly than any of his contemporaries, that what was needed was an entirely new physics, comparable to, yet different from, the preceding **classical physics** in which he and his fellow physicists had been trained. Just as the classical mechanics of Newton provides one of the main pillars of classical physics, so would quantum mechanics provide a foundation for the new coherent subject of quantum physics that was to come.

Fittingly, it was Born's brilliant young research assistant, Werner Heisenberg, who made the key breakthrough that ended the era of ad-hoc quantum rules and highlighted the gulf that would forever separate quantum mechanics from its classical counterpart. The details of Heisenberg's breakthrough need not detain us here, but we should note that the complexities of his analysis were soon eased by the introduction of a simpler and more transparent approach pioneered by the Austrian physicist Erwin Schrödinger. It is this latter approach to quantum mechanics, based on waves and therefore known as **wave mechanics**, that is the subject of this book.

Chapter 1 will introduce you to some of the phenomena that caused so much perplexity in the opening decades of the twentieth century. You will also have your first glimpse of the *probabilistic* ideas of quantum physics that eventually cleared away the confusion. These ideas are studied more systematically in Chapter 2, which introduces *Schrödinger's equation*, the fundamental equation of wave mechanics. Chapter 3 then goes on to apply Schrödinger's equation to some simple situations involving particles in microscopic boxes.

Chapter 4 introduces *Heisenberg's uncertainty principle*, which is one of the most celebrated principles of quantum physics since it so clearly distinguishes quantum mechanics from classical mechanics. In the course of doing this, the chapter will show you how to predict the average value, and the spread in values, of measurable quantities in wave mechanics. Chapter 5 uses these ideas to study the quantum analogue of a classical simple harmonic oscillator, exemplified by the vibrational energies of atoms in diatomic molecules. Up to this point, the focus is on concepts like energy and momentum, but without any direct description of the *motion* of a particle from one point in space to another. This omission is remedied in Chapter 6 which analyzes the motion of particles in terms of *wave packets* and explores the circumstances under which the predictions of quantum physics become practically the same as those of classical physics. This is a significant point, since quantum physics claims to be universally valid, applying to large-scale systems as well as to atoms and nuclei.

The extraordinary quantum phenomena that occur when particles are scattered by obstacles is one of the main subjects of Chapter 7. This chapter brings to an end the physical discussion in the book, concluding on a high note by describing the phenomenon of quantum tunnelling and examining its significance in nuclear decays, stellar astrophysics and the working of the scanning tunnelling microscope. There is, however, one additional chapter; this is devoted to mathematical topics, particularly complex numbers, differential equations and probability, and is designed to be studied, whenever appropriate, alongside the other chapters. Don't forget this final chapter, which constitutes a *Mathematical toolkit*, and don't make the mistake of leaving it until the end!

Although this book covers the basic principles of wave mechanics there are many aspects of the quantum world that it does not address. Some of the deeper aspects of quantum mechanics, especially those involving multi-particle systems, angular momentum and the fascinating phenomenon of entanglement, are explored in the second book in the series, while the applications of quantum mechanics to atoms, molecules and solids are the theme of the third book.

In the second half of the twentieth century, quantum mechanics became increasingly important for technology — in semiconductors and lasers, for example. The British Parliament was told that 'something like 25% of the GDP of the USA is based on physics that is totally derived from quantum mechanics research'. Many physicists have expressed the view that the twenty-first century will be the century of quantum technology and quantum devices. If so, it is the principles contained in these three books that will provide the essential background needed to understand these devices and, perhaps, to participate in their development.

Chapter 1 The quantum revolution

Introduction

This chapter gives an overview of some of the ways in which our understanding of the world has been changed by quantum mechanics. In the process of doing this, it will remind you of some basic physics ideas that are needed throughout the course and allow you to review the mathematics of complex numbers.

The chapter is organized as follows. Section 1 starts with the observation that many systems emit light with characteristic patterns of spectral lines. This is related to the fact the energy levels of these systems are quantized (or discrete). Section 2 introduces one of the most pervasive features of quantum mechanics — the fact that it deals with probabilities rather than certainties. This is illustrated by radioactive decay and by commercial devices that exploit quantum mechanics to produce random numbers. Sections 3 and 4 discuss the perplexing fact that photons and electrons behave in some ways like particles, and in other ways like waves. In the course of describing electron waves, we introduce some concepts that will be needed later in the course, including de Broglie's relationship and wave functions. Section 5 takes a closer look at the quantum-mechanical principles that lie behind the phenomenon of interference and, finally, Section 6 reflects on the quantum mysteries that lie ahead.

1.1 The quantization of energy

In the 1830s the philosopher Auguste Comte gave an example of something that he believed would forever be beyond human knowledge: the composition of the stars. Evidently he was unaware that in 1813 Joseph von Fraunhofer had found **spectral lines** in light from the Sun. Analyzing heavenly substances became possible in the 1860s when Gustav Kirchhoff and Robert Bunsen used a spectroscope to examine the colours emitted by chemical elements heated in flames. They found that the patterns of spectral lines that appear provide a unique signature for each element – its **spectrum** (Figure 1.1).

Figure 1.1 The spectra of some common elements. Top, helium; centre, iron; bottom, neon.

The same patterns of spectral lines are found in stars. So Comte could not have been more wrong: thanks to spectral analysis, we now know in considerable detail

Figure 1.2 Niels Bohr (1885–1961). Bohr accounted for the spectrum of atomic hydrogen by applying simple quantum rules to determine the energy levels of a hydrogen atom. He received the Nobel prize for physics in 1922.

what stars are made of. Much has also been learned about the history of the Universe by studying the abundances of the elements in stars. The spectral lines produced by atoms under given conditions are the same wherever those atoms are in the Universe and they have not changed for billions of years. But how do atoms, indestructible enough to have survived unchanged for billions of years, produce spectral lines?

The first steps towards an explanation were taken by the Danish physicist Niels Bohr (Figure 1.2), who gave a semi-quantum explanation of the spectrum of the simplest atom, hydrogen. A major theme of this course will be the success of quantum mechanics in explaining the spectra of nuclei, atoms, molecules etc., going far beyond what Bohr achieved.

1.1.1 Energy levels

One of Bohr's key discoveries was that atoms possess discrete **energy levels**; their energies are said to be **quantized**. As a result, atoms behave differently from, say, red-hot pokers. The temperature of a cooling poker appears to fall continuously as energy is lost to the environment. An atom, however, is characterized by a series of discrete energy levels, so it cannot lose energy continuously. Instead, it jumps from one energy level to another of lower energy, emitting a packet of electromagnetic radiation, a **photon**, as it does so. After a short time it may emit another photon, jumping to a state of still lower energy. This continues until the atom reaches its state of lowest energy, the **ground state**. Each photon is a packet of electromagnetic radiation with a **frequency** determined by the energy of the photon. If an atom jumps from a state of energy E_j to a state of lower energy E_i, then the energy of the emitted photon, E_{photon}, and the corresponding frequency of the radiation, call it f_{ji}, are determined by the change in energy of the atom:

$$E_{\text{photon}} = hf_{ji} = E_j - E_i, \tag{1.1}$$

where $h = 6.63 \times 10^{-34}$ J s is **Planck's constant**, the fundamental constant that characterizes most quantum phenomena. The emitted radiation can also be described in terms of the **angular frequency** $\omega_{ji} = 2\pi f_{ji}$, and the energy of the photon is then:

$$E_{\text{photon}} = \hbar\omega_{ji} = E_j - E_i, \tag{1.2}$$

where $\hbar = h/2\pi = 1.06 \times 10^{-34}$ J s.

$\hbar$ is pronounced 'h-bar'.

Each kind of atom has its own characteristic series of energy levels, $E_1, E_2, \ldots, E_i, \ldots, E_j, \ldots$, with E_1 corresponding to the ground state. An atom that has been excited to a high energy, maybe in a flame or on the surface of a star, will lose energy by emitting radiation at a characteristic set of frequencies $f_{ji} = |E_j - E_i|/h$. These frequencies account for the spectral lines and provide the atom's unique spectral signature, anywhere in the Universe.

In this course you will see how the energy levels of atoms can be calculated using the principles of quantum mechanics. Not just atoms but also molecules, atomic nuclei and the fundamental constituents of nuclei have states with specific energies (Figure 1.3).

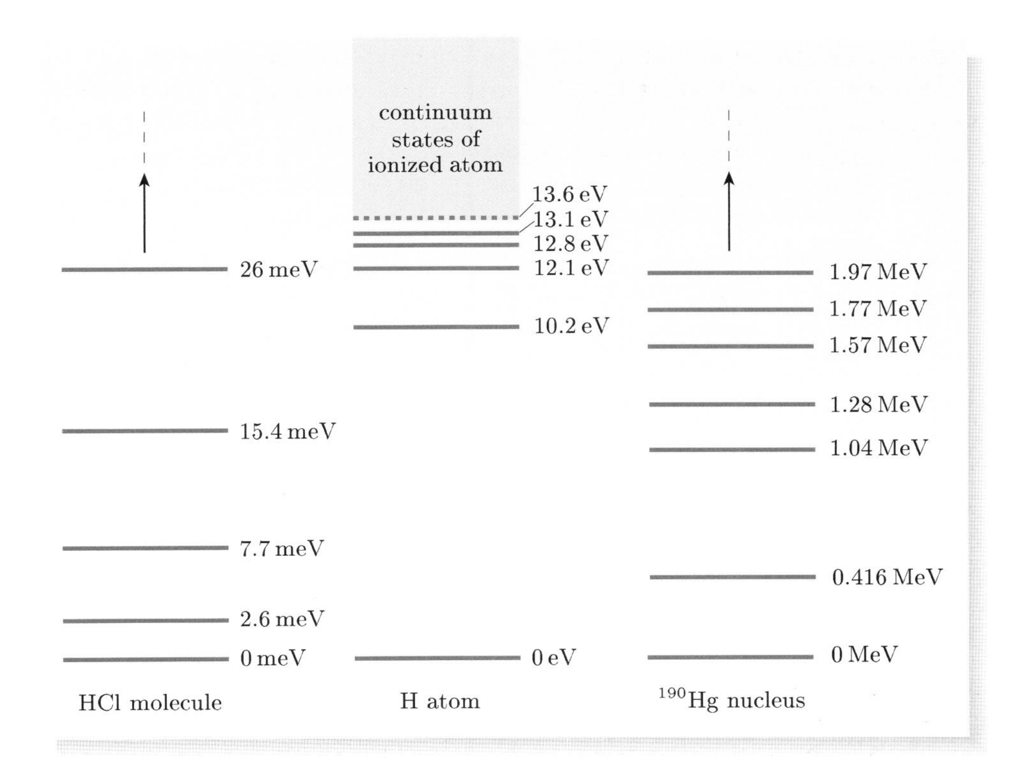

Figure 1.3 Comparing the rotational energy levels of a hydrogen chloride molecule (HCl), the electronic energy levels in a hydrogen atom (H), and the nuclear energy levels of the mercury isotope ^{190}Hg.

Each horizontal line in Figure 1.3 represents the energy of a particular state of the system above the ground state, which has been assigned zero energy in each case. The energy scales are very different. The most convenient unit of measurement for electronic energy levels in atoms is the **electronvolt** (eV), with $1\,\text{eV} = 1.60 \times 10^{-19}\,\text{J}$. Rotational energies of molecules are best measured in millielectronvolts ($1\,\text{meV} = 10^{-3}\,\text{eV}$) and nuclear energies are measured in megaelectronvolts ($1\,\text{MeV} = 10^{6}\,\text{eV}$). The factor of a million or so between atomic and nuclear energies lies behind the fact that electricity generators can get roughly a million times more energy from a lump of uranium than from the same mass of coal. Not all energy levels are shown: for the HCl molecule and the Hg nucleus there are countless energy levels above the top of the scale. The hydrogen atom also has indefinitely many energy levels, but crowded together below the limiting energy of 13.6 eV, above which energy the atom loses its electron and is said to be **ionized**. All these energy levels are known from a study of the photons emitted when the molecule, atom or nucleus jumps from one energy level to another.

Exercise 1.1 Atomic nuclei, like atoms, have discrete energy levels. A nucleus jumps between two energy levels 1 MeV apart emitting a photon; a photon with such an energy is called a gamma-ray photon. What is the ratio of the frequency of a 1 MeV gamma-ray photon to a 1 MHz frequency (corresponding to a photon in the AM radio band)?

Exercise 1.2 Do you think that a red-hot poker *really* cools down continuously, or does it, too, cool in jumps if viewed closely enough? ■

If atoms, molecules, etc. have a ground state of lowest energy, do they also have a state of *maximum* energy? Can their energy be increased without limit? What happens when you pump more and more energy into an atom by bombarding it with photons or by allowing it to collide with other energetic atoms? Such collisions certainly occur in a neon tube, or in a star, for example. The answer is that above a certain energy the atom becomes *ionized*, that is to say, an electron is ejected. The positively-charged remnant of the original atom is referred to as an **ion**. If still more energy is supplied, the ion may lose another electron and become doubly ionized.

When an atom is ionized, the combined system comprising the ion and a detached electron can exist at *any* energy above that at which the electron is ejected. The system is said to be 'in the **continuum**' a term drawing attention to the fact that, unlike the discrete energies for which the electron is bound, a continuous range of energies is now possible. Figure 1.3 shows the continuum of states that exists above 13.6 eV for an ionized hydrogen atom.

- Which element has atoms that cannot be doubly ionized?
- A hydrogen atom cannot be doubly ionized since it has only one electron.
- A helium atom has **atomic number** $Z = 2$. An **alpha particle** is a doubly-ionized atom of helium. How many electrons does it have?
- A neutral helium atom has two electrons, so a doubly-ionized helium atom (an alpha particle) has no electrons.

1.1.2 Democritus had true insight!

The fact that atoms and molecules have particular energy levels is of exceptional scientific importance, and not just because the differences between energy levels provide the foundation for spectroscopy. It explains a deep mystery that sometimes goes unnoticed. Why is any element or compound always exactly the same, no matter what processes it has undergone? You can boil water and condense it, freeze it and melt it and it is still the same old water. It was the attempt to understand such behaviour that led the ancient Greek thinkers Leucippus (480–420 BC) and Democritus (460–370 BC) to put forward the idea of indestructible infinitely-hard atoms, with no internal structure.

We now know that atoms *do* have internal structure and that this accounts for their spectra. But how can atoms with internal structure be virtually indestructible? Only since the advent of quantum mechanics have we been able to resolve this issue. The key point is that all atoms of a given kind have *identical* ground states. Bump them, jostle them hard enough, shine light on them, and they might jump into an excited state of higher energy, but they will eventually lose energy and return to the ground state. Jostle them even harder or bombard them with electrons, as in an arc lamp or on the surface of the Sun, and some of their electrons may get knocked out: an atom may become ionized. But an ionized atom is a charged atom, and charged atoms will pick up electrons, given a chance, so the atom will reassemble and end up back in its ground state again. Unless they undergo nuclear processes, atoms are effectively indestructible and their ground states are always the same. These are ideas of enormous explanatory power. They explain why atoms have survived for billions of years with unchanged spectral signatures. None of this was intelligible before the advent of quantum theory.

1.2 A farewell to determinism

A deeply-held belief of physicists at the beginning of the twentieth century was that the same effect inevitably follows a given cause; that precisely-defined circumstances always lead to the *same* precisely-defined consequences. For example, if the *exact* positions and velocities of colliding snooker balls were known, physicists were confident that they could predict exactly where the balls would go. In this sense, the world was believed to be **deterministic**. But quantum physics changes all that: some things in Nature are governed by chance. Classical physics is deterministic even when it might appear not to be; it is the sheer complexity of a sample of gas, for example, that leads to apparent randomness. The quantum world, however, is irreducibly **indeterministic** and some things cannot be predicted, *even in principle*, with the largest computer imaginable. In this course we shall see how quantum mechanics can predict much about atoms, but some things, such as when a *particular* radioactive nucleus will decay, remain a matter of pure chance.

1.2.1 The exponential law of radioactive decay

Our first example of quantum indeterminism is the exponential law of radioactive decay, discovered by Ernest Rutherford in 1900, before quantum mechanics was established. It was only many years later that the significance of this discovery sank in. What makes it remarkable is the fact that, as we now understand, nuclei can be *absolutely identical*. We cannot have absolutely identical copies of macroscopic objects like white snooker balls. However, all atomic nuclei of a specific kind in their ground states *are* absolutely identical. But though identical, they do *not* decay at identical times; there is *no* hidden property telling a nucleus when to decay.

The fact that atoms or nuclei of a given kind, as well as fundamental particles like electrons, can be *identical* is a uniquely quantum phenomenon, without parallel in the everyday world. It has profound significance for our understanding of matter, as you will see later in the course.

Each nucleus of a specific kind (e.g. the uranium **isotope** $^{238}\mathrm{U}$) has the same probability of decaying in a short time interval δt. We write this probability as $\lambda\,\delta t$, where λ is called the **decay constant** of the given type of nucleus. The decay constant has the same value for all nuclei of the given kind and is independent of time. It describes the statistical behaviour of a large number of nuclei, but does not tell us when an *individual* nucleus will decay. For example, $^{238}\mathrm{U}$ nuclei have a very small decay constant, and typically survive for billions of years, but an *individual* $^{238}\mathrm{U}$ nucleus could decay within the next second. Nuclei do not have internal clocks telling them when to decay, and a nucleus that has been around for a long time is no more likely to decay at a certain instant than one that has just been created.

If at time t there are $N(t)$ nuclei of a given kind then, in the small time interval from t to $t + \delta t$, the decrease in N is the number of nuclei that decay in the given interval. The average number of nuclei decaying in the given interval is equal to the number $N(t)$ of nuclei that were present at the start, multiplied by the probability $\lambda\,\delta t$ that any one of them will decay. Hence, *on average*, the number of nuclei changes over the interval by

$$\delta N = -\lambda N\,\delta t. \tag{1.3}$$

The minus sign shows that the number present, N, *decreases* as nuclei decay. The N on the right-hand side reflects the fact that the number decaying at some time t is proportional to the number present. We stress that if $\lambda\,\delta t$ is the probability of a nucleus decaying in time δt, then $N\lambda\,\delta t$ is the *average* number decaying in time δt when there are N nuclei present, so this is a statistical law. From Equation 1.3 we see that, dividing by δt and taking the limit of a small time interval, the number $N(t)$ of the original kind of nucleus falls at a rate

$$\frac{\mathrm{d}N}{\mathrm{d}t} = -\lambda N(t). \tag{1.4}$$

It is easy to confirm that Equation 1.4 is satisfied by the exponential function

$$N(t) = N(0)\,\mathrm{e}^{-\lambda t}. \tag{1.5}$$

Values of $T_{1/2}$ cover a very wide range. For ^{232}Th, $T_{1/2} = 1.4 \times 10^{10}$ years; for ^{212}Po, $T_{1/2} = 0.3 \times 10^{-6}$ s.

The time taken for half the nuclei to decay, on average, is called the **half-life**, $T_{1/2}$. From Equation 1.5 we have $1/2 = \mathrm{e}^{-\lambda T_{1/2}}$, from which it follows that

$$T_{1/2} = \frac{\ln 2}{\lambda} = \frac{0.693}{\lambda}, \tag{1.6}$$

and hence

$$N(t) = N(0)\mathrm{e}^{-0.693t/T_{1/2}}. \tag{1.7}$$

This equation says that, if there are $N(0)$ nuclei when we start the clock, then when $t = T_{1/2}$, one-half will remain and when $t = 2T_{1/2}$, one-quarter will remain, and so on. Since decay is a statistical process, these numbers of nuclei remaining are true on the average, and are not accurate when N is small, so it is not true that, starting with one nucleus, half a nucleus will remain after time $T_{1/2}$.

The exponential law of Equation 1.6 describes how Rutherford observed nuclei to decay. At the time of its discovery, it would have been natural to suppose that this law is a consequence of some deterministic mechanism in nuclei. If nuclei behaved like complex classical systems, one would suppose that the decay time of any individual nucleus could be predicted if only we knew enough about its internal dynamics. In a complex classical system, such as the jar of numbered balls used in a lottery draw, apparently random behaviour emerges from very complicated dynamics. If we knew the exact positions and velocities of the balls and all the forces acting on them, we could dream of predicting the result of the draw; only the complexity of the calculation would thwart us. But such calculations are impossible for a nucleus, *even in principle*. There are no hidden parameters to distinguish individual nuclei and allow their decay times to be predicted. This is why radioactive decay is so startling. All ^{238}U nuclei that are in their ground state, for example, are *identical* even though they will not decay at identical times. This is utterly beyond explanation by classical physics.

The emission of photons by atoms is also inherently indeterministic; the half-life, $T_{1/2}$, of an atomic state as it waits to decay to a state of lower energy is typically a few nanoseconds. The decay from state to state by photon emission follows the same general rule for all the systems shown in Figure 1.3, although the half-lives vary widely. All hydrogen atoms in the same excited state are identical structures of an electron and a proton, but they decay at random. No internal clocks tell them when to decay. In principle, events in the macroscopic world, such as the switching on of a light, could be *triggered* by random nuclear or atomic processes: in this sense the world is fundamentally *probabilistic*. Nevertheless, there are many entirely predictable things, such as the fact that the half-lives measured for any two large collections of identical nuclei will always be the same wherever and whenever they are measured, that copper will always be reddish in colour and the energy levels of hydrogen will always be the same. This interplay between the absolutely predictable and the absolutely unpredictable is a characteristic feature of quantum physics. We shall see several examples in this chapter.

Exercise 1.3 The *Cassini* spacecraft sent to Saturn was powered by energy released by the radioactive plutonium isotope, ^{238}Pu, which has a half-life of $87.7\,\text{years}$. What fraction of the power produced when *Cassini* was launched would still be produced on arrival at Saturn after $7.30\,\text{years}$? ■

1.2.2 An application of quantum indeterminism

Quantum indeterminism is not confined to decaying atoms and nuclei, but pervades the microscopic world. Practical devices that make essential use of quantum chance include **quantum random number generators**, that have been developed and built by physicists in Vienna, Geneva and elsewhere.

Random numbers are very important in modern science and engineering and are required for a wide range of computational procedures and simulations, for cryptography, and even for slot machines and lottery games. Some of these tasks are very demanding, needing very rapid generation of numbers, and requiring 'true' random numbers rather than the so-called pseudo-random numbers generated by computer algorithms; present-day computers are deterministic and cannot provide truly random numbers. A quantum random number generator provides a rapid source of numbers that are guaranteed to be truly random.

The device is based on a **half-silvered mirror**, a mirror coated in such a way that if many photons fall on any part of it, at an angle of incidence of $45°$, half are transmitted and half are reflected. Such a mirror is an example of a **beam splitter**. In Figure 1.4 photons of light fall one-by-one upon a half-silvered mirror. Half of these photons pass straight through and half are reflected through $90°$. That simple statement conceals a remarkable quantum subtlety. Each photon in some sense actually goes *both* straight through *and* is reflected at $90°$, but is only detected at one of the two detectors, a remarkable and counter-intuitive quantum feature that we shall meet many times in this course.

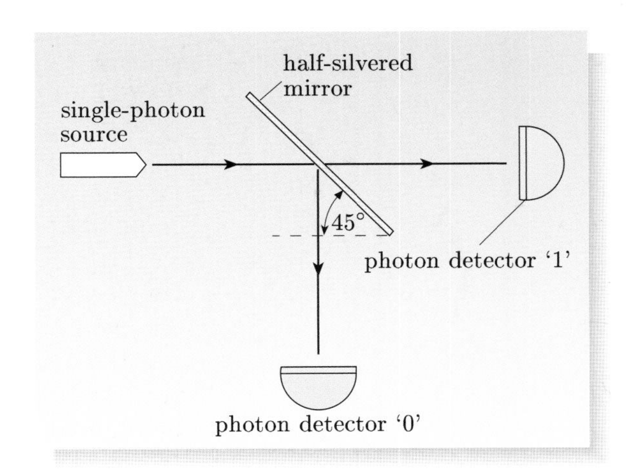

Figure 1.4 Photons from a source emitting one photon at a time fall on a half-silvered mirror and are detected with equal probabilities at detector '1' and detector '0'.

Which of the two detectors fires when a photon is incident on the mirror is completely random, and we can no more say which it will be than we can say when a specific radioactive nucleus will decay. But just as there is a statistical law that describes the decay of a nucleus in terms of its decay constant λ, so we can say that, at least in the ideal case, 50% of the incident photons will be detected by detector '1' and 50% by detector '0'. The probability of a photon being detected by detector '0' is 0.5, and the probability of it being detected by detector '1' is also 0.5. The sum of these probabilities is equal to unity, since the photon is certain to be detected at either '0' or '1'. This behaviour is very different from that of a *classical* electromagnetic field: if an electromagnetic wave is incident on a half-silvered mirror, then, according to classical physics, there would be a field with half the incident intensity at each detector no matter how weak the field. But a photon must be detected in one detector or the other; in this respect photons behave like particles.

The two detectors in Figure 1.4 are labelled '0' and '1' for a good reason: photon counts in these detectors are interpreted as binary digits 0 and 1 respectively. It is not difficult to extract true random numbers from the random sequence of 0s and 1s. A commercial random number generator based on a half-silvered mirror was marketed late in 2004, and is shown in Figure 1.5.

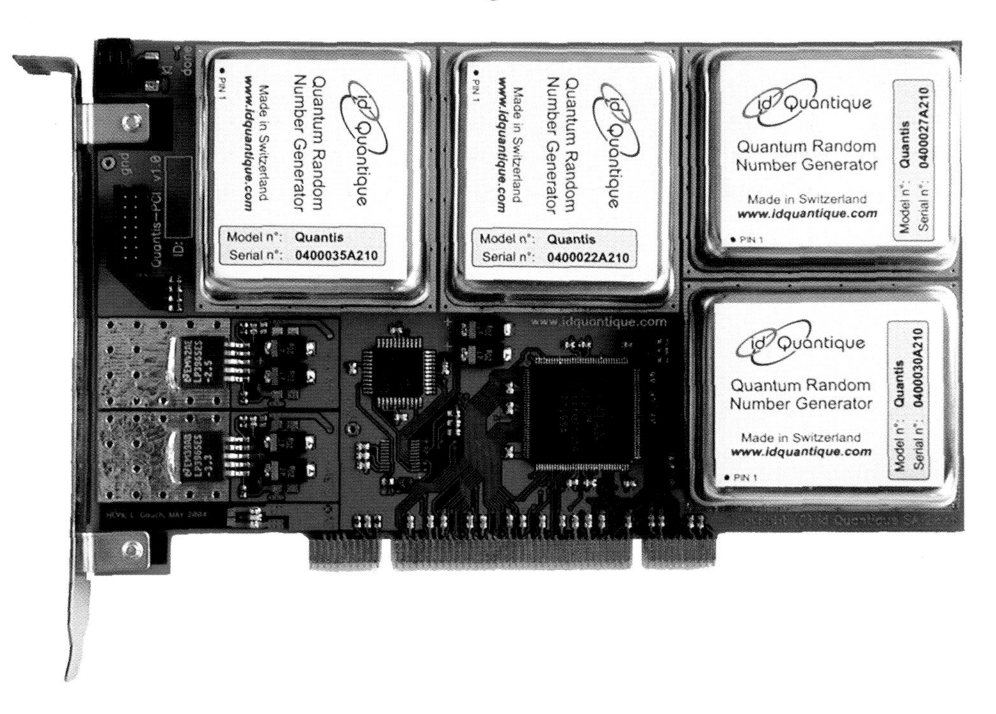

Figure 1.5 Four quantum random number generators ready to slide into a spare slot of a PC. Produced by id Quantique of Geneva.

1.3 Light as waves — light as particles

The photon concept was a radical departure from the wave picture that had become a cornerstone of physics. Although Newton put forward cogent reasons for regarding light as a stream of corpuscles, it was the wave theory developed by Huygens and others that became widely accepted at the beginning of the nineteenth century when optical interference and diffraction were demonstrated by Thomas Young (Figure 1.6) and his successors.

Figure 1.6 Thomas Young (1773–1829), who established the wave nature of light.

Interference describes the effects observed when two or more waves are superposed, leading to a reinforcement in the disturbance in some places and a cancellation in others. **Diffraction** describes the spreading of a wave that occurs when it passes through an aperture or around an obstacle. The two phenomena are linked because the details of diffraction can be explained by the interference of many waves taking different routes. Interference and diffraction are characterized by patterns of intensity which can only be understood in terms of waves. Other experiments showed that light can be polarized, implying that its waves are transverse, rather than longitudinal. In other words, whatever vibrates must vibrate perpendicular to the direction of propagation. This picture was completed in the 1860s, when Maxwell showed that transverse electromagnetic waves, with exactly the speed of light, are a consequence of electromagnetic theory, as summarized by **Maxwell's equations**. The existence of these waves was experimentally verified by Hertz in 1887.

Einstein's radical proposal of 1905 that electromagnetic radiation comes in packets, later known as photons, appeared deeply inconsistent with Maxwell's theory and can still seem unsettling: does light consist of particles or waves? We now understand light to be neither just waves nor just a stream of particles; it behaves like nothing else but itself. Some experiments reveal wave-like properties, some reveal particle-like properties. We say that light exhibits **wave–particle duality**. This is *not* an embarrassed avoidance of the question: 'is light particles or waves?' When we think about it, there is no reason why light should behave like water waves or like bullets; many things in the quantum world have no correspondence with anything in the everyday world. Why, after all, should they?

Describing waves

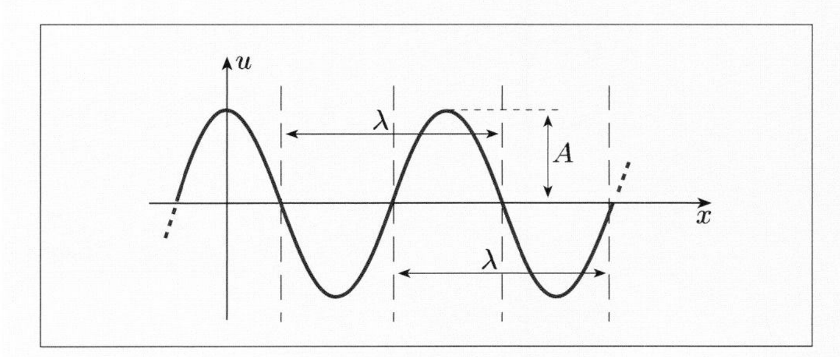

Figure 1.7 A sinusoidal plane wave moving along the x-axis. At a given time, the wavelength λ is the distance between two successive points at which the wave is at the same stage of its cycle. At any given position, the period T is the time taken for the wave to complete one cycle.

It may be helpful to review the definitions needed to describe waves moving in one dimension (see Figure 1.7). The expression

$$u(x,t) = A\cos(2\pi(x/\lambda - t/T)) = A\cos(kx - \omega t) \qquad (1.8)$$

represents a sinusoidal plane wave moving along the x-axis with **wavelength** λ and **frequency** $f = 1/T$, where T is the **period**. We mostly use the second form of the equation, involving the **wave number** $k = 2\pi/\lambda$ and the **angular frequency** $\omega = 2\pi f$. The positive constant A is called the **amplitude** of the wave.

The wave

$$u(x,t) = A\cos(kx - \omega t + \phi) \qquad (1.9)$$

has an additional **phase constant** ϕ. The argument of the cosine, $kx - \omega t + \phi$, is called the **phase** of the wave; this is not to be confused with the phase constant, which is equal to the phase at $x = 0$ and $t = 0$. Phases will be important when we discuss interference effects.

The speed of the wave in Equation 1.8 or 1.9 is given by $v = f\lambda = \omega/k$. For light in free space (not in a medium like glass) this is the speed of light in a vacuum, c.

1.3.1 Light behaving as waves

Figure 1.8 shows examples of the patterns made by light passing through apertures and falling upon a screen. (The screen is really much more distant from the apertures than can be shown in the diagram.) Figure 1.8a shows the **diffraction pattern** produced when light passes through a single narrow slit S_1. The overall size of the diffraction pattern depends on the size of the aperture: in Figure 1.8b the pattern broadens out when we use a narrower slit S_2. Light of a longer wavelength also leads to a broader pattern but this is not shown here. The wave nature of light was not established until around 1800, partly because the wavelength of light is very small compared to the size of everyday objects.

The pattern in Figure 1.8c made Young famous. It occurs when two very narrow slits S_3 and S_4 are spaced so that their broad diffraction patterns become superimposed and interference takes place. The characteristic feature of this **two-slit interference pattern** is the series of dark **interference minima**, separated by bright **interference maxima**. The dark regions would have been bright with either slit alone. The wave picture explains these as the places where the disturbances caused by the wave that originates from S_3 cancel, or nearly cancel, the disturbances caused the wave originating from S_4. The details of the two-slit pattern depend on the width and separation of the slits; widely-spaced narrow slits can lead to a regular sequence of light and dark bands.

1.3.2 Light behaving as particles

In Section 1.1 we referred to photons as 'packets of electromagnetic radiation'. These are the original 'light quanta' of energy hf introduced by Einstein in a famous paper of 1905 in which, among other things, he gave an explanation of the photoelectric effect. The mature concept of a photon with particle-like properties required some 20 more years to develop. Einstein was perplexed that, although electromagnetic theory says that an atom should radiate in all directions like a radio antenna, atoms seemed to radiate in a single direction as if emitting particles. In an important paper in 1916, in which he tried to reconcile these ideas (and in which he uncovered the underlying principle of lasers) he referred to 'radiation bundles'. When in 1923 Compton showed that these radiation bundles had an associated momentum vector and transferred momentum to electrons (the Compton effect), Einstein immediately recommended him for the Nobel prize in physics. We now understand that a photon of frequency f carries momentum of magnitude

$$p_{\text{photon}} = \frac{hf}{c} = \frac{h}{\lambda}, \tag{1.10}$$

where λ is the wavelength. Photons are detected as localized particles when a single grain darkens on a photographic film, or a Geiger counter clicks in response to the passage of a gamma-ray photon.

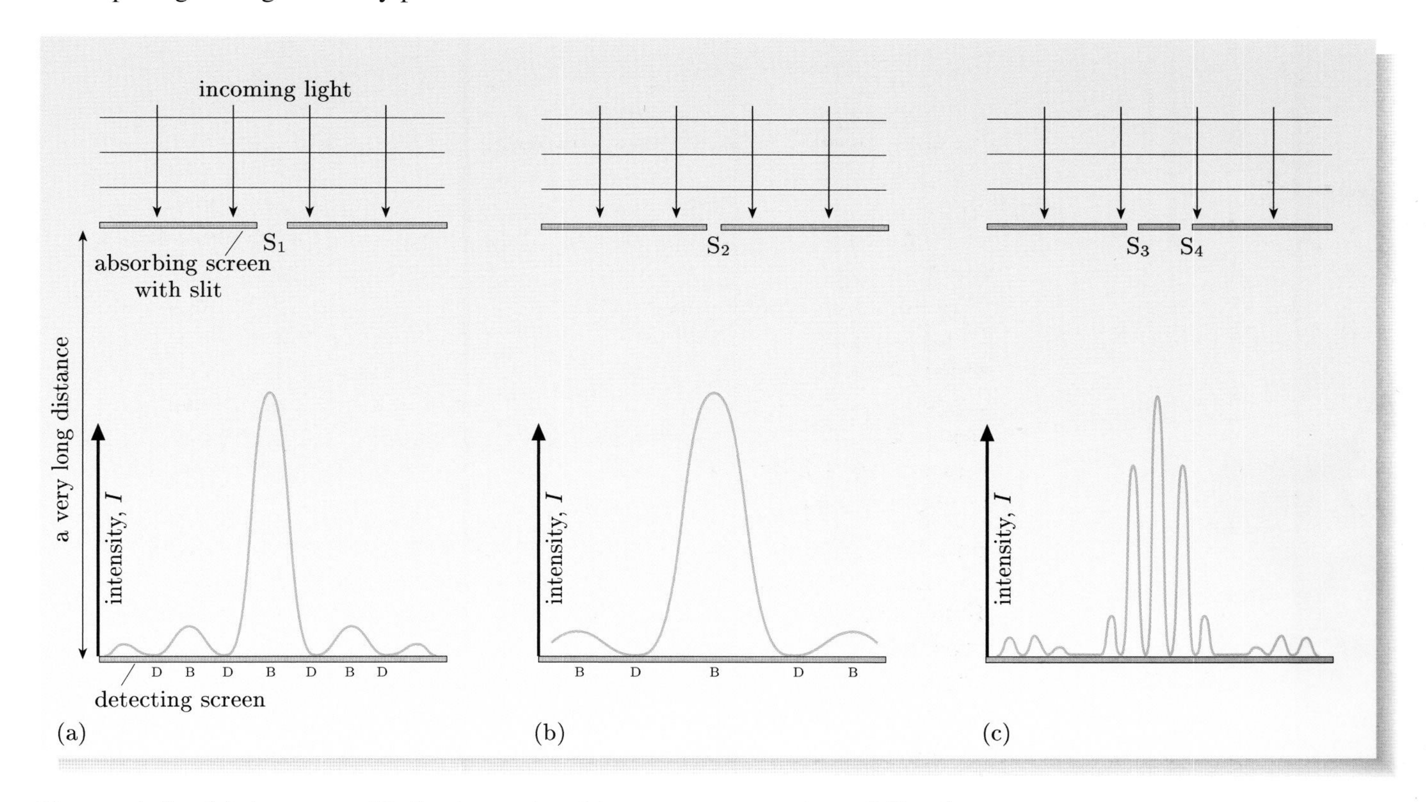

Figure 1.8 (a) A narrow slit, S_1, in an absorbing screen generates a diffraction pattern on a distant detecting screen (not shown to scale). The intensity I of the light on each point on the screen is indicated by a coloured curve. Bright regions are marked B, dark D. The slit extends perpendicular to the plane of the page, as does the pattern on the detecting screen. (b) When a narrower slit S_2 is used, the diffraction pattern becomes more spread out. (c) When there are two very narrow identical slits, S_3 and S_4, separated by a distance much greater than the width of each slit, a characteristic two-slit interference pattern appears. In reality, the slits would be narrow compared to λ, the wavelength of the light.

Exercise 1.4 (a) What is the magnitude of the momentum carried by a single gamma-ray photon of energy $1\,\text{MeV}$? (b) Use the equivalence of force and rate of change of momentum to determine how many such gamma-ray photons per second, falling perpendicularly onto a block of lead and absorbed by it, would exert a force of $1\,\text{N}$. (c) How many AM radio photons per second would exert the same force? (You can use the result of Exercise 1.1.) ■

1.3.3 Light behaving like waves *and* particles

In 1909 G. I. Taylor studied diffraction patterns using light of extremely low intensity: so low that a three-month exposure was required to form a diffraction pattern on a photographic film. Such low intensities mean that after a short time interval, just a few scattered spots appear on grains of the film, apparently at random. But after a long time the accumulated spots form the same diffraction pattern as a short exposure with high intensity light.

There are two important features in what Taylor found:

1. The pattern that eventually forms is that predicted for *waves* passing through the aperture but, when the waves reach the film, they register as just one spot at a time. The diffraction pattern corresponds to a set of waves with a particular wavelength. But it is as if the waves answer the question 'Where are you?' by darkening a single sensitive grain in the film, just as a gamma ray makes a single click in a Geiger counter.
2. There are places in the diffraction pattern with very few spots: interference minima. Since the diffraction pattern depends on the size of the aperture, *increasing* the size of the aperture will *reduce* the density of spots at certain places (and also increase the density of spots where there were previously interference minima).

   From our modern perspective, we can also add:
3. The regions with a low density of spots are *not* the result of different photons interfering with each other. This has been confirmed by modern experiments in which we can be sure that there is only one photon in the apparatus at any given time.
4. The *unpredictability* of where the next spot appears contrasts with the *predictable* diffraction pattern that eventually emerges. This is analogous to the fact that we cannot predict when a particular radioactive nucleus will decay, but a collection of many nuclei does have a well-determined half-life.

Taylor's experiment is an excellent example of *wave–particle duality* in action: the diffraction pattern is that predicted by a wave theory, but the light manifests itself at single points on the screen, like particles.

Exercise 1.5 Look carefully at the first one of the four points listed above. There is a close parallel with an aspect of what happens at the half-silvered mirror in the random number generator. What is it? ■

Figure 1.9 Louis de Broglie (1892–1987), who proposed that particles such as electrons have wave-like properties. He received the Nobel prize for physics in 1929.

1.4 Electrons as particles — electrons as waves

Even before physicists assimilated the idea that light had a particle-like aspect, another bombshell exploded: it was proposed that particles, such as electrons, have wave-like properties. This idea was first put forward by Louis de Broglie in 1924 (Figure 1.9). 'But surely', common sense tells us, 'electrons are not waves! An electron has a definite electric charge. How can a wave have an electric charge? Electrons are indivisible, while waves simultaneously go both ways around an obstacle. Electrons leave spots on a phosphorescent screen like that in an old fashioned TV tube.'

Without doubt, electrons do commonly behave like particles: we can trace their tracks as they bend under the influence of magnetic fields, as in Figure 1.10, and we can hear the clicks made in a suitable Geiger counter by beta particles (electrons created and emitted by certain unstable nuclei) as they arrive one-by-one. But this does not mean that electrons behave *only* like particles.

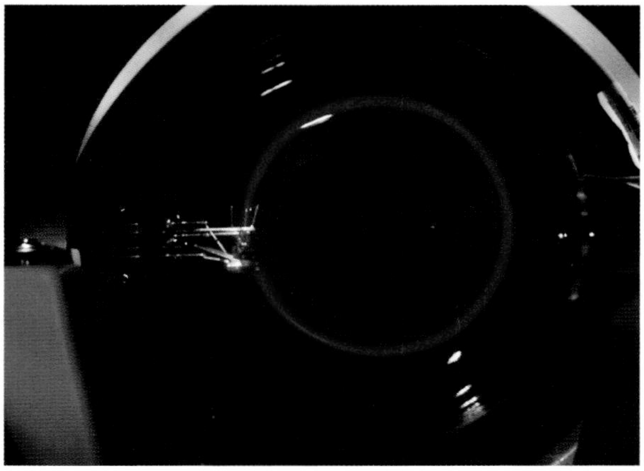

In fact, there is a symmetry in Nature: both photons and electrons are detected as localized particles, but travel through space as waves. The fact that electrons travel like waves was verified by diffraction experiments carried out by C. Davisson and L. Germer in the USA and by G. P. Thomson in the UK. The diffraction of electrons has important practical applications, including electron microscopes. It has now been established experimentally that wave-like behaviour occurs for many kinds of particle: electrons, protons, nuclei and molecules, although experimental difficulties make verification difficult for objects as large as charging elephants.

Figure 1.10 Electrons moving in a magnetic field inside an evacuated glass bulb. The bluish circular path is made visible by a small amount of gas in the glass bulb.

1.4.1 The de Broglie relationship

The scale of a diffraction pattern depends on the wavelength of the wave. So, if we are to explain the diffraction of electron waves, we must understand what determines their wavelength. This issue was settled by Louis de Broglie. In 1924, de Broglie asserted that free electrons (that is, electrons that experience no forces) have a wavelength inversely proportional to the magnitude of their momentum, p. In fact,

$$\lambda_{\text{dB}} = \frac{h}{p}, \qquad (1.11)$$

The subscript dB stands for de Broglie.

where h is Planck's constant. The wavelength λ_{dB} is called the **de Broglie wavelength**, and Equation 1.11 is called the **de Broglie relationship**.

We now know that all particles have a de Broglie wavelength given by Equation 1.11. This includes photons, as you can see from Equation 1.10. The validity of the de Broglie relationship for the motion of nuclei (composite particles of many protons and neutrons) has been known for many years. More recently, Equation 1.11 has been verified for molecules as large as 'bucky balls' (C_{60}) and even heavier molecules.

The de Broglie relationship tells us that, the larger the momentum, the smaller the de Broglie wavelength. For example, the electrons in the beam of a transmission electron microscope have been accelerated to high values of momentum, and have very short de Broglie wavelengths, much smaller than the wavelength of light. This allows electron microscopes to resolve much finer detail than ordinary light microscopes.

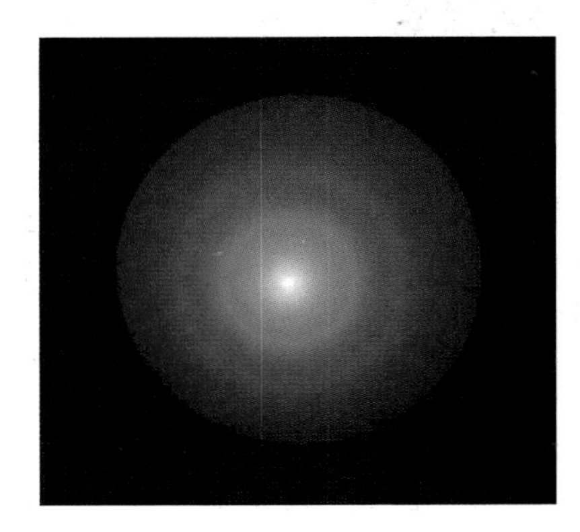

Figure 1.11 The diffraction of high-energy electrons by graphite, producing the rings shown here, can be used to measure the spacing of the carbon atoms in the graphite.

The diffraction of electrons can be exploited to study the structure of matter on an atomic scale and below. The rings in Figure 1.11 are formed when an electron beam diffracts from the regular array of carbon atoms in graphite, and the spacing of the carbon atoms can be deduced from the diameter of these rings. In a similar way, the diffraction of very high-energy electrons by atomic nuclei is the main source of information on nuclear sizes. Very high energies are required to ensure that the de Broglie wavelengths of the electrons are small enough to resolve the nuclear details.

Do not confuse the de Broglie wavelength with the size of a particle. For particles of a given energy, the greater the mass, the greater the momentum, and the smaller the de Broglie wavelength. This means that the de Broglie wavelength can be much larger or much smaller than the particle itself. For example, the de Broglie wavelength of the electrons used by Davisson and Germer was about the size of an atom, although electrons are, as far as we can tell, point particles. By contrast, the de Broglie wavelength of a uranium nucleus that has been accelerated to $1000\,\mathrm{MeV}$ is about one-hundredth the size of the uranium nucleus. Because of their large mass, even slowly-moving macroscopic objects have momenta that are enormous on an atomic scale, with correspondingly small de Broglie wavelengths. The fact that the de Broglie wavelengths of macroscopic bodies are *very* much smaller than the bodies themselves is an important reason why quantum effects are not apparent for everyday objects.

1.4.2 Diffraction and the uncertainty principle

A further detail of electron diffraction is that the diffraction pattern becomes broader as the aperture becomes narrower. This is exactly the same as for light (compare Figures 1.8a and b). It is interesting to note that this feature can be interpreted using the *uncertainty principle*, a fundamental rule of quantum mechanics which will be discussed in Chapter 4. In loose terms, the uncertainty principle tells us that it is impossible to know both the position and momentum of a particle with perfect accuracy. In fact, the more we know about the position of a particle, the less we can know about its momentum. Consider an electron wave *immediately* after it has passed through a narrow slit. At this instant, the electron is somewhere within the width of the slit, so its coordinate in the direction perpendicular to both the slit and to the direction of motion of the beam is known quite accurately. If we make the slit narrower, this coordinate will be known more accurately, and the uncertainty principle tells us that the corresponding component of momentum is known less accurately. The greater uncertainty in transverse momentum means that there is a greater uncertainty in where the electron will arrive on the screen, and this explains why the diffraction pattern is broader.

1.4.3 Electrons behaving like waves *and* particles

Wave–particle duality for *photons* was demonstrated by Taylor's experiment, which produced a diffraction pattern (characteristic of waves) formed from a series of spots (characteristic of particles). Something very similar occurs with electrons and other material particles.

Figure 1.12 shows an experimental arrangement that is analogous to the two-slit interference experiment for light. A beam of electrons is incident on an arrangement called a *biprism*, which mimics the effect of two slits. The electrons arriving at a particular point on the detector can pass either side of a filament, which is analogous to photons passing through either of two slits in a screen.

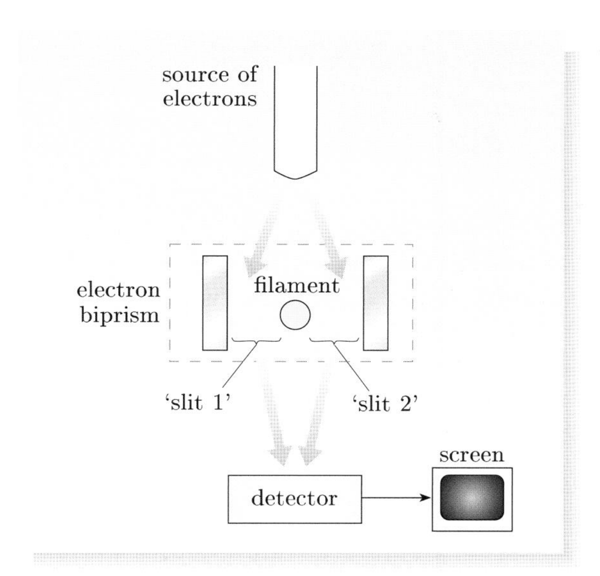

Figure 1.12 Schematic representation of an experimental arrangement used by Tonomura in 1989 to demonstrate the wave–particle duality of electrons. The two paths of the electrons around the filament are analogous to the paths through two slits in an otherwise absorbing screen.

Figure 1.13 shows how single-electron events build up over a 20-minute exposure to form an interference pattern. In (a) just 7 electrons have appeared on the screen and it is impossible to discern a pattern. It is still hard to see a pattern with 270 electrons in (b) though it is discernable in (c) where 2000 electrons have been detected. The pattern is quite clear after a much longer exposure involving 60 000 electrons shown in (d). Here, then, is wave–particle duality for electrons. The electrons appear as indivisible particles, arriving all-or-nothing at points on the detecting screen but, in the long run, the distribution of spots on the screen follows the interference pattern expected for waves.

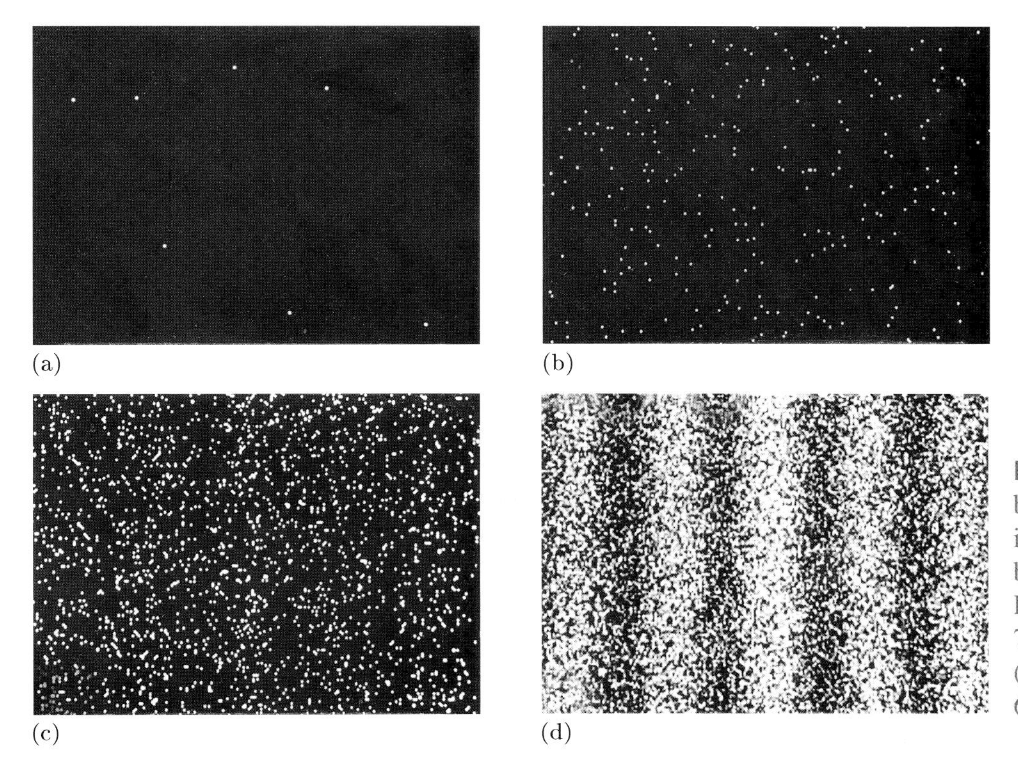

Figure 1.13 The step-by-step build-up of a two-slit interference pattern, each spot being made by a single electron. Patterns formed with: (a) 7 electrons; (b) 270 electrons; (c) 2000 electrons and (d) 60 000 electrons.

A single electron is not predestined to arrive at a particular point on the screen; until a spot irreversibly appears in just one pixel, the electron could have appeared at any point where the electron wave was non-zero. One can send any number of electrons towards a pair of slits, and they will *not* end up at identical points on the screen any more than identical uranium nuclei undergo radioactive decay at the same time. They *will* form a well-defined pattern, just as identical nuclei have a well-defined decay constant, λ.

In summary: an electron passing through two slits provides a good example of the interplay between determinism and indeterminism and, at the same time, the interplay between wave and particle behaviour. It is typical of the quantum world that we cannot say where an individual electron will be detected, but we can predict the pattern produced by many electrons.

1.4.4 Describing electron waves: the wave function

If electrons behave like waves, there must be a mathematical function that describes the value of the wave at each point in space and each instant in time. This function is called the **wave function**, and it is denoted by the symbol Ψ. For a single electron, the wave function depends on the position vector $\mathbf{r} = (x, y, z)$ and time, and so takes the form $\Psi(\mathbf{r}, t)$. However, we are often interested in situations that can be treated in one spatial dimension, allowing the wave function to be expressed as $\Psi(x, t)$.

Ψ is the upper case Greek letter psi, pronounced 'sigh'.

As time passes, the electron waves travel through space, much as the sound waves travel in a concert hall. The equation that governs how the wave of a single electron, or any microscopic system, changes in time is known as **Schrödinger's equation**. It is explained in Chapter 2. Here, we simply note that Schrödinger's equation takes the wave function $\Psi(\mathbf{r}, t_1)$, describing the electron for all $\mathbf{r}$ at time t_1, and carries it forward to give $\Psi(\mathbf{r}, t_2)$, describing the electron for all $\mathbf{r}$ at time t_2. Schrödinger's equation tells us how an electron wave spreads out as it passes around a biprism and on to a detecting screen; this allows us to calculate the final interference pattern.

Leaving aside Schrödinger's equation for the moment, we still have the major task of interpreting the wave function. Because the wave function is defined throughout a region of space, it can be thought of as a field, but is unlike any classical field you have met. An electric field, for example, has values that are real vector quantities. It is easy to measure the electric field at any point by measuring the force exerted on an electric charge placed at that point. By contrast, the wave function Ψ is a complex quantity, so it has both real and imaginary parts. Partly for this reason, the wave function cannot be measured directly. Nevertheless, we can use Ψ to calculate the *probability* of finding the electron in any small region at a particular time. That is clearly appropriate in situations where electrons are diffracted, since the place where a particular electron will register cannot be predicted, but we can predict the diffraction pattern formed by many electrons, and the maxima in that pattern are the places where the electron is most likely to be found. The precise statement of how $\Psi(x, t)$ is used to predict probabilities is given by **Born's rule**, named after Max Born (1882–1970), who belatedly received the Nobel prize for physics in 1954.

Born's rule

The probability of finding the particle at time t in a small volume element δV, centred on position $\mathbf{r}$, is

$$\text{probability} = |\Psi(\mathbf{r}, t)|^2\, \delta V. \tag{1.12}$$

$|\Psi|$ is the modulus of the complex quantity Ψ.

For systems that are modelled in one dimension, the probability of finding the particle at time t in a small interval δx, centred on position x, is

$$\text{probability} = |\Psi(x, t)|^2\, \delta x. \tag{1.13}$$

Note that the probability is proportional to the real quantity $|\Psi|^2$, and is not given by Ψ or Ψ^2. This is because Ψ has complex values, and the probability must be a real number.

For the specific case of an electron diffracted by a slit and striking a screen with sensitive pixels, Born's rule says: the probability of the electron being detected in a given pixel is proportional to the size of the pixel, and to the value of $|\Psi|^2$ at the position of the pixel. We cannot predict where a particular electron will appear, but the pattern formed by many electrons *is* predicted by Schrödinger's equation, which governs the way Ψ spreads out from the slit to the screen.

Because the wave function is complex, you will need to manipulate complex numbers throughout your study of this course. It is worth refreshing your memory of them by reading Section 8.1 of the *Mathematical toolkit* now. You may wish to spend a significant amount of time on this, as it will be a good investment for future chapters.

The rules given in the *Mathematical toolkit* for complex numbers apply equally well to functions with complex values. Here are the most important points to remember.

1. Any complex function, $f(x)$ can be expressed as a sum of real part $u(x)$ and an imaginary part $v(x)$:

   $$f(x) = u(x) + \mathrm{i}v(x), \tag{1.14}$$

   where $u(x)$ and $v(x)$ are real-valued functions.
2. The complex conjugate of $f(x)$ is written as $f^*(x)$ and given by

   $$f^*(x) = u(x) - \mathrm{i}v(x). \tag{1.15}$$

3. The modulus, or magnitude, of $f(x)$ is written as $|f(x)|$. Dropping the argument x for simplicity, this can be calculated as

   $$|f|^2 = f^*f = (u - \mathrm{i}v)(u + \mathrm{i}v) = u^2 + v^2. \tag{1.16}$$

4. Any complex function, f, can also be expressed in terms of its modulus $|f|$ and phase ϕ:

   $$f = |f|\mathrm{e}^{\mathrm{i}\phi}. \tag{1.17}$$

Since

$$e^{i\phi} = \cos\phi + i\sin\phi,$$

the real and imaginary parts of f are, respectively, $\text{Re}(f) = |f|\cos\phi$ and $\text{Im}(f) = |f|\sin\phi$, which checks out since

$$|f|^2 = \left(|f|\cos\phi\right)^2 + \left(|f|\sin\phi\right)^2 = |f|^2.$$

This will often appear in this course in the form:

$$|f|^2 = f^*f = |f|e^{-i\phi} \times |f|e^{+i\phi} = |f|^2e^0 = |f|^2.$$

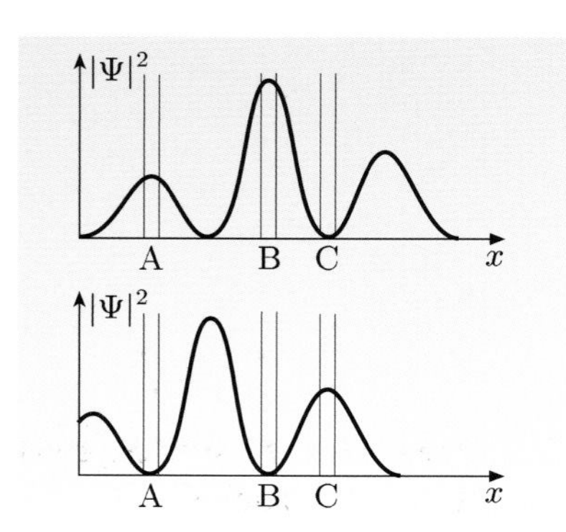

Figure 1.14 The upper graph shows $|\Psi(x,t)|^2$ for an electron at an instant of time and the lower graph shows it at a later instant. Regions A, B and C are referred to in the text.

Exercise 1.6 Figure 1.14 shows $|\Psi(x,t)|^2$ for an electron at two closely spaced instants in time, the lower graph representing the later time. Three regions along the x-axis, A, B and C are marked. In which of these regions in the upper graph would the electron be most likely to be found, least likely to be found, and somewhere in between? Remembering that the lower graph represents a later instant in time, would you say that the electron is moving in the direction of increasing or decreasing x?

Exercise 1.7 In the last exercise, we concluded that the electron was moving in the negative x-direction. Does this mean that the electron would inevitably be found at a smaller value of x at the later time? ■

1.4.5 Particles and the collapse of the wave function

What does it really mean to say that an electron, or any other entity, is a *particle*? Here is a reasonable definition, but one with far reaching consequences in the context of quantum mechanics.

In quantum mechanics, a *particle* is an entity which is found in only one place when its position is measured.

The point is illustrated by the phenomenon of electron diffraction. When a single electron leaves a slit, there is a certain volume within which the wave function $\Psi(\mathbf{r},t)$ is non-zero, and outside which the wave function is practically zero. The electron may be found anywhere in the volume where $|\Psi|^2 > 0$. The motion of this volume follows from Schrödinger's equation, which governs the time development of $\Psi(\mathbf{r},t)$. By the time the electron wave reaches the screen, it is spread out over a wide area.

The screen is covered with pixels (arrangements of atoms that can record the arrival of an electron). The arrival of the electron wave at the screen is marked by a permanent change in *one* of these pixels. Crucially, only one of the pixels responds and the others remain unchanged. It is as if the electron is asked by the screen where it wishes to appear. The electron responds by triggering a reaction in *one* of the pixels, and automatically avoids reacting with any other. If we try to interpret this in terms of news spreading out from the responding pixel, telling other pixels not to respond, the news must spread out instantly. This is an example of *quantum non-locality*, in which what happens at one place affects what happens at other places in a way that cannot be explained by communication at the speed of light, the maximum speed at which any information can be transmitted. The detection of the particle at one and only one location implies that the wave function has somehow 'collapsed' onto a single pixel in a sudden and abrupt process, called the **collapse of the wave function**. This is yet another aspect of the quantum revolution that will be discussed in detail later in the course.

1.4.6 The de Broglie wave function

Although de Broglie's ideas preceded the development of Schrödinger's equation and Born's rule, it is convenient to call the wave function $\Psi(\mathbf{r},t)$ for a free particle a **de Broglie wave function**. We shall consider the case of a free particle moving with momentum p and energy E in the positive x-direction. To simplify the description, we ignore any motion in the y- and z-directions, and describe the particle by a wave function $\Psi_{\text{dB}}(x,t)$. What would we guess for the form of this wave function? At first sight, we might suppose that

Remember that the subscript dB stands for 'de Broglie'.

$$\Psi_{\text{dB}}(x,t) = A\cos(kx-\omega t) \qquad \text{(FIRST GUESS!)}, \tag{1.18}$$

as in Equation 1.8.

In fact, this first guess is not quite right since wave functions, including Ψ_{dB}, are complex. Equation 1.18 is just the real part. Instead we have:

One may use either of the equivalent expressions, $e^{\text{i whatever}}$ or $\exp(\text{i whatever})$ according to typographical taste.

$$\begin{aligned}\Psi_{\text{dB}}(x,t) &= A[\cos(kx-\omega t) + \mathrm{i}\sin(kx-\omega t)] \\ &= A\exp[\mathrm{i}(kx-\omega t)] = A\mathrm{e}^{\mathrm{i}(kx-\omega t)}.\end{aligned} \tag{1.19}$$

The constants k and ω that characterize the de Broglie wave can be related to properties of the corresponding free particle. The wave number k is given by $k = 2\pi/\lambda$. Combining this with the de Broglie relationship, $\lambda = h/p$, we see that

$$k = \frac{2\pi p}{h} = \frac{p}{\hbar}.$$

Remember, $\hbar = h/2\pi$.

For light, we know that the angular frequency is related to the photon energy by $E = \hbar\omega$. We shall assume that a similar relationship applies to de Broglie waves, so that

$$\omega = \frac{E}{\hbar}.$$

Because de Broglie waves describe free particles, with no potential energy, the energy E can be identified with the kinetic energy, E_{kin} of the particle. Thus:

> The de Broglie wave function in Equation 1.19 represents a free particle of momentum $\hbar k$ and kinetic energy $\hbar\omega$.

The constant A is complex, and may be written in exponential form as $A = |A|\mathrm{e}^{\mathrm{i}\phi}$. This means that the de Broglie wave function can also be expressed as

$$\Psi_{\text{dB}}(x,t) = |A|\mathrm{e}^{\mathrm{i}(kx-\omega t+\phi)}. \tag{1.20}$$

The following exercise shows that this wave function describes a particle that is equally likely to have any value of x.

Exercise 1.8 Show that the wave function of Equation 1.19 represents a particle that has the same probability of being found in any small interval of fixed length δx, no matter where this interval is placed along the x-axis. ■

The wave function Ψ_{dB} is clearly an idealization — we generally deal with particles that are more likely to be found on Earth than in another galaxy! Even so, idealizations have a role to play in physics. Later in the book you will see how more localized wave functions describing free particles can be constructed, using de Broglie waves as building blocks.

1.5 The origin of quantum interference

This section takes another look at the diffraction and interference patterns produced by electrons and photons. This time, we will concentrate on describing such phenomena using the language of quantum physics. In the case of electron diffraction, this means interpreting the diffraction pattern in terms of the *wave function*. In the case of interference by photons, our description will involve a new concept called a *probability amplitude*. The section will also introduce a fundamental rule, which we shall call the *interference rule*, which lies at the heart of quantum physics.

1.5.1 Electron interference caused by a double slit

Let's return to Figure 1.8c, which shows the interference pattern due to waves passing through a double slit. There is nothing mysterious about this pattern if one thinks of it in terms of water waves. At some places, the wave coming from one slit tends to produce the same disturbance as the wave coming from the other slit. At these points the waves reinforce one another and produce a maximum (**constructive interference**). In the case of water waves, if you measured the strength of the wave by the up-and-down displacement of a floating cork from its mean position, then the displacement would be greatest at these points. At another series of points, the waves coming from the two slits would produce disturbances that cancel, and a minimum would be observed (**destructive interference**). In the case of water waves, a cork placed at one of these points would remain at rest. Notice two things.

1. Nothing would change if the waves falling on the slits were very weak except that the periodic displacement of the cork would also be very weak.
2. The displacement at any point is just the sum of the displacement that would be there if only one slit was open, plus the displacement that would be there if just the other slit was open.

More precisely, if $D_1(t)$ is the displacement of the cork from its mean position at time t due to a wave coming through slit 1, and $D_2(t)$ is the displacement of the cork from its mean position at time t due to a wave coming through slit 2, then when both slits are open the displacement is

$$\text{total displacement of cork at time } t = D_1(t) + D_2(t). \tag{1.21}$$

Now consider what happens when electrons fall on a pair of slits. The previous section described an experiment which simulated the effect of two slits by using a biprism (Figure 1.12), and the overall pattern produced by many electrons was shown in Figure 1.13d. This is very similar to the interference pattern produced by water waves, but there is one crucial difference. The electrons appear as indivisible particles, arriving all-or-nothing at points on the screen. This is *unlike* the behaviour of classical waves such as water waves. Water waves of very low intensity would produce a weak but *continuous* interference pattern that would cause a floating cork to oscillate everywhere except at the interference minima.

The observation of discrete spots produced by individual electrons might lead you to think that the electrons travel from the slits to the screen as discrete particles, but this cannot be true either. It does not begin to explain why the pattern of spots follows the outline of an interference pattern. If the electrons travelled to the screen as particles, the number of electrons arriving at any given point on the screen would be the sum of those arriving from slit 1 and those arriving from slit 2. Closing off one of the slits would always reduce the number of electrons arriving at the screen. But this is not what is observed. Closing off one of the slits converts a two-slit interference pattern into a single-slit diffraction pattern. These two patterns have minima in different places, so a point that was a minimum for the two-slit interference pattern will no longer be a minimum for the single-slit diffraction pattern. Closing off one of the slits therefore causes an *increase* in the number of electrons arriving at this point. Such behaviour cannot be understood in terms of a particle model, but is explained by the interference of waves. This, of course, brings us back to the conclusion of a previous section — electrons propagate as waves and are detected as particles — they display wave–particle duality.

How are these facts explained in quantum mechanics? Fortunately, the concept of a wave function, together with Born's rule, allows us to make a consistent interpretation. The wave function propagates through space much like water waves, though there are some differences in detail. Water waves obey a certain partial differential equation while the wave function Ψ obeys another (the Schrödinger equation). Such differences do not matter here; the only important point is that the wave function propagates as a wave. The positions of the maxima and minima in its interference pattern depend on geometric factors, such as the spacing between the slits and their distance from the screen. However, the wave function itself is not directly measurable. There is no 'bobbing cork' that can be used to find the value of the complex wave function. Instead, we have Born's rule, which tells us that the probability of finding an electron in a small region is proportional to the square of the modulus of the wave function in that region. Each electron appears as a discrete spot at a random location but, in the long run, the overall pattern of spots follows the probability distribution set by $|\Psi|^2$. In this way, the wave function, and its interpretation, combine both wave- and particle-like aspects.

To give a more detailed account of electron interference, let us denote by $\Psi(\mathbf{r},t)$ the wave function that emerges from the two slits. It turns out that this wave function can be expressed as the sum of two terms:

$$\Psi(\mathbf{r},t) = \Psi_1(\mathbf{r},t) + \Psi_2(\mathbf{r},t), \tag{1.22}$$

where $\Psi_1(\mathbf{r}, t)$ is the part of the wave function associated with propagation through slit 1 and $\Psi_2(\mathbf{r}, t)$ is the part of the wave function associated with propagation through slit 2. The terms are scaled in such a way that the complete wave function, Ψ, describes one electron that passes through either of the slits, although we cannot say which one.

It follows from Born's rule that the probability of finding an electron in a small volume element δV, located at $\mathbf{r}$, is

$$\text{probability} = |\Psi(\mathbf{r}, t)|^2\,\delta V = |\Psi_1(\mathbf{r}, t) + \Psi_2(\mathbf{r}, t)|^2\;\delta V. \tag{1.23}$$

Omitting the arguments of the functions for clarity, this can be rewritten as

$$\begin{aligned}\text{probability} &= (\Psi_1^* + \Psi_2^*)(\Psi_1 + \Psi_2)\,\delta V \\ &= (\Psi_1^*\Psi_1 + \Psi_2^*\Psi_2 + \Psi_1^*\Psi_2 + \Psi_2^*\Psi_1)\,\delta V \\ &= (|\Psi_1|^2 + |\Psi_2|^2 + \Psi_1^*\Psi_2 + \Psi_2^*\Psi_1)\,\delta V.\end{aligned}$$

If we had squared the moduli first, and then added the squares, we would have obtained $|\Psi_1|^2 + |\Psi_2|^2$, which represents the sum of two *single*-slit diffraction patterns. By adding the wave functions first, and then taking the square of the modulus of the sum, we get an additional term $(\Psi_1^*\Psi_2 + \Psi_2^*\Psi_1)$, known as the *interference term*. The interference term can be *positive or negative*, corresponding to constructive or destructive interference. When it is included, the contributions from the two slits cancel at points where $\Psi_1(\mathbf{r}, t)$ and $\Psi_2(\mathbf{r}, t)$ have the same magnitude and opposite signs. This is like the water waves from two slits cancelling in Equation 1.21, except that here we are dealing with a wave function that represents the state of a *single electron*. In summary, we can make the following statement:

> To allow for the interference of two terms in a wave function, it is essential to add the terms *before* taking the square of the modulus.

Exercise 1.9 Consider two functions: $f(x) = \sin x$ and $g(x) = \cos x$. Which of the functions $p(x) = |f(x) + g(x)|^2$ and $q(x) = |f(x)|^2 + |g(x)|^2$ is zero for certain values of x and which is never zero? ■

In the case of two-slit electron interference, the presence of interference maxima and minima implies that waves come from both slits. These waves describe a single electron, so we are unable to say which slit the electron passed through. In other words, *we must abandon the classical idea of a trajectory.* An electron might originate within a source, be accelerated towards the pair of slits, and end up being detected in a particular pixel, but we absolutely *cannot* say that it follows a particular path from the source to the place where it was detected. We *cannot* say which slit it went through.

If we did somehow determine which slit the electron went through, and this left a record of some kind (referred to as 'which-way' information), it would be wrong to add the two wave functions before taking the square of their modulus. If an electron is known to pass through slit 1, its wave function will be proportional to Ψ_1, and there will be no interference with Ψ_2. The pattern produced by such electrons is that for a single slit: *the two-slit interference pattern is destroyed when information is collected about which slit the electron passes through.* This has been verified experimentally.

Abandoning trajectories might make you feel uneasy. How do we reconcile it with the fact that electrons and other subatomic particles sometimes do appear to have trajectories, as revealed by cloud-chamber tracks (Figure 1.15)? The answer is that the cloud chamber provides which-way information. If a charged particle emerged from a double slit into a cloud chamber, you would see just one track and certainly not a two-slit interference pattern. As soon as the particle enters the chamber and induces a water droplet to form, the wave function collapses. The water droplet records which slit the particle passed through and this 'decides' the direction of motion of the particle. The particle maintains substantially the same direction of motion in the medium until it slows down or is deflected by a close encounter with a nucleus.

Figure 1.15 Alpha-particle tracks in a cloud chamber. Each track was made by a single alpha particle from a radioactive source, which ionizes a trail of molecules in air containing supersaturated water vapour. The ions become seeds upon which water droplets form.

1.5.2 Probability amplitudes

Our interpretation of electron diffraction in terms of wave functions can be used to describe other processes, but there are some limitations. For example, Schrödinger's equation applies to non-relativistic material particles, such as electrons, but it does not apply to photons. Fortunately, there is another way of calculating probabilities, but to explain it we need to introduce a new concept, that of a *probability amplitude*.

The term *probability amplitude* is part of the language of quantum physics. It should not be confused with the *amplitude* of a wave.

You will recall that Born's rule tells us that the probability of finding a particle in a given region is proportional to the square of the modulus of the wave function in that region. The idea of calculating a probability by taking the square of the modulus of something is deeply embedded in quantum mechanics. The 'something' we need to take the modulus squared of is called a **probability amplitude**. So, we have:

$$\text{probability} = |\text{probability amplitude}|^2.$$

More formally, if a process has a number of possible outcomes, and we associate a probability amplitude a_i with outcome i, the probability of this outcome is $|a_i|^2$.

Probability amplitudes can be used in situations where wave functions are difficult to use. If we wished to calculate the decay probability λ for a nucleus, we would first calculate a probability amplitude α, and then calculate $\lambda = |\alpha|^2$. When photons are reflected from half-silvered mirrors, the probability amplitudes for going through the mirror and for being reflected both have magnitudes equal to $1/\sqrt{2}$, so the corresponding probabilities are both equal to $1/2$.

The rule for adding wave functions before taking the square of their modulus has its counterpart for probability amplitudes. In fact, we have the following general rule, which we shall informally call the **interference rule**.

The interference rule

If a given process, leading from an initial state to a final state, can proceed in two or more alternative ways, and no information is available about which way is followed, the probability amplitude for the process is the sum of the probability amplitudes for the different ways.

When calculating the probability of the process, it is essential to add all the contributing probability amplitudes before taking the square of the modulus.

In the next section, you will see how this rule helps us to explain the interference of photons in the laboratory.

1.5.3 Photon interference in the laboratory

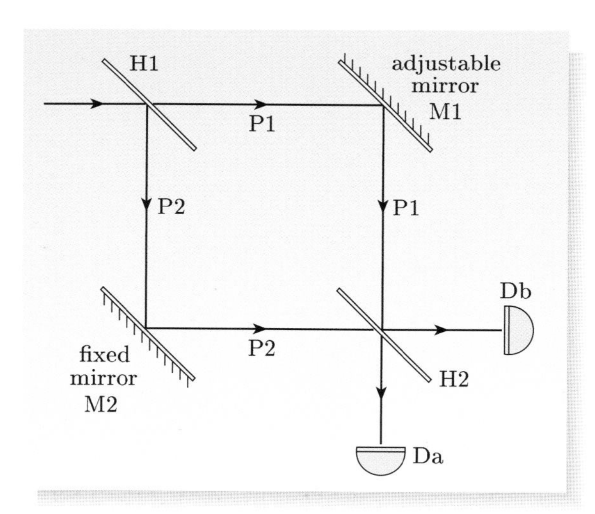

Figure 1.16 Light enters from the left, striking a half-silvered mirror, H1. It can then take two possible paths P1 and P2, being reflected off mirrors M1 and M2 before passing through a second half-silvered mirror, H2, to be detected at either of the detectors Da or Db.

The adjustment of mirror M1 is very sensitive: changing the difference in path length by half a wavelength changes constructive interference to destructive interference.

In Section 1.2.2 we saw a half-silvered mirror in action, producing random numbers. But the fun with half-silvered mirrors really starts with *two* of them and they provide a nice example of the principle of adding probability amplitudes before you square. As shown in Figure 1.16, a **Mach–Zehnder interferometer** is an arrangement of two half-silvered mirrors, H1 and H2, which act as beam splitters and two fully-reflecting mirrors, M1 and M2. This figure shows the two possible light paths, P1 and P2, leading from H1 to H2. Note that paths P1 and P2 do not have exactly identical lengths, and their relative length can be adjusted by making tiny adjustments to the position of mirror M1. It would be natural to suppose that, when light following both paths reaches H2, half the light is detected at one detector Da, and half is detected at the other detector Db. Indeed, this is exactly what happens when either one of the paths is blocked. If only path P1 is open, then the light on this path is split by H2 to be detected at detector Da or Db, with half going each way. The same thing happens when only path P2 is open.

With paths P1 and P2 both open, we find that, as the length of path P1 (say) is continuously adjusted, the light intensity measured at Da oscillates from zero to a maximum value that corresponds to the detection of all incident photons. The light measured at Db does the same, but in such a way that the maxima occur for the same path lengths that give minima at Da. This behaviour can be understood for classical light as an interference effect. If the two path lengths are such that the waves coming via P1 and P2 are exactly in phase (wave maxima matching) when emerging towards Da then the two waves reinforce one another and all the light is detected at Da. If this happens, there will be no signal at Db. All intermediate cases are covered by varying one of the path lengths, as shown in Figure 1.17.

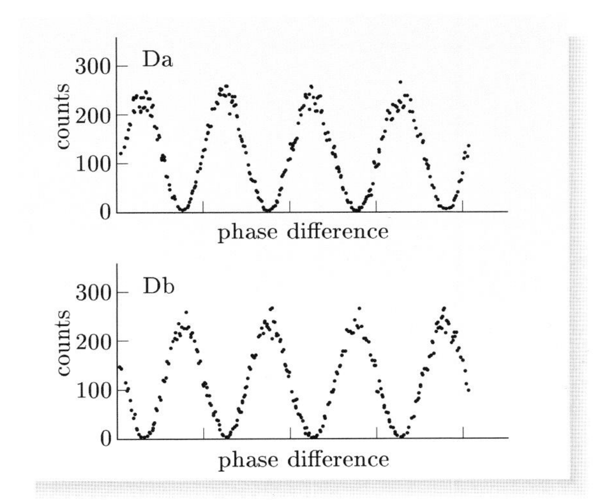

Figure 1.17 Light signals at detectors Da and Db vary as shown when the phase difference between the waves travelling along P1 and P2 is continuously changed by varying a path length. The maxima in one counter coincide with minima in the other. The dots indicate the statistical nature of the counts in both counters. (Experimental results due to Grangier *et al.*, 1986.)

With a weak light source sending just one photon into the interferometer at a time, exactly the behaviour predicted by the classical wave picture occurs if the photons are counted for long enough. That is how Figure 1.17 was obtained, with each detector Da and Db counting individual photons. So let us now turn to the quantum description in terms of probability amplitudes.

Each photon has some probability amplitude a_1 for reaching detector Da via P1, and a probability amplitude a_2 for reaching the same detector via P2. These depend on the lengths of paths P1 and P2 respectively. *Since there is no way of determining which path was taken by the photon*, we must use the interference rule for probability amplitudes: we add the probability amplitudes before squaring the modulus. The total probability of detection at Da is therefore

$$p_{\text{a}} = |a_1 + a_2|^2. \tag{1.24}$$

In general $a_1 = |a_1|\text{e}^{\text{i}\theta_1}$ and $a_2 = |a_2|\text{e}^{\text{i}\theta_2}$ are complex numbers. If the path lengths are such that $\theta_1 = \theta_2$, these probability amplitudes add constructively and p_{a} is a maximum; if $\theta_1 = \theta_2 + \pi$, the probability amplitudes add destructively and p_{a} is a minimum.

A very similar description applies at the other detector. Let the probability amplitudes for being detected at Db via P1 be b_1 and via P2 be b_2. Then the probability of detection at Db is

$$p_\mathrm{b} = |b_1 + b_2|^2.$$

This turns out to be small when p_a is large, and vice versa, as shown in Figure 1.17.

Exercise 1.10 Assume that the path lengths are such that $p_\mathrm{b} = 0$. What will be found at Db if, in an effort to determine which path a photon takes, path P1 is blocked? What will p_b be in this case?

Exercise 1.11 Imagine that a Mach–Zehnder interferometer has one path blocked, and an incident photon can appear at either of the detectors. Explain why un-blocking the blocked path can, if the path lengths are appropriate, *stop* photons being detected by one of the two detectors. Explain both for classical waves and for photons entering the interferometer one at a time. ■

We now have a more satisfactory way of speaking about particles that encounter a beam splitter or a double slit. Instead of saying rather vaguely, as we did in Section 1.2.2, that '... each photon in some sense actually goes *both* straight through *and* is reflected at 90°...' we can say that the photon is in a *superposition of two states*: one in which it travels one way, and one in which it travels the other. If this still sounds mysterious, it is because it matches *nothing* in our everyday experience of the world. One could say of superposition, like other key quantum notions, that it is really more a matter of getting used to it than of understanding it ...especially if, by 'understanding', you mean having a mental model involving everyday things that exactly parallels what is going on at the quantum level. This course will give you plenty of opportunity to get used to the idea of quantum superposition.

1.6 Quantum physics: its scope and mystery

Quantum physics is the foundation of our understanding of the material world around us, of the energy source of stars, and of the existence of galaxies. Why, for example is copper reddish and gold yellowish? The answer lies in the way electrons in metals obey quantum rules, and the way light interacts with these electrons according to quantum rules.

When do we need quantum physics? There are many phenomena where classical physics gives practically the right answers. For example, we can explain the reflection of radio waves from a sheet of copper using classical physics, but quantum physics is required to explain why light reflected from copper appears reddish and why gold has its characteristic lustre. A photon of visible light has much more energy, $E = hf$, than a radio-wave photon, so many more photons are involved when a given amount of energy is transferred with radio waves than with light. In general, the more photons that are involved in a process, the less a quantum description is required. For example, quantum effects are not important for designing radio antennas, but quantum effects are important in the interactions of gamma rays. The emission by a nucleus of a single gamma-ray photon, and its interaction with matter, are entirely quantum phenomena, the gamma-ray photon having a frequency billions of times higher than an AM radio-wave photon.

It is important to realize that quantum mechanics, rather than classical mechanics, is the fundamental theory. *In principle quantum mechanics itself can give an account of where and when classical concepts are valid.* Marconi did not need quantum mechanics, and quantum mechanics itself tells us why. But the devices within a modern radio depend on quantum mechanics in an essential way, and the reason for that is understood too.

At the heart of the strange new quantum world are wave functions and probability amplitudes that, for alternative (but unrecorded) routes, must be added before taking the square of the modulus. From these, *probabilities* can be calculated; we simply must face the fact that Nature is not deterministic. In Chapter 2 we begin the story of how wave functions and probability amplitudes are calculated for a wide range of cases.

If you find what you have just seen profoundly strange, even disturbing, then that is an entirely appropriate response! I end with the words of one of the principal creators of quantum mechanics, Werner Heisenberg:

> 'I remember discussions with Bohr which went on through many hours till very late at night and ended almost in despair; and when at the end of the discussion I went alone for a walk in the neighbouring park I repeated to myself again and again the question: Can nature possibly be as absurd as it seemed to us in these atomic experiments?'
>
> *Physics and Philosophy*, 1958

Summary of Chapter 1

Section 1 Microscopic systems on the scale of molecules, atoms, nuclei and subnuclear particles have discrete energy levels. Transitions between these energy levels involving the emission or absorption of photons are a prime source of information concerning the structure of matter. Systems such as ionized atoms have a continuum of energy levels above the highest discrete level of the bound system.

Section 2 The world is not deterministic; observations involving photons, electrons, etc. are probabilistic. Radioactive decay provides an example: identical nuclei of a specific kind have a well-defined probability of decaying, but the instant of decay of an individual nucleus is unpredictable. A practical application of quantum indeterminism is the quantum random number generator based on the random appearance of photons on either side of a half-silvered mirror. A characteristic feature of the quantum world, which reappears in every chapter in different forms, is this: quantum physics predicts a fixed overall pattern of some kind, but individual events occur randomly within that pattern.

Section 3 Light exhibits 'wave–particle duality' i.e. it exhibits both wave-like and particle-like behaviour. Photons are not simply packets of electromagnetic radiation, but have momentum and energy, and are detected in localized regions.

Section 4 Electrons and indeed all material particles also exhibit wave–particle duality, displaying either wave or particle aspects depending on the type of experiment to which they are subjected. The de Broglie wavelength is related to momentum by $\lambda = h/p$. The propagation of a particle as a wave is described by a complex-valued wave function which obeys an equation called Schrödinger's equation. In one dimension, the wave function is usually represented by $\Psi(x,t)$. Its significance is understood as follows: the probability of detecting a particle at time t in a small interval δx, centred on position x, is $|\Psi(x,t)|^2\,\delta x$.

Section 5 To allow for the interference of two terms in a wave function, it is essential to add the two terms *before* taking the square of the modulus. More generally, squaring the modulus of a probability amplitude gives a probability. The interference rule states that when a given process can happen in two or more alternative (but unrecorded) ways, the probability amplitudes for the different ways must be added before taking the square of the modulus.

Section 6 Quantum physics has fundamentally changed our understanding of the world. Do not be surprised if quantum mechanics seems shocking; its creators found it so!

Achievements from Chapter 1

After studying this chapter, you should be able to:

1.1 Explain the meanings of the newly defined (emboldened) terms and symbols, and use them appropriately.

1.2 Give a general account of the way quantum physics has changed our understanding of the world.

1.3 Write an account of, and answer questions concerning, the existence and significance of discrete energy levels in atoms and other microscopic systems.

1.4 Explain what is meant by quantum indeterminism, with reference to real examples.

1.5 Explain what is meant by wave–particle duality, drawing on experiments involving photons and material particles such as electrons.

1.6 Explain how quantum physics accounts for wave–particle duality, using wave functions and probability amplitudes.

1.7 State and use the de Broglie relationship for free particles.

1.8 State and use Born's rule for the wave function.

1.9 State and use the interference rule.

1.10 Explain when it is appropriate to add wave functions or probability amplitudes before squaring the modulus, and when it is not appropriate.

1.11 Explain why the concept of a particle trajectory cannot be taken for granted in quantum mechanics.

After studying Chapter 8.1 you should also be able to:

1.12 Manipulate complex numbers, recognizing the significance of modulus and phase and appreciating their relationship to the real and imaginary parts.

Chapter 2 Schrödinger's equation and wave functions

Introduction

The story is told that the physicist Peter Debye attended a talk given by de Broglie concerning his electron waves. Debye asked de Broglie what equation his waves obeyed, without receiving a satisfactory response. In the audience was Erwin Schrödinger ... and the rest is history! In the first half of 1926, Schrödinger (Figure 2.1) presented his famous wave equation which provides the basis for the wave-mechanical approach to explaining and predicting quantum phenomena.

Figure 2.1 Erwin Schrödinger (1887–1961), who was awarded the Nobel prize for physics in 1933.

Let us try to understand the problem Schrödinger was trying to solve. In Chapter 1 we saw that the propagation of an electron through a slit and on to a detecting screen is described by a time-dependent wave function, $\Psi(\mathbf{r}, t)$. The wave function contains information about where the electron is likely to be found. At time t, the probability of finding the electron in any small region centred on a position $\mathbf{r}$ is proportional to the value of $|\Psi(\mathbf{r}, t)|^2$.

As time passes, the wave function changes, and the region where the electron is likely to be found also changes — in ordinary speech, we say that the electron moves. In order to understand this motion, we must know how the wave function varies in time at each point in space. Such a problem is like many others in physics. It is like predicting how water waves, sound waves or light waves propagate through space. In each case, there is a particular partial differential equation that links a rate of change in time to a rate of change in space — a *wave equation*. We need the wave equation that governs the dynamics of the quantum wave function: this equation is called *Schrödinger's equation*, and it is the main subject of this chapter.

A second major problem must also be tackled. You saw in Chapter 1 that systems have definite energy levels, and that transitions between these energy levels are accompanied by the emission of photons with characteristic frequencies. The resulting spectra can be very intricate and provide 'fingerprints' of nuclei, atoms and molecules. Of course, we would like to know where these energy levels come from.

Rather surprisingly, the second problem is closely related to the first. This is because Schrödinger's equation has a special set of solutions describing what are called *stationary states*. Each stationary state has a definite energy, and corresponds to an energy level of the system. It turns out that stationary-state wave functions have a very simple time-dependence; they describe standing waves that oscillate without going anywhere, rather like the vibrations of a plucked guitar string. Allowing for this fact, Schrödinger's equation leads to a simpler condition, called the *time-independent Schrödinger equation*, which determines the spatial variation of the standing wave. In practice, the energy levels of a quantum system are determined by solving the time-independent Schrödinger equation, and you will see many examples of this throughout the course. One can imagine Schrödinger's thoughts when he solved the time-independent Schrödinger equation for a system of one proton and one electron and found the energy levels of the hydrogen atom. Although Bohr had calculated these with a semi-quantum model, everyone knew that Bohr's model was far from the whole truth — it could not even be generalized to helium, the next simplest atom. Within months of Schrödinger's success, Heisenberg had used the time-independent Schrödinger equation to explain the helium spectrum for the first time, and the assumptions behind Schrödinger's ideas were vindicated.

This chapter introduces Schrödinger's equation and the associated time-independent Schrödinger equation. We will not derive Schrödinger's equation here, but we will show that this equation is plausible, with its roots in Newtonian mechanics. Ultimately, Schrödinger's equation is justified by the fact that it *predicts the result of experiments*.

Section 2.1 begins by considering free particles, which are described by de Broglie waves. It is possible to guess the form of Schrödinger's equation in this case, but it is not obvious how to generalize the result to include the effects of forces. In Section 2.2 we introduce a number of mathematical concepts, including that of a linear operator, which will help us to make this transition. In Section 2.3, we set out the three steps that lead to Schrödinger's equation for any system that is understood in classical terms. Section 2.4 then describes some general properties of Schrödinger's equation and its solutions. Finally, Section 2.5 introduces the time-independent Schrödinger equation which determines the energy levels of a quantum system.

The *Mathematical toolkit* reviews differential equations (in Section 8.2) and partial differential equations (in Section 8.3). You may find it helpful to treat this material as an essential part of your study of this chapter.

2.1 Schrödinger's equation for free particles

We start with the simplest case: a free particle, (that is, a particle free from any forces), moving in one dimension, in the positive x-direction. In Chapter 1 you saw that such a particle is described by the de Broglie wave function

$$\Psi_{\text{dB}}(x,t) = A\mathrm{e}^{\mathrm{i}(kx-\omega t)}, \tag{2.1}$$

where A is a complex constant. You also saw that this wave function corresponds to a free particle with a momentum of magnitude

$$p = \frac{h}{\lambda} = \hbar k, \tag{2.2}$$

and an energy

$$E = hf = \hbar\omega. \tag{2.3}$$

Although Ψ_{dB} gives a complete description of a de Broglie wave, it is helpful to take a step backwards, and ask what partial differential equation this wave function satisfies. This will give us a valuable clue for writing down Schrödinger's equation in more complicated situations.

We shall simply write down a suitable partial differential equation and check that Ψ_{dB} in Equation 2.1 satisfies it. The appropriate equation is

$$\mathrm{i}\hbar\frac{\partial\Psi_{\text{dB}}}{\partial t} = -\frac{\hbar^2}{2m}\frac{\partial^2\Psi_{\text{dB}}}{\partial x^2}. \tag{2.4}$$

It is not difficult to see that Ψ_{dB} satisfies this equation provided that a certain relationship between ω and k is satisfied. The following worked example gives the details.

Worked Example 2.1

Verify that the de Broglie wave function $\Psi_{\text{dB}}(x,t) = A\mathrm{e}^{\mathrm{i}(kx-\omega t)}$ satisfies Equation 2.4, provided that ω and k obey the condition $\hbar\omega = (\hbar k)^2/2m$.

Solution

To answer a question like this, we simply insert the given function into both sides of the partial differential equation. Substituting Equation 2.1 into the left-hand side of Equation 2.4 gives

$$\begin{aligned}\mathrm{i}\hbar\frac{\partial \Psi_{\text{dB}}}{\partial t} &= \mathrm{i}\hbar\frac{\partial}{\partial t}\left(A\mathrm{e}^{\mathrm{i}(kx-\omega t)}\right)\\ &= \mathrm{i}\hbar\left(-\mathrm{i}\omega A\mathrm{e}^{\mathrm{i}(kx-\omega t)}\right) = \hbar\omega\,\Psi_{\text{dB}}(x,t).\end{aligned}$$

Similarly, substituting it into the right-hand side gives

$$\begin{aligned}-\frac{\hbar^2}{2m}\frac{\partial^2\Psi_{\text{dB}}}{\partial x^2} &= -\frac{\hbar^2}{2m}\frac{\partial^2}{\partial x^2}\left(A\mathrm{e}^{\mathrm{i}(kx-\omega t)}\right)\\ &= -\frac{\hbar^2}{2m}\left((\mathrm{i}k)^2A\mathrm{e}^{\mathrm{i}(kx-\omega t)}\right) = \frac{(\hbar k)^2}{2m}\,\Psi_{\text{dB}}(x,t).\end{aligned}$$

Equation 2.4 is satisfied if these two expressions are equal for all x and t. This is true if

$$\hbar\omega = \frac{(\hbar k)^2}{2m}, \tag{2.5}$$

which is the condition specified in the question.

Essential skill

Verifying that a given function satisfies a given partial differential equation

The relationship that ω and k must obey, Equation 2.5, is highly significant. Using the de Broglie relationship $p = \hbar k$, and the Newtonian expression for the magnitude of momentum, $p = mv$, the right-hand side of Equation 2.5 becomes

$$\frac{(\hbar k)^2}{2m} = \frac{p^2}{2m} = \frac{(mv)^2}{2m} = \tfrac{1}{2}mv^2, \tag{2.6}$$

which is the kinetic energy of the particle. Since we are dealing with a de Broglie wave, describing a free particle that experiences no forces, the potential energy may be taken to be zero. It follows that $(\hbar k)^2/2m$ can be interpreted as the energy of the particle, E, and we know from Equation 2.3 that this is equal to $\hbar\omega$. Consequently, Equation 2.5 is automatically satisfied, and we see that Ψ_{dB} is a solution of Equation 2.4.

Let's take stock. We are looking for a partial differential equation that governs the dynamics of the wave function. We have seen that Equation 2.4 accomplishes this for Ψ_{dB}. To avoid clutter, we now write this equation without the subscript:

$$\mathrm{i}\hbar\frac{\partial\Psi}{\partial t} = -\frac{\hbar^2}{2m}\frac{\partial^2\Psi}{\partial x^2} \quad \text{(for a free particle).} \tag{2.7}$$

In fact, this is Schrödinger's equation for a free particle with zero potential energy.

It is worth noting two points:

1. Equation 2.7 contains the first partial derivative $\partial\Psi/\partial t$ on the left-hand side. This has an important consequence: if we know the wave function everywhere at time t, we can evaluate the right-hand side of Equation 2.7, and hence find $\partial\Psi/\partial t$, the rate of change of the wave function at time t. This allows us to find Ψ at a slightly later time $t + \delta t$ and, continuing the process, to find Ψ at *any* later time. In other words, future values of the wave function are fully determined by its values *now*, and by the form of Schrödinger's equation.
2. Equation 2.7 involves the imaginary number $\mathrm{i} = \sqrt{-1}$. It may seem unusual for a fundamental equation in physics to involve complex numbers in this way. In electromagnetism for example, complex numbers appear for convenience, but in quantum physics they are essential. Indeed, you have already seen that the wave function is a complex quantity.

So far, so good, but not all particles are free. In realistic situations, such as an electron and a proton in a hydrogen atom, the particles are subject to forces. The main task in the first half of this chapter is to find an equation, more general than Equation 2.7, that governs the time-dependence of the wave function for a particle that is not free from forces.

2.2 Mathematical preliminaries

The task of generalizing Equation 2.7 is not trivial and involves some bold leaps of faith. This section will prepare the way by introducing three crucial mathematical concepts: *operators*, *linear operators* and *eigenvalue equations*. For the moment, these will be presented as pure mathematics, but you will soon see how these concepts are woven deeply into the language of wave mechanics.

2.2.1 Operators

We first introduce the notion of an *operator* acting on a function. For present purposes, we can say that

> An **operator** is a mathematical entity that transforms one function into another function.

An example that will occur many times in this course is the differential operator $\mathrm{d}/\mathrm{d}x$ that transforms the function $f(x)$ into its derivative $\mathrm{d}f/\mathrm{d}x$. For example, $5x$ is transformed into the constant function 5, and $3x^3 + \sin(4x)$ is transformed into $9x^2 + 4\cos(4x)$. When dealing with functions of more than one variable, we shall also encounter operators like $\partial/\partial x$ and $\partial^2/\partial x^2$ which take the first and second partial derivatives (with respect to x) of any function that they act on.

The special case $\alpha = 1$ gives the **identity operator**, written $\widehat{\mathrm{I}}$.

Two other operators are very straightforward. One is simply multiplying a function by a constant: if α is a number, then multiplication by α can be thought of as an operator. Applied to a function $f(x)$, it gives a new function $g(x) = \alpha f(x)$ which is just α times $f(x)$. The other simple operator is multiplication by a function $\beta(x)$. Applying this to a function $f(x)$, produces a new function $g(x) = \beta(x)f(x)$. This might seem trivial, but we shall often meet such operators.

It is sometimes helpful to represent operators by symbols. Just as we can talk about a function $f(x)$, so we can talk about an operator O. It is then useful to have a notation that indicates that O is an operator. The convention used throughout quantum mechanics is to place a hat above the symbol for the operator. So, instead of O, we write $\widehat{\mathrm{O}}$. *You are expected to do this in your written work.* In print, operators are also set in roman (i.e. erect, rather than sloping) type. This is a detail you may notice, but should not try to reproduce in handwriting.

The hat is more properly called a caret, but physicists generally pronounce the symbol $\widehat{\mathrm{O}}$ as O-hat.

Notation for operators

When an operator is denoted by a symbol, the symbol must always wear a hat. We write: $\widehat{\mathrm{O}}$, $\widehat{\mathrm{H}}$, $\widehat{\mathrm{p}}_x$, etc.

Worked Example 2.2

Essential skill

Acting with operators on functions

Consider the operators $\widehat{\mathrm{O}}_1$, $\widehat{\mathrm{O}}_2$ and $\widehat{\mathrm{O}}_3$:

$$\widehat{\mathrm{O}}_1 = \frac{\mathrm{d}}{\mathrm{d}x}; \quad \widehat{\mathrm{O}}_2 = 3\frac{\mathrm{d}}{\mathrm{d}x} + 3x^2; \quad \widehat{\mathrm{O}}_3 = \frac{\mathrm{d}^2}{\mathrm{d}x^2} + 5;$$

and the function $f(x) = 4x^2$. Find the new functions obtained by operating on $f(x)$ with each of these operators.

Solution

Applying each of the operators in turn:

$$\widehat{\mathrm{O}}_1 f(x) = \frac{\mathrm{d}}{\mathrm{d}x}(4x^2) = 8x,$$

$$\widehat{\mathrm{O}}_2 f(x) = \left(3\frac{\mathrm{d}}{\mathrm{d}x} + 3x^2\right)(4x^2) = 24x + 12x^4,$$

$$\widehat{\mathrm{O}}_3 f(x) = \left(\frac{\mathrm{d}^2}{\mathrm{d}x^2} + 5\right)(4x^2) = 8 + 20x^2.$$

Exercise 2.1 The operators $\widehat{\mathrm{O}}_1$, $\widehat{\mathrm{O}}_2$ and $\widehat{\mathrm{O}}_3$, are defined by:

$$\widehat{\mathrm{O}}_1 = \frac{\partial}{\partial x}; \qquad \widehat{\mathrm{O}}_2 = 3\frac{\partial}{\partial x} + 3x^2; \qquad \widehat{\mathrm{O}}_3 = \frac{\partial^2}{\partial x^2} + 5.$$

Find the new functions obtained by acting with each of these operators on

(a) $g(x, t) = 3x^2t^3$; (b) $h(x, t) = \alpha \sin(kx - \omega t)$. ■

2.2.2 Linear operators

Nearly all the operators you will meet in the context of quantum mechanics have the special property of being *linear*.

Linear operators

An operator $\widehat{\mathrm{O}}$ is said to be **linear** if it satisfies the equation

$$\widehat{\mathrm{O}}(\alpha f_1 + \beta f_2) = \alpha(\widehat{\mathrm{O}} f_1) + \beta(\widehat{\mathrm{O}} f_2) \tag{2.8}$$

for all functions f_1 and f_2 and all complex constants α and β.

Complex, of course, includes *real*.

A good example is provided by the differential operator $\mathrm{d}/\mathrm{d}x$. We can demonstrate that this is a linear operator by using the rules of differentiation since

$$\frac{\mathrm{d}}{\mathrm{d}x}\big(\alpha f_1(x) + \beta f_2(x)\big) = \frac{\mathrm{d}}{\mathrm{d}x}\big(\alpha f_1(x)\big) + \frac{\mathrm{d}}{\mathrm{d}x}\big(\beta f_2(x)\big) = \alpha\frac{\mathrm{d}f_1}{\mathrm{d}x} + \beta\frac{\mathrm{d}f_2}{\mathrm{d}x}$$

for all functions $f_1(x)$ and $f_2(x)$ and all complex constants α and β.

However, not all operators are linear. For example, the operator defined by

$$\widehat{\mathrm{O}} f(x) = \mathrm{e}^{f(x)}$$

is *not* linear because the left-hand side of Equation 2.8 gives

$$\widehat{\mathrm{O}}\big(\alpha f(x) + \beta g(x)\big) = \mathrm{e}^{\alpha f(x) + \beta g(x)} = \mathrm{e}^{\alpha f(x)} \times \mathrm{e}^{\beta g(x)},$$

while the right-hand side gives an entirely *different* result:

$$\alpha(\widehat{\mathrm{O}} f(x)) + \beta(\widehat{\mathrm{O}} g(x)) = \alpha \mathrm{e}^{f(x)} + \beta \mathrm{e}^{g(x)}.$$

Exercise 2.2 Which of the following operators is linear?

(a) $\widehat{\mathrm{Q}}$, defined by $\widehat{\mathrm{Q}} f(x) = \log f(x)$.

(b) $\widehat{\mathrm{R}}$, defined by $\widehat{\mathrm{R}} f(x) = \mathrm{d}^2 f/\mathrm{d}x^2$.

(c) $\widehat{\mathrm{S}}$, defined by $\widehat{\mathrm{S}} f(x) = [f(x)]^2$. ■

2.2.3 Eigenvalue equations

The next mathematical concept we need is that of an *eigenvalue equation*. Suppose that we let an operator $\widehat{\mathrm{A}}$ act on a function $f(x)$. This will produce a new function, $g(x) = \widehat{\mathrm{A}} f(x)$, which in general is not closely related to the original function, $f(x)$. However it may be possible to find a particular function, $f_1(x)$ with the property that

$$\widehat{\mathrm{A}} f_1(x) = \lambda_1 f_1(x),$$

where λ_1 is a constant. The effect of the operator $\widehat{\mathrm{A}}$ on $f_1(x)$ is then especially simple; it just returns the original function, multiplied by a constant. In this case, the function $f_1(x)$ is said to be an *eigenfunction* of the operator $\widehat{\mathrm{A}}$, and the constant λ_1 is said to be the *eigenvalue* corresponding to the eigenfunction $f_1(x)$.

You may also have met eigenvalue equations in a different form: $\mathbf{Av} = \lambda\mathbf{v}$, where $\mathbf{A}$ is a square matrix and $\mathbf{v}$ is a vector, i.e. a column matrix. In this case, λ is said to be an eigenvalue of the matrix $\mathbf{A}$, and $\mathbf{v}$ is an eigenvector.

Generally, an operator may have several different eigenfunctions, each of which has a corresponding eigenvalue. The following box summarizes the situation.

Eigenvalue equations, eigenfunctions and eigenvalues

Given an operator $\widehat{\mathrm{A}}$, the **eigenvalue equation** for that operator is

$$\widehat{\mathrm{A}}f(x) = \lambda f(x), \tag{2.9}$$

where λ is a complex constant.

The eigenvalue equation is generally satisfied by a particular set of functions $f_1(x)$, $f_2(x)$, ..., and a corresponding set of constants λ_1, λ_2, ... These are the **eigenfunctions** and the corresponding **eigenvalues** of the operator $\widehat{\mathrm{A}}$.

Eigenvalue equations occur throughout physics and applied mathematics. For example, the modes of vibration of strings and membranes are determined by eigenvalue equations. Usually, neither the eigenfunctions nor the eigenvalues are known at the outset, and the mathematical challenge is to find them. In some contexts, the search for eigenfunctions is limited to a particular class of physically-sensible functions, for example functions that are continuous, or that do not diverge at infinity; when restrictions like this are applied, they will be made explicit.

Worked Example 2.3

Essential skill

Checking whether a given function is an eigenfunction

Confirm that $\sin(qx)$ is an eigenfunction of $\mathrm{d}^2/\mathrm{d}x^2$ and determine its eigenvalue. Is $\sin(qx)$ an eigenfunction of $\mathrm{d}/\mathrm{d}x$?

Solution

We let the operators act on the given function, and see what emerges.

$$\frac{\mathrm{d}^2}{\mathrm{d}x^2}\big(\sin(qx)\big) = \frac{\mathrm{d}}{\mathrm{d}x}\big(q\cos(qx)\big) = -q^2\sin(qx),$$

so $\sin(qx)$ is an eigenfunction of $\mathrm{d}^2/\mathrm{d}x^2$ with eigenvalue $-q^2$. However,

$$\frac{\mathrm{d}}{\mathrm{d}x}\big(\sin(qx)\big) = q\cos(qx),$$

so $\sin(qx)$ is not an eigenfunction of $\mathrm{d}/\mathrm{d}x$.

Exercise 2.3 Show that $\mathrm{e}^{\mathrm{i}\alpha x}$ is an eigenfunction of $\mathrm{d}/\mathrm{d}x$ and $\mathrm{d}^2/\mathrm{d}x^2$ and determine the corresponding eigenvalues. ■

2.3 The path to Schrödinger's equation

With the mathematical preliminaries out of the way, we now return to the task of writing down Schrödinger's equation for a particle that is subject to external forces.

2.3.1 Observables and linear operators

Although we are aiming to generalize de Broglie waves, it is helpful to consider them once more, using the mathematical language of eigenfunctions. Omitting the dB subscript for simplicity, the de Broglie wave function is $\Psi(x,t) = A\mathrm{e}^{\mathrm{i}(kx-\omega t)}$. It is easy to see that this is an eigenfunction of both $\partial/\partial x$ and $\partial^2/\partial x^2$. Partially differentiating once, and then again, gives results similar to those you found in Exercise 2.3:

$$\frac{\partial}{\partial x}\Psi(x,t) = \mathrm{i}k\Psi(x,t), \tag{2.10}$$

$$\frac{\partial^2}{\partial x^2}\Psi(x,t) = (\mathrm{i}k)^2\Psi(x,t) = -k^2\Psi(x,t). \tag{2.11}$$

The eigenvalues, $\mathrm{i}k$ and $-k^2$, do not seem to have much physical significance, but we can scale both sides of the above equations by constants to obtain something more striking. Multiplying Equation 2.10 by $-\mathrm{i}\hbar$ and Equation 2.11 by $-\hbar^2/2m$, we obtain

$$-\mathrm{i}\hbar\frac{\partial}{\partial x}\Psi(x,t) = \hbar k\Psi(x,t) \tag{2.12}$$

$$-\frac{\hbar^2}{2m}\frac{\partial^2}{\partial x^2}\Psi(x,t) = \frac{(\hbar k)^2}{2m}\Psi(x,t). \tag{2.13}$$

The eigenvalues $\hbar k$ and $(\hbar k)^2/2m$ are now highly significant. Using the de Broglie relation $p = \hbar k$, and the fact that $E = p^2/2m$, we see that they are the momentum and kinetic energy associated with the de Broglie wave. Put another way, the de Broglie wave function $\Psi(x,t) = A\mathrm{e}^{\mathrm{i}(kx-\omega t)}$ is an eigenfunction of *both* $-\mathrm{i}\hbar\partial/\partial x$ *and* $-(\hbar^2/2m)\partial^2/\partial x^2$, and the corresponding eigenvalues are the momentum and kinetic energy associated with this wave. This could be a curiosity, but it turns out to be far more than that.

One new term is needed before we state a fundamental quantum-mechanical principle. Quantum mechanics is unusual in dealing with concepts like the wave function that cannot be measured directly. It is therefore worth having a term that describes all the other types of quantity — position, momentum, energy, and so on — that *can* be measured. Such measurable quantities are called **observables** in quantum mechanics. We can now state a principle that pervades the whole of quantum mechanics:

Operators, eigenvalues, observables and measurement outcomes

Each observable O is associated with a linear operator $\widehat{\mathrm{O}}$.
As a general rule, the eigenvalues of the operator $\widehat{\mathrm{O}}$ are the only possible outcomes of a measurement of the observable O.

Applying this principle to Equations 2.12 and 2.13, we see that the quantum-mechanical operators corresponding to momentum and kinetic energy are $-\mathrm{i}\hbar\partial/\partial x$ and $-(\hbar^2/2m)\partial^2/\partial x^2$. Using the convention of placing hats on operators, the transition from the momentum p_x to the **momentum operator** $\widehat{\mathrm{p}}_x$, and from the kinetic energy E_{kin} to the **kinetic energy operator** $\widehat{\mathrm{E}}_{\text{kin}}$, can be expressed as follows:

$$p_x \Longrightarrow \widehat{\mathrm{p}}_x = -\mathrm{i}\hbar\frac{\partial}{\partial x} \tag{2.14}$$

$$E_{\text{kin}} \Longrightarrow \widehat{\mathrm{E}}_{\text{kin}} = -\frac{\hbar^2}{2m}\frac{\partial^2}{\partial x^2}. \tag{2.15}$$

The symbol $\Longrightarrow$ is used to indicate the transition from a classical variable to a quantum-mechanical operator.

A subscript x has been placed on the momentum observable p_x and its associated operator $\widehat{\mathrm{p}}_x$ to remind us that the momentum we are talking about is that in the direction of increasing x. The inclusion of a subscript will help avoid problems later, when we deal with particles moving in the negative x-direction or with particles moving in three dimensions.

Two other important observables are represented by very simple operators. The position coordinate x is associated with a **position operator** $\widehat{\mathrm{x}}$, which simply tells us to multiply by the variable x, so

$$\widehat{\mathrm{x}}\Psi(x,t) = x\Psi(x,t). \tag{2.16}$$

Also, any function $f(x)$ of the position coordinate is an observable, and the corresponding operator simply tells us to multiply by that function of x. In particular, we will be concerned with the *potential energy function*, $V(x)$. The corresponding **potential energy operator** tells us to multiply by the potential energy function. Thus,

$$\widehat{\mathrm{V}}(x)\Psi(x,t) = V(x)\Psi(x,t). \tag{2.17}$$

Note that neither Equation 2.16 nor Equation 2.17 is an eigenvalue equation since x and $V(x)$ are both functions of x and not constants.

Exercise 2.4 A certain free particle is described by the wave function $\Psi(x,t) = A\mathrm{e}^{\mathrm{i}(-kx-\omega t)}$. This describes a plane wave propagating along the negative x-axis. What is the result of operating with $\widehat{\mathrm{p}}_x$ on this wave function? How do you interpret your answer? ■

2.3.2 Guessing the form of Schrödinger's equation

The idea that observables are represented by operators provides the key to extending Schrödinger's equation beyond the uneventful world of free particles. Using the kinetic energy operator $\widehat{\mathrm{E}}_{\text{kin}} = (-\hbar^2/2m)\partial^2/\partial x^2$ introduced in the last section, Schrödinger's equation for a free particle moving in one dimension takes the form

$$\mathrm{i}\hbar\frac{\partial\Psi(x,t)}{\partial t} = \widehat{\mathrm{E}}_{\text{kin}}\Psi(x,t). \tag{2.18}$$

How do you think this equation might generalize to a particle that is not free?

A particle that is not free has both kinetic and potential energy. It is therefore reasonable to suppose that the appropriate generalization is

$$\mathrm{i}\hbar\frac{\partial\Psi(x,t)}{\partial t} = \left(\widehat{\mathrm{E}}_{\text{kin}} + \widehat{\mathrm{V}}(x)\right)\Psi(x,t).$$

Using the explicit form of the kinetic energy operator, and remembering that the potential energy operator simply tells us to multiply by the potential energy function, we obtain

$$\mathrm{i}\hbar \frac{\partial \Psi(x,t)}{\partial t} = -\frac{\hbar^2}{2m} \frac{\partial^2 \Psi(x,t)}{\partial x^2} + V(x)\Psi(x,t). \tag{2.19}$$

This is **Schrödinger's equation** for a particle of mass m moving in one dimension, with potential energy function $V(x)$. We have used a fair amount of guesswork and intuition to motivate it, but revolutionary science always involves a leap of the imagination. Here, we have taken the Schrödinger equation for a free particle and included the effects of interactions by adding the potential energy function in the simplest possible way. The left-hand side of the equation has been left unchanged, and still involves $\partial\Psi/\partial t$, the first-order partial derivative of Ψ with respect to time.

2.3.3 A systematic recipe for Schrödinger's equation

We have guessed the form of Schrödinger's equation for a single particle moving in one dimension. However, a more general procedure is needed. We now present a three-step recipe that is guaranteed to produce Schrödinger's equation for any system with a well-defined classical description. The recipe applies to systems in three dimensions and to systems that contain many particles but, to begin with, we will continue to discuss a single particle in one dimension. The recipe is based on the ideas that led to Equation 2.19. The first step is to give a suitable *classical* description of the system.

Step 1: Specify the Hamiltonian function

In classical physics, the motion of a single particle in one dimension (along the x-axis) is governed by Newton's second law:

$$ma_x = F_x, \tag{2.20}$$

which relates the acceleration a_x of a particle of mass m to the force F_x acting on it. In quantum mechanics however, the key concepts are not acceleration and force, but momentum and potential energy. It is possible to rewrite Newton's second law in terms of these concepts.

Firstly, we note that the force is related to the gradient of the **potential energy function**, $V(x)$. For example, a particle moving along the x-axis with a potential energy function $V(x)$ experiences the force $F_x = -\partial V/\partial x$. The significance of this relationship between force and potential energy is illustrated in Figure 2.2. Secondly, we can write $ma_x = \mathrm{d}(mv_x)/\mathrm{d}t = \mathrm{d}p_x/\mathrm{d}t$. So the x-component of Newton's second law becomes

$$\frac{\mathrm{d}p_x}{\mathrm{d}t} = -\frac{\partial V}{\partial x}. \tag{2.21}$$

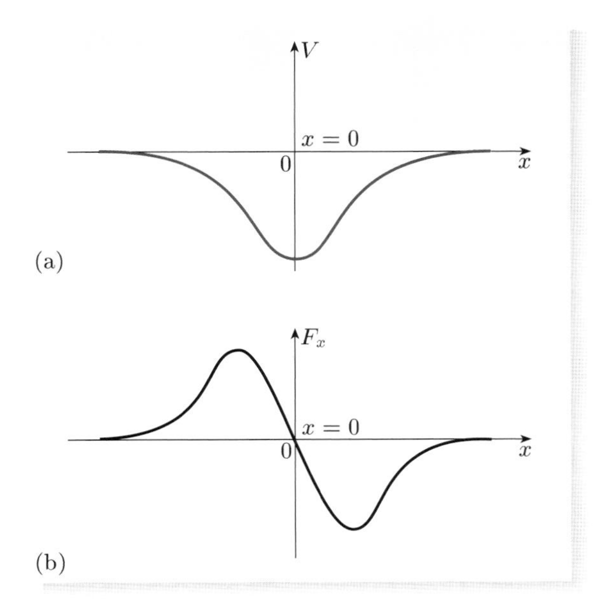

Figure 2.2 (a) An example of a potential energy function $V(x)$ in one dimension. This potential energy function has a minimum at $x = 0$ and tends to zero far from the origin. A classical particle with total energy $E < 0$ would be trapped by this potential energy well, and oscillate to and fro about the origin. (b) The corresponding force, F_x. The magnitude of the force is greatest where the potential energy function has the steepest slope, and is equal to zero where the potential energy function has zero slope (i.e. at $x = 0$). The force F_x is negative for $x > 0$ and positive for $x < 0$, and so tends to pull the particle back towards the origin.

The Irish mathematician William Hamilton (Figure 2.3) took these ideas one step further. He introduced a quantity now called the **Hamiltonian function**, H. This is the total energy of the system, *but with the kinetic energy expressed in terms of the momenta of the particles*. For a particle of mass m moving along the x-axis, subject to a potential energy $V(x)$, the Hamiltonian function is

$$H = \frac{p_x^2}{2m} + V(x). \tag{2.22}$$

Because the kinetic energy term $p_x^2/2m$ does not depend explicitly on x, we have $\partial H/\partial x = \partial V/\partial x$, so Equation 2.21 can be written as

$$\frac{\mathrm{d}p_x}{\mathrm{d}t} = -\frac{\partial H}{\partial x}, \tag{2.23}$$

which is one of Hamilton's equations of classical mechanics. Each classical system is characterized by its own Hamiltonian function, and it is the *form* of this function (rather than the value) that really matters, as this is what determines how the system develops in time. It is the Hamiltonian function that provides the bridge to quantum mechanics. Writing down the Hamiltonian function for the system of interest is the first step in our recipe for constructing Schrödinger's equation.

Figure 2.3 Sir William Rowan Hamilton (1805–1865), who reformulated Newtonian mechanics to simplify the calculation of planetary orbits. Hamilton's form of mechanics was later found to be highly suitable for making the transition to quantum mechanics.

Exercise 2.5 Write down the Hamiltonian function for a free particle of mass m, confined to the x-axis.

Exercise 2.6 A particle of mass m, confined to the x-axis, is subject to *Hooke's law*, i.e. it experiences a force that is directed towards the origin and is proportional to the particle's displacement from the origin, so $F_x = -Cx$, where C is called the *force constant*. Verify that this force is consistent with the potential energy function $V(x) = \frac{1}{2}Cx^2$, and write down the Hamiltonian function for the system. ■

The force $F_x = -Cx$ is exactly what is needed for simple harmonic motion in one dimension. Hence $V(x) = \frac{1}{2}Cx^2$ is the potential energy function for a one-dimensional harmonic oscillator.

Step 2: From function to operator

The second step in our recipe is to replace the Hamiltonian function by the corresponding **Hamiltonian operator**. Incidentally, since the Hamiltonian function describes an observable quantity — the total energy of the classical system — the Hamiltonian operator will be the *energy operator* for the system. Some rules for converting observables into operators were given in Section 2.3.1. For example, we made the correspondence

$$p_x \implies \widehat{\mathrm{p}}_x = -\mathrm{i}\hbar\frac{\partial}{\partial x},$$

from which it follows that the kinetic energy operator is

$$\frac{\widehat{\mathrm{p}}_x^2}{2m} = \frac{1}{2m}\left(-\mathrm{i}\hbar\frac{\partial}{\partial x}\right)^2 = -\frac{\hbar^2}{2m}\frac{\partial^2}{\partial x^2}, \tag{2.24}$$

in agreement with Equation 2.15. Applying these rules, the Hamiltonian operator takes the form

$$\widehat{\mathrm{H}} = \frac{\widehat{\mathrm{p}}_x^2}{2m} + \widehat{\mathrm{V}}(x) = -\frac{\hbar^2}{2m}\frac{\partial^2}{\partial x^2} + V(x). \tag{2.25}$$

Use the answer to Exercise 2.6.

Exercise 2.7 Write down the Hamiltonian operator for a particle of mass m, confined to the x-axis, and subject to Hooke's law with force $F_x = -Cx$. ■

Step 3: Assembling Schrödinger's equation

The final step is to assemble Schrödinger's equation, which always take the same general form

$$\mathrm{i}\hbar\frac{\partial\Psi(x,t)}{\partial t} = \widehat{\mathrm{H}}\Psi(x,t),$$

where $\widehat{\mathrm{H}}$ is the Hamiltonian operator. Of course, this operator depends on the system under study. For a particle of mass m moving along the x-axis, $\widehat{\mathrm{H}}$ is given by Equation 2.25 and Schrödinger's equation becomes

$$\mathrm{i}\hbar\frac{\partial\Psi(x,t)}{\partial t} = -\frac{\hbar^2}{2m}\frac{\partial^2\Psi(x,t)}{\partial x^2} + V(x)\Psi(x,t).$$

which agrees with Equation 2.19.

Exercise 2.8 Write down Schrödinger's equation for a particle of mass m confined to the x-axis for the following cases:

(a) a free particle,

(b) a particle subject to Hooke's law with force constant C. ■

Our recipe for constructing Schrödinger's equation is very general. So far, we have applied it to a single particle in one dimension, but it also works for many-particle systems in three dimensions, including nuclei, atoms, molecules and metals. As the system becomes more complicated, so does its Hamiltonian function, and hence its Schrödinger equation, but the same principles apply.

To illustrate the point, consider a system of two particles of masses m_1 and m_2, interacting with one another and moving in three dimensions. The electron and proton in a hydrogen atom would be a good example. The Hamiltonian function of such a system takes the form

$$H = \frac{p_{1x}^2 + p_{1y}^2 + p_{1z}^2}{2m_1} + \frac{p_{2x}^2 + p_{2y}^2 + p_{2z}^2}{2m_2} + V(r) \tag{2.26}$$

where p_{1x} is the x-component of the momentum of particle 1, and so on. The potential energy function $V(r)$ expresses the interaction between the particles and is assumed to depend only on the distance r between them. Now, we can make the transition to quantum mechanics by replacing each momentum component by the appropriate linear operator. For example, we replace p_{1x} by $\widehat{\mathrm{p}}_{1x} = -\mathrm{i}\hbar\partial/\partial x_1$, where x_1 is the x-coordinate of particle 1. This leads to the Hamiltonian operator

$$\widehat{\mathrm{H}} = -\frac{\hbar^2}{2m_1}\left[\frac{\partial^2}{\partial x_1^2} + \frac{\partial^2}{\partial y_1^2} + \frac{\partial^2}{\partial z_1^2}\right] - \frac{\hbar^2}{2m_2}\left[\frac{\partial^2}{\partial x_2^2} + \frac{\partial^2}{\partial y_2^2} + \frac{\partial^2}{\partial z_2^2}\right] + V(r).$$

The Schrödinger equation for the two-particle system is obtained by substituting this operator into

$$\mathrm{i}\hbar\frac{\partial \Psi}{\partial t} = \widehat{\mathrm{H}}\Psi.$$

In this case, the wave function is a function of all the coordinates of both particles, and also of time: $\Psi = \Psi(x_1, y_1, z_1, x_2, y_2, z_2, t)$. This book will avoid such complications, but you will see how they are dealt with later in the course, when we discuss hydrogen, helium and more complicated atoms.

2.4 Wave functions and their interpretation

Now that we have a recipe for writing down Schrödinger's equation, you might suppose that the next step would be to solve this equation in simple cases. However, the solution of Schrödinger's equation is a **wave function**, $\Psi(x, t)$, which has a much less obvious physical significance than water waves or sound waves, for example. We therefore begin by discussing the *interpretation* of the wave function.

The first point to make is that the wave function is complex. Given the factor of $\mathrm{i} = \sqrt{-1}$ that explicitly appears in Schrödinger's equation, and the fact that potential energies, masses and so on, are real, a purely real function Ψ, with $\mathrm{Im}(\Psi) = 0$, cannot satisfy Schrödinger's equation.

The complex nature of the wave function is a nuisance when it comes to drawing graphs. To give a complete picture of the wave function, we need to show both the real and imaginary parts. However, for some purposes, this would be too much information. As we noted in Chapter 1, the wave function itself cannot be measured. It is the square of the modulus of the wave function that tells us where a particle is likely to be found. This is the content of Born's rule, proposed in 1926 by Max Born (Figure 2.4).

Figure 2.4 Max Born (1882–1970), was the first to see that the wave function must be interpreted in terms of probability. He was awarded the 1954 Nobel prize for physics.

2.4.1 Born's rule and normalization

You met Born's rule in Chapter 1, so we give only a brief reminder here. For our purposes, it is sufficient to consider the version that applies to a single particle in one dimension. In this case, Born's rule states that the probability of finding the particle within a small interval δx centred on the point x is

$$\text{probability} = |\Psi(x,t)|^2\,\delta x. \tag{2.27}$$

The quantity $|\Psi(x,t)|^2$ is called the **probability density**; in one dimension this represents the probability *per unit length* of finding the particle around a given point. In general, there will be a region where $|\Psi(x,t)|^2$ is appreciably non-zero, and this is the region where the particle is likely to be found. However, the spatial extent of this region should not be confused with the size of the particle whose behaviour the wave function describes. For example, the wave function describing an electron in a hydrogen atom is appreciable over the volume occupied by the atom, whereas the electron itself has no discernible size.

Equation 2.27 presumes that the interval δx is small enough for $\Psi(x,t)$ to be effectively constant over that interval. If this is not the case, the probability of finding the particle between $x = a$ and $x = b$ is given by the integral:

$$\text{probability} = \int_{x=a}^{x=b} |\Psi(x,t)|^2\,\mathrm{d}x.$$

Of course, we know that, if the position of the particle is measured, it must be found *somewhere*. Since certainty is characterized by a probability of 1, we require that

$$\int_{-\infty}^{\infty} |\Psi(x,t)|^2\,\mathrm{d}x = 1. \tag{2.28}$$

This is called the **normalization condition**, and wave functions that satisfy it are said to be **normalized**.

If we have a technique for solving Schrödinger's equation, we have no reason to expect that the solution that emerges will be normalized. Unless we are very lucky, it will not be normalized. Normalization is something that is imposed *after* solving Schrödinger's equation in order to ensure that the wave function is consistent with Born's rule.

Suppose that we are given a function $\Phi(x,t)$ that satisfies Schrödinger's equation, but is not normalized. How should we proceed? The first thing to note is that if $\Phi(x,t)$ is a solution of Schrödinger's equation then so is $A\Phi(x,t)$, where A is any complex constant. To obtain a normalized wave function $\Psi(x,t)$, we write $\Psi(x,t) = A\Phi(x,t)$ and then impose normalization by requiring that

$$1 = \int_{-\infty}^{\infty} |\Psi(x,t)|^2\,\mathrm{d}x = \int_{-\infty}^{\infty} |A\,\Phi(x,t)|^2\,\mathrm{d}x = |A|^2 \int_{-\infty}^{\infty} |\Phi(x,t)|^2\,\mathrm{d}x.$$

The wave function $\Psi(x,t)$ will then be normalized *provided that* we choose

$$|A| = \left(\int_{-\infty}^{\infty} |\Phi(x,t)|^2\,\mathrm{d}x\right)^{-1/2}. \tag{2.29}$$

Any complex number A that satisfies this condition will achieve normalization but, for simplicity, we generally choose A to be real and positive. The constant A is called the **normalization constant**.

Although Schrödinger's equation does not guarantee that wave functions are normalized, it does have an important property that underpins Born's rule. If a wave function $\Psi(x, t_0)$ is normalized at a particular instant t_0, and we use Schrödinger's equation to find $\Psi(x, t)$ at any other time t, then it can be shown that the wave function $\Psi(x, t)$ remains normalized. This will be proved much later in the Course.

2.4.2 Wave functions and states

The wave function can be combined with Born's rule to find out where a particle is most likely to be found. But there is much more to physics than this, and the wave function has a much wider role to play. That wider role concerns the **state** of a system.

In classical physics the state of a system can be completely specified by giving the values of a set of measurable quantities. For example, in the case of a free particle of mass m, it would be sufficient to specify the position and velocity of the particle at some particular time. Thanks to Newton's laws, this would determine the position and velocity of the particle at any other time.

In quantum physics the situation is very different. The idea that a system may exist in a variety of states is still of crucial importance, but the state of a system cannot generally be specified by listing the values of measurable quantities, such as position and velocity. Due to quantum indeterminism, even a free particle does not possess precise values of position and velocity at any particular time. *In wave mechanics, the state of a quantum system at a given time is specified by a wave function at that time.* In fact, according to wave mechanics:

The wave function describing a particular state of a system provides the *most complete description* that it is possible to have of that state.

We emphasize that the wave function contains all the information we can possibly have about the state of a quantum system. For example, in the case of a free particle, it contains all the information we can have about its energy and momentum. (Later chapters will explain how this information is extracted in more complicated cases.) However, that information is generally probabilistic; it provides only the *possible* outcomes of measurements and the *probability* of each of those outcomes. This is a deep fact about the way the world works, rather than a failure of the wave function to specify the state of the system in a more precise way: there is an intrinsic and unavoidable indeterminism in the quantum world.

As emphasized in Chapter 1, the indeterminism of quantum physics limits the scope of predictions about the future, but it does not prevent all predictions. Given the wave function that describes the state of a quantum system at some particular time, Schrödinger's equation determines how that wave function changes with time and hence allows us to obtain probabilistic information about the system at later times. This will always be the case as long as the system remains undisturbed. However, when a measurement is made, and out of all possible outcomes one alone becomes actual, the state suddenly changes and the wave function undergoes a correspondingly abrupt change, which is called a 'collapse'. This *collapse of the wave function* is *not* described by Schrödinger's equation.

Finally, it is important to note that the wave function that describes a state is *not* unique. If a wave function is multiplied by a complex number of the form $e^{i\alpha}$, where α is real, the state described by this modified wave function is unchanged, and all the predictions about the possible outcomes of measurements and their probabilities are unchanged. For example, since $|e^{i\alpha}| = 1$, it is clear that

$$|e^{i\alpha}\Psi(x,t)|^2 = |e^{i\alpha}|^2|\Psi(x,t)|^2 = |\Psi(x,t)|^2,$$

so Born's rule and the normalization of the wave function are unaffected. Complex constants of the form $e^{i\alpha}$, where α is real, are called **phase factors**. When we describe the state of a system by a wave function, the choice of the phase factor multiplying the whole wave function is arbitrary. However, it is important not to misunderstand this point: if a wave function is the sum of two parts, the *relative* phase of the parts *does* matter, since it determines whether the parts will interfere constructively or destructively. You saw this effect in the two-slit interference of electrons described in Chapter 1.

2.4.3 The superposition principle

Wave functions have another important property: they obey the following principle.

The superposition principle

If Ψ_1 and Ψ_2 are solutions of Schrödinger's equation, then so is the linear combination $a\Psi_1 + b\Psi_2$, where a and b are arbitrary complex numbers.

Recalling the definition of a linear operator (Equation 2.8), and writing Schrödinger's equation in the compact form

$$i\hbar\frac{\partial}{\partial t}\Psi = \widehat{H}\Psi,$$

we see that the superposition principle is valid provided that $i\hbar\partial/\partial t$ and $\widehat{H}$ are both linear operators. It should come as no surprise that they are, since partial differentiation is certainly linear and all observables in quantum mechanics (including the Hamiltonian) are represented by linear operators. You can give a more explicit proof in the following exercise.

Exercise 2.9 Verify that $a\Psi_1(x,t) + b\Psi_2(x,t)$ is a solution of Schrödinger's equation (Equation 2.19), provided that $\Psi_1(x,t)$ and $\Psi_2(x,t)$ are both solutions of Schrödinger's equation.

Exercise 2.10 Is it true to say that any linear combination of normalized wave functions represents a state of a system? ■

The superposition principle is important for many reasons. First, it underlies interference effects, like the two-slit interference of electrons described in Chapter 1. You may recall that this was explained by representing the wave function as the sum of two parts, one associated with passage through slit 1, and the other associated with passage through slit 2. The justification for such a wave function is provided by the superposition principle: each part of the wave function obeys Schrödinger's equation, and so does their sum.

Using the superposition principle, we can combine wave functions in countless ways. For example, we can form linear combinations of different de Broglie waves. The de Broglie waves themselves extend over the whole of space and are therefore idealizations. But some linear combinations of de Broglie waves only extend over a finite region, and provide much more realistic descriptions of real particles.

The superposition principle also leads to some of the deepest puzzles of quantum physics. If a state described by the wave function Ψ_1 has some property (say, energy E_1) and a state described by the wave function Ψ_2 has another property (say, energy E_2), what can be said about the state described by $\Psi_1 + \Psi_2$, suitably normalized? The remarkable answer is that, in some sense, this state has *both* energy E_1 and energy E_2; its energy is *uncertain*. Schrödinger never felt comfortable with this idea, and dramatized the situation by dreaming up wave functions in which a cat is both dead and alive. As you can imagine, the course will have more to say about this later. You will see that the superposition principle underpins quantum teleportation and quantum computers.

2.5 The time-independent Schrödinger equation

You saw in Chapter 1 that some quantum systems have discrete energy levels, and that transitions between those levels produce characteristic patterns of spectral lines. We now turn to the second major topic of this chapter, which is to explain in general terms where these energy levels come from, and to show how their values can be calculated.

In wave mechanics, the state of a system is described by a wave function, and that wave function must satisfy Schrödinger's equation for the system. In this section, for the sake of simplicity, we shall mainly restrict our attention to a one-dimensional system consisting of a particle of mass m subject to a potential energy function $V(x)$. Schrödinger's equation then takes the form of a partial differential equation in two variables, x and t:

$$i\hbar\frac{\partial\Psi(x,t)}{\partial t} = -\frac{\hbar^2}{2m}\frac{\partial^2\Psi(x,t)}{\partial x^2} + V(x)\Psi(x,t), \tag{2.30}$$

We now focus on solutions that describe states of definite energy. It turns out that these solutions can be found by a standard approach to tackling partial differential equations — the method of *separation of variables*. This method is described in the next subsection.

2.5.1 The separation of variables

The key step in the method of **separation of variables** is to assume that the wave function $\Psi(x,t)$ can be written as a product of a function of x and a function of t:

$$\Psi(x,t) = \psi(x)T(t). \tag{2.31}$$

The symbol ψ on the right is the lower-case Greek letter psi. Make sure you can distinguish it from the upper case psi (Ψ) on the left.

There is a mathematical motive behind this assumption: it leads to greatly simplified equations. The simplification is achieved at the expense of only finding *some* of the solutions to Schrödinger's equation, but this does not matter for our purposes since all the energy levels can be accounted for by product wave functions of the form given in Equation 2.31.

When we insert the product wave function into Schrödinger's equation we get:

$$\mathrm{i}\hbar\frac{\partial}{\partial t}\psi(x)T(t) = \left[-\frac{\hbar^2}{2m}\frac{\partial^2}{\partial x^2} + V(x)\right]\psi(x)T(t). \tag{2.32}$$

Since $\partial/\partial t$ does not affect $\psi(x)$ and $\partial^2/\partial x^2$ does not affect $T(t)$, this may be rewritten as

$$\psi(x)\,\mathrm{i}\hbar\frac{\partial T(t)}{\partial t} = T(t)\left[-\frac{\hbar^2}{2m}\frac{\partial^2\psi(x)}{\partial x^2} + V(x)\psi(x)\right].$$

The functions $T(t)$ and $\psi(x)$ both depend on a single variable, so we can replace the partial derivatives by ordinary derivatives. Doing this, and dividing both sides by $\psi(x)T(t)$ we obtain,

$$\frac{1}{T(t)}\,\mathrm{i}\hbar\frac{\mathrm{d}T(t)}{\mathrm{d}t} = \frac{1}{\psi(x)}\left[-\frac{\hbar^2}{2m}\frac{\mathrm{d}^2\psi(x)}{\mathrm{d}x^2} + V(x)\psi(x)\right].$$

Now, the left-hand side of this equation is independent of x and the right-hand side is independent of t. How can two sides of an equation be guaranteed to be equal when each depends on a *different* independent variable? The answer is simple: each side of the equation must be equal to the *same* constant. We shall denote this constant by E, which is not an arbitrary choice of symbol, as you will soon see. Thus,

$$\frac{1}{T(t)}\,\mathrm{i}\hbar\frac{\mathrm{d}T(t)}{\mathrm{d}t} = E = \frac{1}{\psi(x)}\left[-\frac{\hbar^2}{2m}\frac{\mathrm{d}^2\psi(x)}{\mathrm{d}x^2} + V(x)\psi(x)\right].$$

This gives us *two* ordinary differential equations

$$\mathrm{i}\hbar\frac{\mathrm{d}T(t)}{\mathrm{d}t} = ET(t), \tag{2.33}$$

$$-\frac{\hbar^2}{2m}\frac{\mathrm{d}^2\psi(x)}{\mathrm{d}x^2} + V(x)\psi(x) = E\psi(x). \tag{2.34}$$

The first equation involves only the variable t and the second involves only the variable x, so the two variables have been separated from one another. The constant E that appears in both equations is called a **separation constant**.

The method of separation of variables is useful because ordinary differential equations are usually much easier to solve than partial differential equations. Equation 2.33 has a particularly simple solution:

$$T(t) = \mathrm{e}^{-\mathrm{i}Et/\hbar}. \tag{2.35}$$

Exercise 2.11 Verify that $T(t) = \mathrm{e}^{-\mathrm{i}Et/\hbar}$ is a solution of Equation 2.33. ■

The solution for $T(t)$ is universal in the sense that it applies to all systems, irrespective of the potential energy function $V(x)$. However, the second differential equation

$$-\frac{\hbar^2}{2m}\frac{\mathrm{d}^2\psi(x)}{\mathrm{d}x^2} + V(x)\psi(x) = E\psi(x) \qquad \text{(Eqn 2.34)}$$

is characteristic of the system under study. This equation will be of major importance in this course. It is called the **time-independent Schrödinger equation** for the given system.

If we know the potential energy function $V(x)$, we can look for a function $\psi(x)$ that satisfies the time-independent Schrödinger equation for a given value of E. If we can find such a function, we can put everything back together and write down the product wave function

$$\Psi(x,t) = \psi(x)T(t) = \psi(x)\mathrm{e}^{-\mathrm{i}Et/\hbar}. \tag{2.36}$$

By virtue of its construction, starting from Equation 2.32, this wave function satisfies Schrödinger's equation. The wave function involves the separation constant E, and we now turn to consider the significance of this quantity.

2.5.2 An eigenvalue equation for energy

In order to interpret the separation constant, E, we write the time-independent Schrödinger equation in the form

$$\left[-\frac{\hbar^2}{2m}\frac{\partial^2}{\partial x^2} + V(x)\right]\psi(x) = E\psi(x) \tag{2.37}$$

where we have reverted to using partial derivatives. This makes no difference because the function $\psi(x)$ only depends on a single variable. However, it helps us to recognize the term in square brackets as the Hamiltonian operator for the system. It follows that the time-independent Schrödinger equation can be written in the compact form

$$\widehat{\mathrm{H}}\psi(x) = E\psi(x). \tag{2.38}$$

This is an eigenvalue equation of the kind introduced in Section 2.2.3. In this case, the operator concerned is the Hamiltonian operator, which is the operator that corresponds to the total energy of the system, so:

The time-independent Schrödinger equation of a system is the energy eigenvalue equation of that system.

In general, the eigenvalues of a quantum-mechanical operator $\widehat{\mathrm{O}}$ are the possible measured values of the corresponding observable, O. Hence, the separation constant, E can be interpreted as the energy of the system. The wave function in Equation 2.36 is associated with a particular value of E, so it is natural to assume that:

The product wave function $\Psi(x,t) = \psi(x)\mathrm{e}^{-\mathrm{i}Et/\hbar}$, where $\psi(x)$ is an energy eigenfunction with eigenvalue E, describes a state of definite energy E. If the energy of the system is measured in this state, the value E will certainly be obtained.

For reasons that will be explained later, the states of definite energy that are described by product wave functions are called *stationary states*.

2.5.3 Energy eigenvalues and eigenfunctions

The time-independent Schrödinger equation is the eigenvalue equation for energy. When eigenvalue equations arise in specific physical contexts, it is sometimes agreed to restrict the search for eigenfunctions to a certain class of 'physically-sensible function'. This is the case with the time-independent Schrödinger equation.

> We insist that the energy eigenfunctions $\psi(x)$ should not diverge as $x \to \pm\infty$.

It is easy to see that eigenfunctions that diverge as $x \to \pm\infty$ are never going to be of any physical interest. If we were to try to use such eigenfunctions in product wave functions (stationary states) we would find that these stationary states, and their linear combinations, would also diverge at infinity. Born's rule (Equation 2.27) forbids such solutions since it requires that the wave function be normalized (Equation 2.28).

In the next chapter, you will see how the time-independent Schrödinger equation is solved in specific cases. However, it is important to have an overview of the type of result that emerges. A typical potential energy function is shown in Figure 2.5.

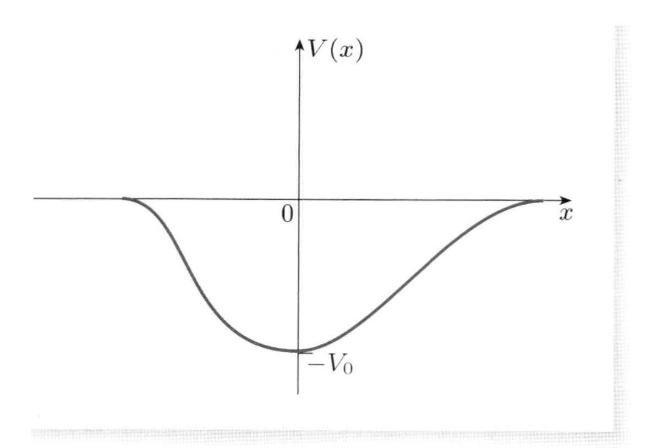

Figure 2.5 A finite well. The potential energy function, $V(x)$, has a local minimum, is finite for all values of x, and remains finite as x approaches $\pm\infty$.

This potential energy function has a minimum and then levels off, approaching flatness at infinity. It could describe the potential energy that arises when one particle is attracted to another fixed particle. Potential energy functions of this type are called **finite wells**. Finite wells may be drawn in many ways. In our example we have chosen to identify the top of the well with zero energy and the bottom with energy $-V_0$, which you can see is negative.

We shall concentrate, to begin with, on energies below the top of the well (that is, energies that are negative in Figure 2.5). If we try to solve the time-independent Schrödinger equation in this case, bearing in mind that the solutions must not diverge at infinity, something quite remarkable happens. Physically acceptable solutions are only found for certain specific values of E. These are the energy eigenvalues below the top of the well, and they form a discrete set — that is to say, they can be labelled E_1, E_2, E_3, ... Corresponding to each energy eigenvalue, there is an energy eigenfunction, so the eigenfunctions also form a discrete set, labelled $\psi_1(x)$, $\psi_2(x)$, $\psi_3(x)$, ... Equation 2.37 for the eigenfunctions becomes

$$\left[-\frac{\hbar^2}{2m}\frac{\partial^2}{\partial x^2} + V(x)\right]\psi_n(x) = E_n\psi_n(x), \tag{2.39}$$

where the label n (which is called a **quantum number**) is useful in picking out a particular eigenvalue and eigenfunction.

Now, the energy eigenvalues give us the possible energies of the system. So the discrete set of energy eigenvalues implies a discrete set of energy levels, two of which are shown in Figure 2.6. In other words, quantization of energy arises directly from the time-independent Schrödinger equation *and the conditions that identify its physically acceptable solutions*. Transitions between the energy levels then account for the characteristic pattern of spectral lines observed for the system. An example is shown in Figure 2.7.

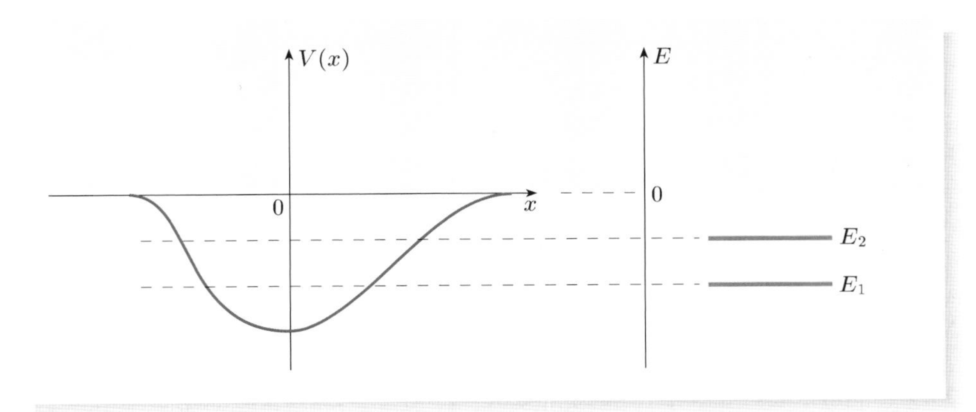

Figure 2.6 For a given finite well, the energy eigenvalues between the top and bottom of the well are restricted to discrete values, the energy levels of the system, two of which are shown. (Energy levels for real nuclei, atoms and molecules were given in Figure 1.3.)

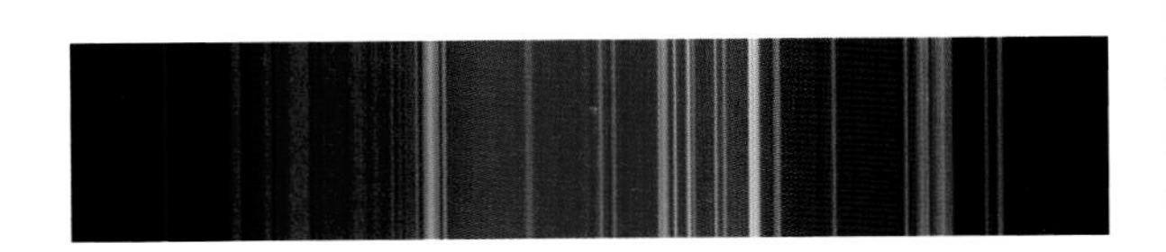

Figure 2.7 The optical spectrum of molecular nitrogen (N_2). Each line is due to photons having an energy that is the difference between two energy eigenvalues of the Hamiltonian operator for N_2.

2.5.4 Stationary states and wave packets

Corresponding to the energy eigenvalue E_n and the energy eigenfunction, $\psi_n(x)$, we have the stationary-state wave function

$$\Psi_n(x,t) = \psi_n(x)\mathrm{e}^{-\mathrm{i}E_nt/\hbar}. \tag{2.40}$$

If the system is in this state, and the energy is measured, the value E_n is certain to be obtained. This is a state of definite energy. We can now explain why these particular solutions, $\Psi_n(x,t)$, of Schrödinger's equation are said to describe **stationary states**.

Remember Born's rule, which tells us that the probability of finding the particle in a small interval of length δx, centred on x, is for any Ψ_n

$$\text{probability} = |\Psi_n(x,t)|^2\,\delta x = \Psi_n^*(x,t)\Psi_n(x,t)\,\delta x. \tag{2.41}$$

Let us see what this implies for the stationary-state wave function of Equation 2.40. Substituting this wave function into Born's rule gives

$$|\Psi_n(x,t)|^2\,\delta x = \left[\psi_n(x)\mathrm{e}^{-\mathrm{i}E_nt/\hbar}\right]^*\psi_n(x)\mathrm{e}^{-\mathrm{i}E_nt/\hbar}\,\delta x.$$

Now, we know that E_n is an energy, so it follows that it is a *real* quantity, equal to its own complex conjugate ($E_n^* = E_n$). Using this fact, we see that

$$|\Psi_n(x,t)|^2\,\delta x = \psi_n^*(x)\psi_n(x)\mathrm{e}^{\mathrm{i}E_nt/\hbar}\mathrm{e}^{-\mathrm{i}E_nt/\hbar}\,\delta x.$$

Noting that the product of the two exponentials is unity, we conclude that

$$|\Psi_n(x,t)|^2\,\delta x = |\psi_n(x)|^2\,\delta x. \qquad (2.42)$$

It follows that for the state described by $\Psi_n(x,t)$ the probability of finding the particle in any small interval of length δx is independent of time. Although we shall not prove it yet, anything that we can measure is also independent of time in this state: that is why it is described as being stationary.

It is not quite true to think of the state as frozen; there is a time-dependent phase factor, $\mathrm{e}^{\mathrm{i}E_nt/\hbar}$ so the real and imaginary parts of the wave function change with time, even though the modulus of the wave function does not. All parts of the wave oscillate in phase with one another, rather like a plucked guitar string, but with the added complication that the wave function is complex.

Stationary states correspond to nothing that is familiar in the world of human-sized objects. Before the introduction of wave mechanics it might have been thought that the lines seen in atomic spectra had something to do with electrons moving in closed orbits around atomic nuclei and occasionally jumping from one orbit to another. Such a picture is inconsistent with wave mechanics. In the stationary states that wave mechanics uses to explain the energy levels of atoms, nothing can really be said to 'move' at all. The probability of detecting the electron in any given region never changes.

Of course, there are objects that do move in nature and quantum physics must have a way of describing them. It is clear that stationary states cannot do this. However, it must be remembered that stationary states are a *special type* of solution to Schrödinger's equation. There are other types of solution, not described by product wave functions, that can describe motion. You will see in Chapter 6 how these non-stationary states are constructed. The key idea will turn out to be the superposition principle. By adding together different stationary-state wave functions in an appropriate way, we can produce a normalized wave function which is not itself a stationary state, and can describe motion. Such linear combinations of stationary states are called *wave packets*.

The concept of a wave packet will also help us to get around another difficulty. So far, we have said nothing about the energies above the top of the finite well in Figure 2.5. This was quite deliberate because the situation here is more subtle. For any value of E above the top of the well, it is possible to find an energy eigenfunction $\psi(x)$ that satisfies the time-independent Schrödinger equation, and does not *diverge* as $x \to \pm\infty$. Thus, in addition to the discrete energy levels below the top of the well there is also a *continuum* of levels above the top of the well (see Figure 2.8). However, it turns out that eigenfunctions above the top of the well *do not approach zero* as $x \to \pm\infty$, so neither they, nor the corresponding stationary-state wave functions $\Psi(x,t)$, can be normalized. Born's rule tells us that these wave functions are physically unacceptable and cannot really describe states. But we know that there are states that correspond to the continuum above the top of the well — atoms after all can be ionized into states of positive energy. How are these states to be described? It turns out that certain linear combinations of positive-energy stationary-state wave functions *can* be normalized, and it is these *wave packets* that are used to describe systems in the continuum. We cannot do full justice to this point here, but more extensive discussions will be given in Chapter 6.

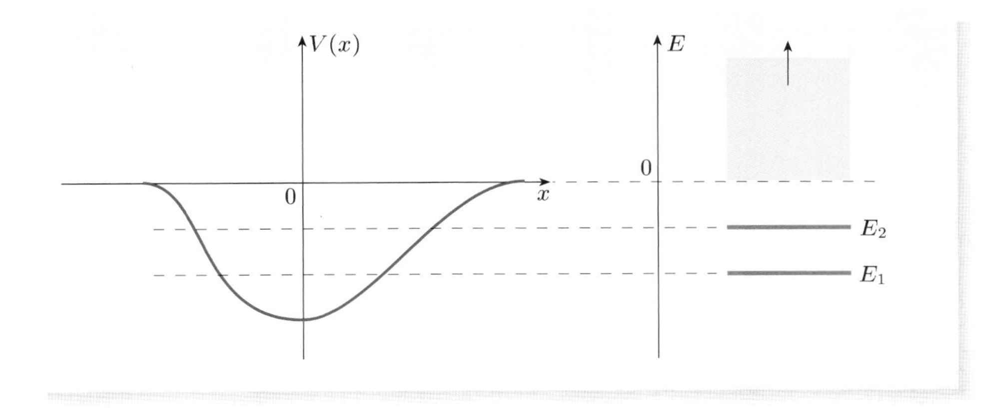

Figure 2.8 For a given finite well, there are energy eigenvalues for all values of energy above the top of the well. These are the continuum energy levels of the system, corresponding to ionized states of atoms.

2.6 Schrödinger's equation, an overview

Schrödinger's equation,

$$\mathrm{i}\hbar\frac{\partial\Psi}{\partial t} = \widehat{\mathrm{H}}\Psi, \tag{2.43}$$

is the equation that governs the time-development of the wave function Ψ that describes the state of a system for which $\widehat{\mathrm{H}}$ is the Hamiltonian operator. If, at a particular time t_0, we are given the wave function $\Psi(x, t_0)$, then we can calculate $\Psi(x, t)$ for all later times, $t > t_0$. To this extent the world is deterministic: the way the world is *now* determines the way the world will be *in the future.* However, the world is also indeterministic; the wave function provides the most complete description we can have of the state of a system, yet it is limited to providing probabilistic information about that state. It may predict the outcome of a particular measurement with probability 1, i.e. with certainty, as in the case of the energy corresponding to a stationary state, but there will inevitably be other aspects of that state that are indeterminate. Schrödinger's equation allows us to predict with great accuracy the *probability* of an electron appearing in a specified region of a screen or the *probability* of a nucleus decaying in the next second, but it will not allow us to predict exactly where a particular electron will be seen or exactly when a particular nucleus will decay.

This element of uncertainty in no way detracts from the great power of Schrödinger's equation. It is the means to understand much (in principle, just about *all*) about atoms, molecules, nuclei, solids, liquid helium, white dwarf stars, neutron stars, the production of the elements.... The problem is how to solve Schrödinger's equation for complex systems. In the following chapters we shall present solutions for some simple systems in order to illustrate general principles that also apply in more complex cases.

As you work through these chapters you will be introduced to all of the basic principles of wave mechanics. However, as you start that journey you may find it useful to keep in mind the following *preliminary version* of these principles:

Preliminary principles of wave mechanics

1. The state of a system at time t is represented by a wave function $\Psi(x, t)$.
2. An observable, such as energy or momentum, is represented by a linear operator, such as $-i\hbar\partial/\partial x$ for the momentum component p_x.
3. As a general rule, the only possible outcomes of a measurement of an observable are the eigenvalues of the associated operator.
4. The time-evolution of a system in a given state is governed by Schrödinger's equation.
5. A measurement will cause the collapse of the wave function — a sudden and abrupt change that is not described by Schrödinger's equation.

Summary of Chapter 2

Section 1 A key task is to establish the equation that governs the time-development of the wave function. This is Schrödinger's equation. It is possible to guess the form of Schrödinger's equation for a de Broglie wave describing a free particle of energy $\hbar\omega$ and momentum $\hbar k$.

Section 2 To extend Schrödinger's equation to particles subject to forces, several mathematical concepts are needed, including operators (specifically linear operators) and eigenvalue equations, which take the form $\widehat{\mathrm{A}} f(x) = \lambda f(x)$.

Section 3 In wave mechanics each observable is associated with a linear operator, and the only possible outcomes of a measurement of an observable are the eigenvalues of the corresponding operator. For example, the momentum component p_x is represented by the linear operator $\widehat{\mathrm{p}}_x = -i\hbar\partial/\partial x$. A de Broglie wave is an eigenfunction of both the kinetic energy and the momentum operators.

Schrödinger's equation for a system can be written down using a three-step recipe. Step 1: write down the Hamiltonian function of the system. This is the energy of the system, including kinetic and potential energies, but with the kinetic energy written in terms of momentum. Step 2: replace all classical observables by the corresponding operators to obtain the Hamiltonian operator. Step 3: write down Schrödinger's equation in the compact form:

$$i\hbar\frac{\partial\Psi}{\partial t} = \widehat{\mathrm{H}}\Psi,$$

and expand the right-hand side by applying the appropriate Hamiltonian operator to a wave function that depends on the coordinates of all the particles, and on time.

Section 4 The wave function $\Psi(x, t)$, a solution of Schrödinger's equation, is complex and cannot be measured. It provides the most complete specification possible of the state of a system. The probability of finding the particle in that state in a small interval δx, centred on x, is $|\Psi(x,t)|^2\,\delta x$ (Born's rule), provided that the wave function has been normalized. Wave functions which differ by an overall multiplicative phase factor describe the same state. Since Schrödinger's equation involves only linear operators, a linear combination of two solutions is also a solution. This is the superposition principle.

Section 5 Schrödinger's equation for a particle moving in one-dimension with potential energy $V(x)$ is satisfied by product wave functions of the form $\Psi(x,t) = \psi(x)T(t)$. The method of separation of variables shows that $T(t)$ has the universal form $\mathrm{e}^{-\mathrm{i}Et/\hbar}$, while $\psi(x)$ satisfies the time-independent Schrödinger equation

$$\left[-\frac{\hbar^2}{2m}\frac{\partial^2}{\partial x^2} + V(x)\right]\psi(x) = E\psi(x).$$

If the energy of the system is measured in a state described by the product wave function $\psi(x)\,\mathrm{e}^{-\mathrm{i}Et/\hbar}$, the value E will be observed with certainty.

The time-independent Schrödinger equation is an eigenvalue equation for energy:

$$\widehat{\mathrm{H}}\psi(x) = E\psi(x).$$

Insisting that the energy eigenfunctions $\psi(x)$ should not diverge as x approaches $\pm\infty$ ensures that, if $V(x)$ is a finite well, the energy eigenvalues between the bottom and the top of the well are discrete while those above the top of the well form a continuum. The product wave functions are usually called stationary-state wave functions because they lead to probability distributions that are independent of time. Those corresponding to the continuum are not physically acceptable since they cannot be normalized. However, some linear combinations of stationary-state wave functions, known as wave packets, can be normalized and can be used to describe the non-stationary states associated with the continuum.

Achievements from Chapter 2

After studying this chapter, you should be able to:

2.1 Explain the meanings of the newly defined (emboldened) terms and symbols, and use them appropriately.

2.2 Recognize the action of simple operators; identify those that are linear; determine whether given functions are eigenfunctions and if so determine the corresponding eigenvalues.

2.3 Write an expression for the operator representing momentum, $\widehat{\mathrm{p}}_x$.

2.4 Give an account of the relationship between operators, their eigenvalues and the possible outcomes of measurements.

2.5 Write down the Hamiltonian function, the corresponding Hamiltonian operator and Schrödinger's equation for simple systems of particles interacting through potential energy functions.

2.6 Write a brief account of the significance of the wave function $\Psi(x,t)$ including its interpretation in terms of position measurements and probabilities; explain the requirement for $\Psi(x,t)$ to be normalized.

2.7 Explain why the linearity of $\widehat{\mathrm{H}}$ and the superposition principle as applied to Schrödinger's equation are relevant to understanding two-slit interference.

2.8 Use the method of separation of variables to show that stationary-state wave functions have the universal form $\Psi(x,t) = \psi(x)\mathrm{e}^{-\mathrm{i}Et/\hbar}$, where $\psi(x)$ satisfies the time-independent Schrödinger equation. Describe the general properties of stationary states.

2.9 For a one-dimensional system consisting of a particle in a finite well, describe the way in which the time-independent Schrödinger equation accounts for the discrete and continuum energy levels.

2.10 State a preliminary version of the principles of wave mechanics.

2.11 Recognize that multiplying $\Psi(x,t)$ by a phase factor $\mathrm{e}^{\mathrm{i}\alpha}$, where α is real, makes no difference to any probabilities extracted from $\Psi(x,t)$.

Chapter 3 Particles in boxes

Introduction

In Chapter 2, we introduced Schrödinger's equation and stressed its importance in wave mechanics, but did not solve it for any specific system. This chapter will set up and solve Schrödinger's equation for a number of situations involving a *particle in a box*. By this, we mean that the particle is more or less confined to a limited region of space but that, within this limited region, the particle feels no forces. The region of space may be in one, two or three dimensions.

Technological devices that exploit the quantum behaviour of particles in boxes are becoming increasingly important. Such devices are generally constructed from **semiconducting materials**, such as silicon and gallium, that have electrical conductivities intermediate between those of good conductors (such as copper), and insulators (such as glass). At the heart of many of these devices is a tiny structure called a **quantum dot** that consists of a speck of one semiconductor embedded in a larger sample of another semiconductor (Figure 3.1). The enclosed speck may be as small as one nanometre across, and contain only a few hundred atoms. Some of the electrons in the speck can become detached from their parent atoms, but are inhibited from entering the surrounding semiconductor; such electrons behave very much like particles in a tiny three-dimensional box.

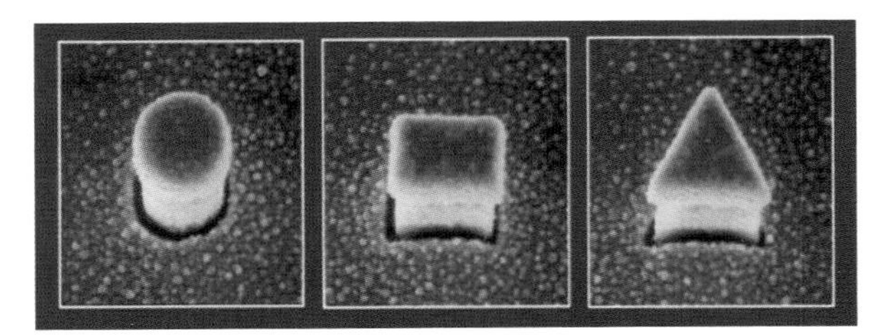

Figure 3.1 Three quantum dots of various shapes. Each consists of a small 'box' of one semiconductor entirely surrounded by another semiconductor.

The trapped electrons have wave-like properties and occupy energy levels, just as the electrons in an atom occupy energy levels. Indeed, quantum dots are sometimes referred to as *artificial atoms*. In contrast to real atoms, however, quantum dots can be tailored to produce a range of properties that atoms and molecules fail to provide. The light-emitting properties of quantum dots are particularly important; quantum dots are used in solid-state lasers (for CD and DVD players), lighting systems, solar cells, and even as fluorescent markers for a range of biomedical applications (Figure 3.2).

Figure 3.2 Quantum dots emitting light after exposure to ultraviolet radiation. The colour of the light depends on the size of the quantum dot, which increases from left to right.

Quantum dots, in which electrons are trapped in a tiny three-dimensional box, are just one example of a type of structure that uses microscopic arrangements of semiconductors to confine electrons. Some other examples are shown in Figure 3.3. In the case of a **quantum wire**, where a thin thread of one

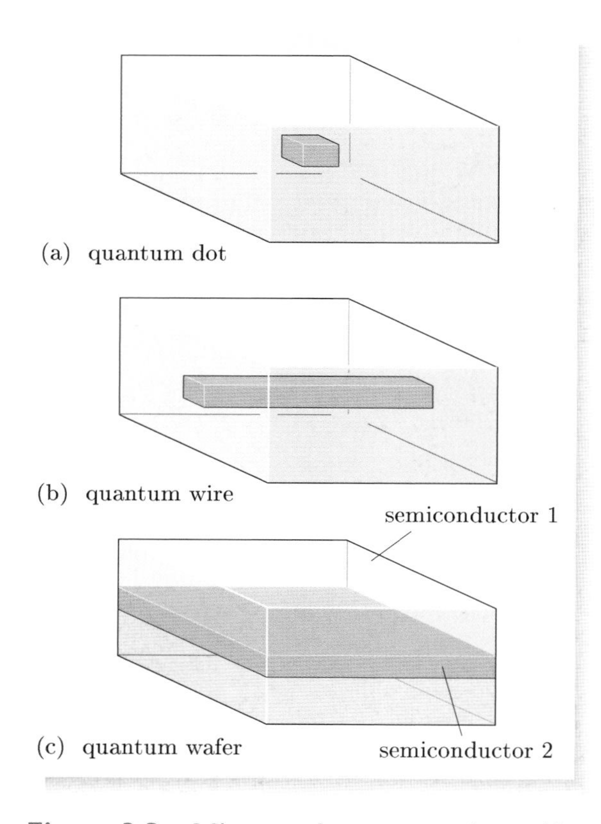

Figure 3.3 Microscopic structures formed by embedding one semiconductor inside another: (a) a quantum dot, (b) a quantum wire and (c) a quantum wafer.

semiconductor is embedded in another, the electrons are free to move along the thread, while being confined in the other two dimensions. These electrons can be thought of as being trapped in a *two-dimensional box* while being unconfined in the third dimension. In a **quantum wafer**, where a very thin layer of one semiconductor is sandwiched between broader layers of another, the electrons are free to move in the plane of the wafer while being confined in the dimension perpendicular to the plane of the wafer. These electrons can be thought of as being trapped in a *one-dimensional box* while forming a sort of freely-moving two-dimensional gas in the plane of the wafer.

We shall not describe the detailed physics of semiconductor devices here, but will concentrate on the underlying quantum behaviour of particles in boxes. The focus will be on finding the energy levels and stationary-state wave functions in these model systems, and on interpreting the results. However, you should bear in mind that our models are closely related to real systems of increasing technological importance.

In the sections that follow we shall consider boxes in one, two and three dimensions, though we shall often refer to them as *wells*, since in each case our starting point will be a classical potential energy function describing the 'well' in which the particle is trapped. Section 3.1 discusses the idealized case of a particle in a one-dimensional infinitely-deep well. Section 3.2 extends the discussion to infinitely-deep wells in two and three dimensions. Finally, Section 3.3 considers a particle in a one-dimensional well of finite depth. You will see how sets of discrete energy levels emerge from Schrödinger's equation (Figure 3.4) and how particles can sometimes be found in regions that would be forbidden according to classical physics (Figure 3.5).

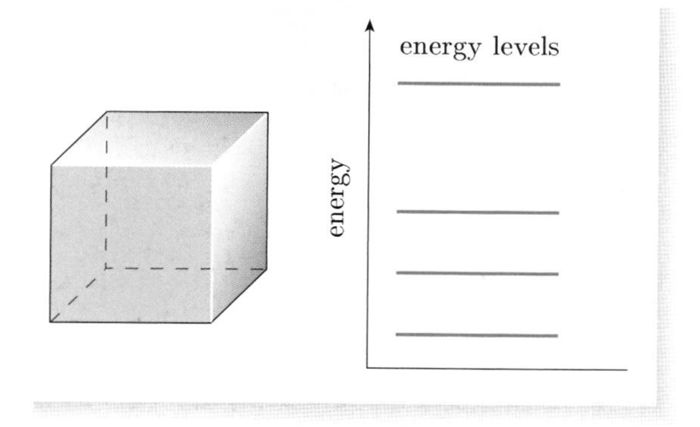

Figure 3.4 Schrödinger's equation shows that a particle confined to an infinitely deep well has a set of discrete energy levels.

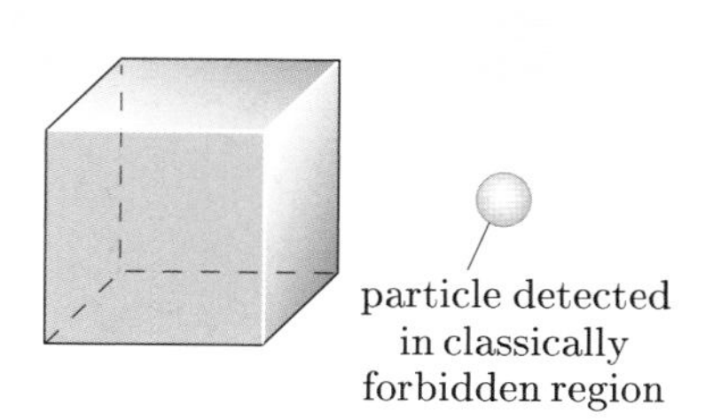

Figure 3.5 In quantum physics there is some chance of a particle being found in regions that are forbidden according to classical physics.

3.1 The one-dimensional infinite square well

The first 'box' we are going to investigate is called a *one-dimensional infinite square well* . This is the kind of box that confines the electrons of a quantum wafer in the dimension perpendicular to the plane of the wafer. We start by considering the classical behaviour of a particle trapped inside such a box and then move on to consider the behaviour of the analogous quantum system by setting up and solving the appropriate Schrödinger equation.

3.1.1 The classical system

The classical system that is our starting point consists of a particle of mass m, confined to a one-dimensional region of length L. The potential energy well responsible for trapping the particle has abrupt, infinitely-high walls, but between the walls, the particle feels no forces. Such a well is called a **one-dimensional infinite square well**. It can be represented by the potential energy function

$$V(x) = 0 \quad \text{for } 0 \leq x \leq L, \tag{3.1}$$

$$V(x) = \infty \quad \text{for } x < 0 \text{ and } x > L, \tag{3.2}$$

which is illustrated in Figure 3.6.

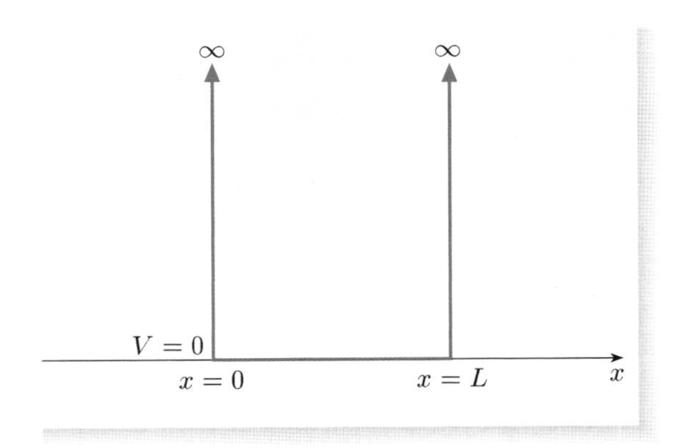

Figure 3.6 The potential energy function for a one-dimensional infinite square well.

This specification of the potential energy function involves some arbitrary choices; we have chosen to locate the walls of the well at $x = 0$ and $x = L$, and have taken the potential energy function to be equal to zero inside the well. We might have chosen differently; by locating the walls of the well at $x = -L/2$ and $x = +L/2$ for example, or by assigning a constant positive value V_0 to the potential energy inside the well. Such choices would alter the potential energy function of the system, and would therefore affect the way we *describe* the particle's behaviour, but they wouldn't influence the *behaviour* itself. We shall return to this point later.

According to *classical physics*, the total energy of a particle trapped in a one-dimensional infinite square well can have any finite positive value. Since the particle's potential energy inside the well is zero, the total energy of the particle is equal to its kinetic energy. A non-zero kinetic energy allows the particle to bounce back and forth between the impenetrable walls with constant speed. Increasing the energy would simply increase the speed. But no matter how great the energy, there is no possibility of the particle being found *outside* the well. The region outside the well is therefore said to be **classically forbidden**.

Exercise 3.1 How would you modify Equations 3.1 and 3.2 to represent a one-dimensional infinite square well with walls at $x = -L/2$ and $x = +L/2$? ■

3.1.2 Setting up Schrödinger's equation

The procedure for writing down Schrödinger's equation was given in Chapter 2. It consists of three steps:

1. Write down the Hamiltonian function $H(p_x, x)$, which is the classical energy of the system expressed in terms of momentum and position.
2. Convert $H(p_x, x)$ into an operator, $\widehat{\mathrm{H}}$, using the replacement rules

$$\widehat{\mathrm{x}} \Longrightarrow x, \quad \widehat{\mathrm{p}}_x \Longrightarrow -\mathrm{i}\hbar\frac{\partial}{\partial x} \quad \text{and} \quad \frac{\widehat{\mathrm{p}}_x^2}{2m} \Longrightarrow -\frac{\hbar^2}{2m}\frac{\partial^2}{\partial x^2}.$$

3. Write down Schrödinger's equation in the form

$$\mathrm{i}\hbar\frac{\partial\Psi(x,t)}{\partial t} = \widehat{\mathrm{H}}\Psi(x,t).$$

For a system consisting of a particle of mass m in a one-dimensional potential energy well $V(x)$, this prescription leads to a Schrödinger equation of the form

$$\mathrm{i}\hbar\frac{\partial\Psi(x,t)}{\partial t} = -\frac{\hbar^2}{2m}\frac{\partial^2\Psi(x,t)}{\partial x^2} + V(x)\Psi(x,t). \tag{3.3}$$

We are interested in the stationary states of this system — states of definite energy. These are special solutions of Schrödinger's equation that are products of separate functions of x and t:

$$\Psi(x,t) = \psi(x)T(t).$$

Following exactly the same steps as in Chapter 2, we replace $\Psi(x,t)$ in Equation 3.3 by the product $\psi(x)T(t)$ and then separate variables to obtain two ordinary differential equations

$$\mathrm{i}\hbar\frac{\mathrm{d}T}{\mathrm{d}t} = ET(t) \tag{3.4}$$

$$-\frac{\hbar^2}{2m}\frac{\mathrm{d}^2\psi(x)}{\mathrm{d}x^2} + V(x)\psi(x) = E\psi(x). \tag{3.5}$$

Equation 3.4 is satisfied by a function of the form

$$T(t) = \mathrm{e}^{-\mathrm{i}Et/\hbar}, \tag{3.6}$$

while Equation 3.5 is the *time-independent Schrödinger equation*. This is the eigenvalue equation for energy; its solutions $\psi(x)$ are the energy eigenfunctions and the corresponding values of E are the energy eigenvalues.

In the case of the one-dimensional infinite square well, $V(x) = 0$ inside the box, so Equation 3.5 becomes

$$-\frac{\hbar^2}{2m}\frac{\mathrm{d}^2\psi}{\mathrm{d}x^2} = E\psi(x) \quad \text{for } 0 \le x \le L, \tag{3.7}$$

while outside the box, where $V(x)$ is infinite, the only solution to Equation 3.5 is

$$\psi(x) = 0 \qquad \text{for } x < 0 \text{ and } x > L. \tag{3.8}$$

The task that now confronts us is that of solving Equation 3.7, subject to appropriate boundary conditions. As you will soon see, the energy eigenfunctions and eigenvalues that emerge form discrete sets, so we can talk about the nth eigenfunction, $\psi_n(x)$, and the nth eigenvalue, E_n. The corresponding stationary-state wave function is

$$\Psi_n(x,t) = \psi_n(x)\,\mathrm{e}^{-\mathrm{i}E_nt/\hbar}. \tag{3.9}$$

This describes a state of definite energy E_n: if you measure the energy of the system in this state, you will certainly get the value E_n. So, by solving Equation 3.7, and obtaining all the energy eigenvalues, we will find all the possible energy levels of the system, together with the stationary-state wave functions, which give a complete description of the state of the system in each of these energy levels.

Exercise 3.2 Explain the distinction between an energy eigenfunction $\psi(x)$ and a wave function $\Psi(x,t)$. ■

3.1.3 Solving the time-independent Schrödinger equation

Solving the time-independent Schrödinger equation inside the infinite square well is not difficult. Equation 3.7 is a second-order differential equation, so its general solution contains two arbitrary constants. Moreover, the general solution $\psi(x)$ must be such that, when differentiated twice, the result is $\psi(x)$ multiplied by the constant $(-2mE/\hbar^2)$. This immediately suggests that the solutions will be of the form

$$\psi(x) = A\sin(kx) + B\cos(kx) \quad \text{for } 0 \le x \le L, \tag{3.10}$$

where A and B are arbitrary constants and k is a constant that depends on the properties of the system.

Exercise 3.3 Show that Equation 3.10 is indeed a solution of Equation 3.7, provided that

$$k = \frac{\sqrt{2mE}}{\hbar}. \tag{3.11}$$

Exercise 3.4 In deriving Equation 3.11, we took the *positive* square root of the equation $k^2 = 2mE/\hbar^2$. It would be equally valid to take the negative square root, to obtain $k = -\sqrt{2mE}/\hbar$. Explain why this case can be ignored without any loss of generality. ■

3.1.4 Boundary conditions for the eigenfunctions

We have found the general solution of the time-independent Schrödinger equation inside the well, but this is not sufficient. The eigenfunctions we are looking for must also satisfy some boundary conditions relating to finiteness and continuity, as we shall now explain.

First, we always require that energy eigenfunctions should remain finite as $x \to \pm\infty$. This is not a problem in the present case because we already know that $\psi(x)$ is equal to zero outside the well (Equation 3.8).

More importantly, there are conditions that refer to the continuity of the eigenfunction and its derivative. Because these conditions are important throughout this chapter, we shall introduce them in a form that applies to all one-dimensional situations. Then, we will specialize to the case of present interest — a one-dimensional infinite square well.

The main point is that the time-independent Schrödinger equation must be satisfied *everywhere*, without any exceptions at individual points. If $V(x)$ is finite, Equation 3.5 implies that $\mathrm{d}^2\psi/\mathrm{d}x^2$ must also be finite. It then follows that $\psi(x)$ and $\mathrm{d}\psi/\mathrm{d}x$ must both be continuous (since any jump in these functions would prevent $\mathrm{d}^2\psi/\mathrm{d}x^2$ from having a finite value).

The potential energies of real systems are always finite, but this may not be true in some model systems. The one-dimensional infinite square well, for example, has a potential energy function that becomes infinite at each wall of the well. At any point where the potential energy function is infinite, Equation 3.5 implies that: either $\psi(x)$ is equal to zero, or $\mathrm{d}^2\psi/\mathrm{d}x^2$ is infinite (or both these things are true). Bearing in mind that infinities only appear in artificial models, it is sensible to take the function $\psi(x)$ to be as smooth as possible (subject to the above considerations). We therefore assume that any infinity in $\mathrm{d}^2\psi/\mathrm{d}x^2$ is associated with a *finite* discontinuity in $\mathrm{d}\psi/\mathrm{d}x$, but that $\psi(x)$ remains continuous *everywhere*. This leads to the following boundary conditions:

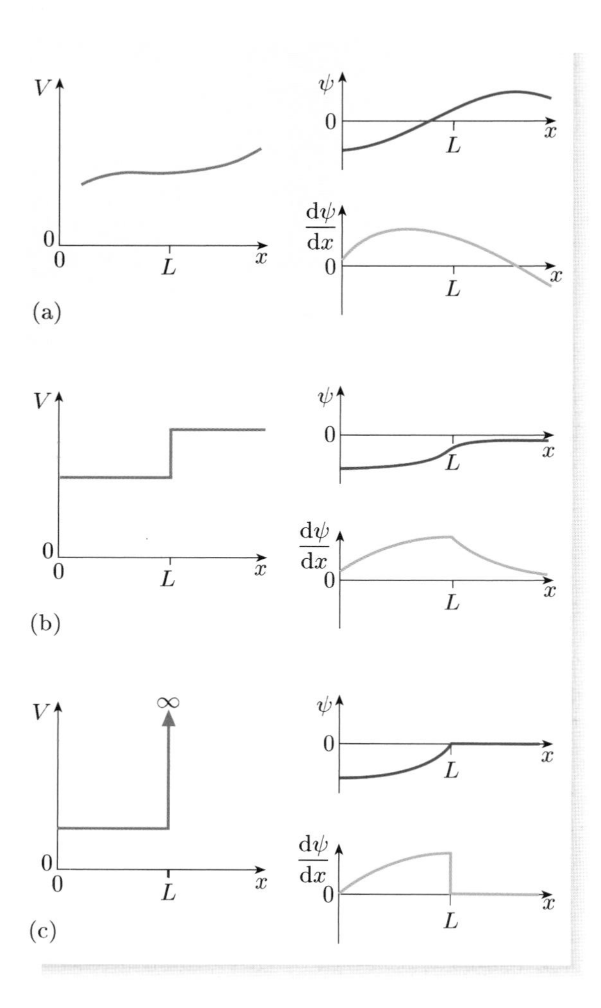

Figure 3.7 A schematic illustration of the continuity boundary conditions: (a) $\psi(x)$ and $\mathrm{d}\psi/\mathrm{d}x$ are both continuous functions provided that $V(x)$ remains finite; (b) this conclusion is unaffected by finite jumps in $V(x)$; (c) an infinite discontinuity in $V(x)$ produces a finite discontinuity in $\mathrm{d}\psi/\mathrm{d}x$, which implies that $\psi(x)$ has a sharp kink, but remains continuous.

Continuity boundary conditions

The eigenfunction $\psi(x)$ is always continuous. The first derivative $\mathrm{d}\psi/\mathrm{d}x$ is continuous in regions where the potential energy function is finite; it need not be continuous at points where the potential energy function becomes infinite.

These conditions are illustrated schematically in Figure 3.7. It is worth noting that they depend on whether the potential energy function is finite, but not on whether it is continuous. The functions $\psi(x)$ and $\mathrm{d}\psi/\mathrm{d}x$ are continuous at points where the potential energy function has a *finite* jump; $\mathrm{d}\psi/\mathrm{d}x$ can only be discontinuous at points where $V(x)$ becomes *infinite*.

The continuity boundary conditions apply at all points in space, but they become particularly useful at points where we need to match expressions for $\psi(x)$ in two neighbouring regions. In the case of a one-dimensional infinite square well, this is necessary at the points $x = 0$ and $x = L$, which mark the walls of the well. Since the potential energy function goes to infinity at these points, we only insist that the eigenfunction $\psi(x)$ remains continuous; there is no restriction on $\mathrm{d}\psi/\mathrm{d}x$. Since $\psi(x)$ is equal to zero outside the well, the continuity boundary conditions give

$$\psi(0) = 0 \quad \text{and} \quad \psi(L) = 0. \tag{3.12}$$

In the next section we shall impose these conditions on the general solutions we found earlier, and examine their consequences.

Technical note: continuity boundary conditions apply to *energy eigenfunctions* and so to *stationary-state* wave functions; they do not necessarily apply to *all* wave functions.

3.1.5 Applying the boundary conditions: energy quantization

Returning to the general solution $\psi(x) = A\sin(kx) + B\cos(kx)$ of the time-independent Schrödinger equation inside the well, and applying the boundary condition $\psi(0) = 0$, it is immediately apparent (since $\sin 0 = 0$ and $\cos 0 = 1$) that

$$B = 0.$$

Then, applying the other boundary condition, $\psi(L) = 0$, it follows that

$$A\sin(kL) = 0. \tag{3.13}$$

We are not interested in cases where $A = 0$, since requiring that $A = 0$ *and* $B = 0$ would leave us with no solution at all. So what Equation 3.13 really tells us is that $\sin(kL) = 0$. The only solutions to this equation are $kL = n\pi$, where n is an integer (i.e. a whole number). We can rule out the possibility $n = 0$, since this would give $\psi(x) = 0$ everywhere. We can also restrict attention to positive n, since we have taken k to be positive (see Exercise 3.4). It therefore follows that $kL = n\pi$ for $n = 1, 2, 3, \ldots$.

Of course, this is actually a condition on the acceptable values of k, since L is a fixed parameter of the system, describing the width of the well. Thus, even though the time-independent Schrödinger equation has solutions corresponding to all positive values of k, the only ones consistent with the continuity boundary conditions correspond to certain discrete values of k that we can denote $k_1, k_2, k_3, \ldots$ etc. where

$$k_n = \frac{n\pi}{L} \quad \text{for } n = 1, 2, 3, \ldots \text{ etc.} \tag{3.14}$$

This is very significant, since it leads to restrictions on the energy eigenfunctions and energy eigenvalues. In particular, it follows from Equation 3.10 that the allowed energy eigenfunctions inside the well are of the form

$$\psi_n(x) = A_n \sin\left(\frac{n\pi x}{L}\right) \quad \text{for } n = 1, 2, 3, \ldots \text{ etc.,} \tag{3.15}$$

where $A_1, A_2, A_3, \ldots$ etc. are constants. Similarly, Equation 3.11 implies that the corresponding energy eigenvalues $E_1, E_2, E_3, \ldots$ etc. are given by

$$E_n = \frac{n^2\pi^2\hbar^2}{2mL^2} = \frac{n^2h^2}{8mL^2} \quad \text{for } n = 1, 2, 3, \ldots \text{ etc.} \tag{3.16}$$

Recall that $\hbar = h/2\pi$.

This is a momentous result. Because the **quantum number** n is restricted to integer values, it shows that the acceptable energy eigenvalues are discrete — i.e. separate and distinct — in this case taking the values $E_1 = \pi^2\hbar^2/2mL^2$, $E_2 = 4\pi^2\hbar^2/2mL^2$, $E_3 = 9\pi^2\hbar^2/2mL^2$, and so on. These are the possible energy levels of the system. The lowest energy level is E_1. According to quantum mechanics, no particle in a one-dimensional infinite square well can have less than this amount of energy. In particular, there is no state of zero energy. The fact that the energy levels are discrete is an example of **energy quantization**.

Quantization does not come from the time-independent Schrödinger equation itself, but rather from the imposed boundary conditions and the fact that we are dealing with a confined system of limited size, L. If we allowed the box to get larger and larger by increasing L, the difference in energy between any two neighbouring levels would approach zero as L approached infinity. Under these circumstances our particle in a box would behave increasingly like a free particle, which can have any positive energy. Energy quantization is an immensely important feature of confined quantum systems. Even so, you should not make the mistake of thinking that it is somehow the heart of quantum physics. The example of the free particle makes it clear that there are quantum systems that do not have discrete energy levels. So, despite its importance, the essence of quantum physics is not to be found in the discreteness of energy.

Exercise 3.5 Show that the difference between any two neighbouring energy levels in a one-dimensional infinite square well of width L does indeed approach zero as L approaches infinity. ■

3.1.6 The normalization condition

Another condition must be imposed in order to obtain sensible wave functions. The physical requirement that the particle must be found somewhere, combined with Born's rule for the probability distribution of position, leads to the *normalization condition* for the wave function $\Psi(x,t)$ that was discussed in Section 2.4.1. We require that

$$\int_{-\infty}^{\infty} |\Psi(x,t)|^2 \, dx = 1. \tag{3.17}$$

We are interested in stationary-state wave functions of the form $\Psi_n(x,t) = \psi_n(x)e^{-iE_nt/\hbar}$. In this case,

$$|\Psi_n(x,t)|^2 = |\psi_n(x)e^{-iE_nt/\hbar}|^2 = |\psi_n(x)|^2|e^{-iE_nt/\hbar}|^2 = |\psi_n(x)|^2,$$

which leads to a normalization condition for the energy eigenfunctions $\psi_n(x)$:

Eigenfunction normalization condition

$$\int_{-\infty}^{\infty} |\psi_n(x)|^2 \, dx = 1. \tag{3.18}$$

In the case of a one-dimensional infinite square well with walls at $x = 0$ and $x = L$, we know that the energy eigenfunction $\psi_n(x)$ must be zero outside the well, so the eigenfunction normalization condition of Equation 3.18 becomes

$$\int_0^L |\psi_n(x)|^2 \, dx = 1. \tag{3.19}$$

Applying this condition to the eigenfunctions of Equation 3.15 we obtain

$$\int_0^L |\psi_n(x)|^2 \, dx = \int_0^L \left| A_n \sin\left(\frac{n\pi x}{L}\right) \right|^2 dx = 1,$$

which may be rewritten as

$$|A_n|^2 \int_0^L \sin^2\left(\frac{n\pi x}{L}\right) dx = 1. \tag{3.20}$$

Most of the integrals in this course can be evaluated by changing the variable of integration and using the list of standard integrals printed inside the back cover of the book. In the present case, we substitute $y = n\pi x/L$, from which it follows that $x = (L/n\pi)y$ and so $dx = (L/n\pi)\,dy$. We must also transform the limits of integration, noting that $x = 0$ corresponds to $y = 0$ and that $x = L$ corresponds to $y = n\pi$. Hence Equation 3.20 becomes

$$|A_n|^2 \frac{L}{n\pi} \int_0^{n\pi} \sin^2 y \, dy = 1.$$

According to the table of integrals inside the back cover, the definite integral in this expression has the value $n\pi/2$, so we get

$$|A_n|^2 \frac{L}{n\pi} \frac{n\pi}{2} = 1.$$

It follows that $|A_n| = \sqrt{2/L}$. Remembering that each A_n may be complex, this means that

$$A_n = \sqrt{\frac{2}{L}}\, e^{i\delta},$$

where δ is a real constant. Phase factors, such as $e^{i\delta}$, which multiply entire eigenfunctions and wave functions, never have any physical consequences, and can be chosen arbitrarily (as discussed in Chapter 2). We shall make the simplest choice by setting $\delta = 0$ so that $e^{i\delta} = 1$. Hence we can say that for each value of n

$$A_n = \sqrt{\frac{2}{L}}. \tag{3.21}$$

Substituting this value into Equation 3.15, and remembering that the eigenfunctions vanish outside the well, we see that the normalized energy eigenfunction $\psi_n(x)$ corresponding to the eigenvalue E_n is

$$\psi_n(x) = \sqrt{\frac{2}{L}} \sin\left(\frac{n\pi x}{L}\right) \quad \text{for } 0 \le x \le L \tag{3.22}$$

$$\psi_n(x) = 0 \quad \text{elsewhere.} \tag{3.23}$$

Figure 3.8 shows the first four of these energy eigenfunctions ψ_1 to ψ_4, together with the corresponding energy levels E_1 to E_4. Note that, as expected, the allowed energy eigenfunctions are continuous but sharply kinked at the walls of the well due to the infinite discontinuity in the potential energy function at these points. Also note that the separation between neighbouring energy levels increases with increasing n due to the factor of n^2 in the expression for E_n (Equation 3.16).

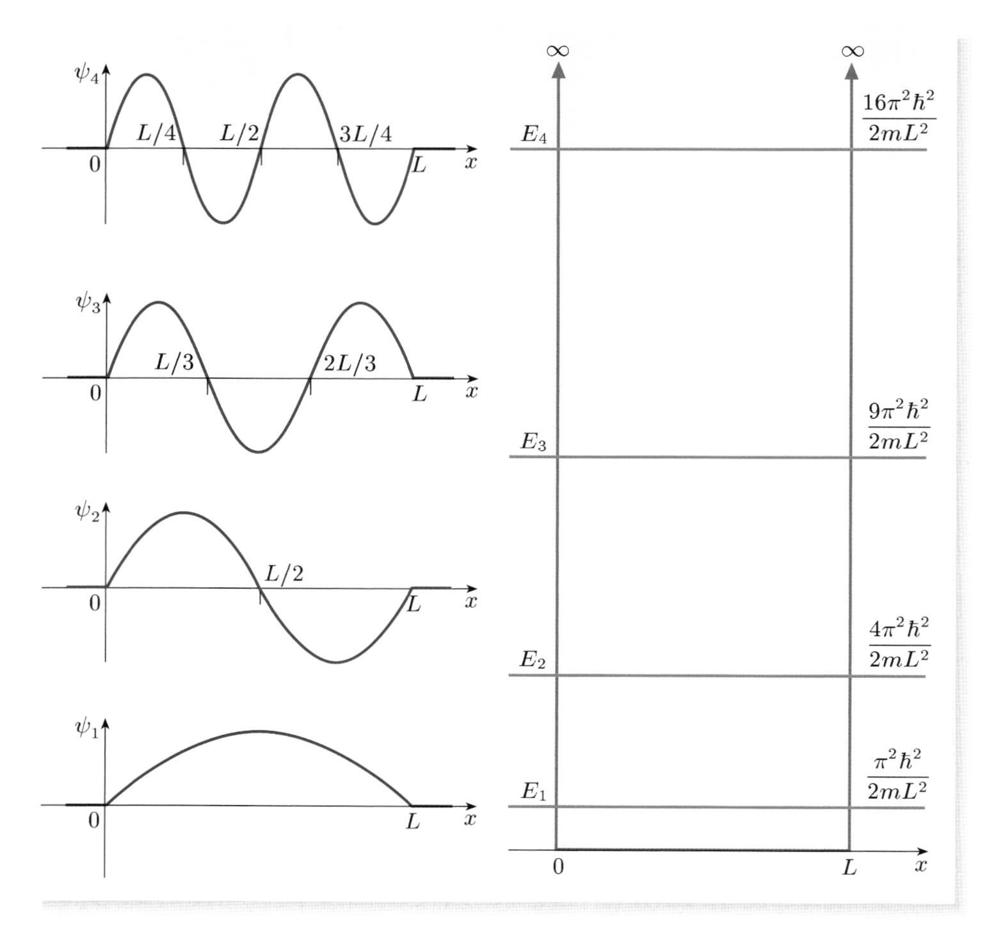

Figure 3.8 Plots of the normalized energy eigenfunctions $\psi_1(x)$, $\psi_2(x)$, $\psi_3(x)$ and $\psi_4(x)$, together with the corresponding energy eigenvalues E_1, E_2, E_3 and E_4, which are the first four energy levels in a one-dimensional infinite square well. We are treating the eigenfunctions as real functions, though any of them could be multiplied by a complex phase factor of the form $e^{i\delta}$ that would not affect the calculation of any measurable quantity.

3.1.7 The stationary-state wave functions

Having found the energy eigenvalues E_n, and energy eigenfunctions $\psi_n(x)$, we can now write down the stationary-state wave functions. For any positive integer,

n, the stationary-state wave function is

$$\Psi_n(x,t) = \sqrt{\frac{2}{L}}\,\sin\left(\frac{n\pi x}{L}\right)\mathrm{e}^{-\mathrm{i}E_nt/\hbar} \quad \text{for } 0 \le x \le L \tag{3.24}$$

$$\Psi_n(x,t) = 0 \quad \text{elsewhere.} \tag{3.25}$$

The constant $\sqrt{2/L}$ ensures that the wave function is normalized. This wave function provides a complete description of the state in which the particle has a definite energy $E_n = n^2\pi^2\hbar^2/2mL^2$. Due to the nature of $\psi_n(x)$ (see Figure 3.8), $\Psi_n(x,t)$ is always equal to zero at the ends of the box. This wave function is also equal to zero at $(n-1)$ points *inside* the box. These $(n-1)$ points are called the **nodes** of the wave function (the zeroes at the ends of the box are not counted as nodes). Thus, the ground state wave function, with $n = 1$, has no nodes.

Each stationary-state wave function, of the kind specified by Equation 3.24, describes a complex *standing wave*. A **standing wave** is one that oscillates without propagating through space. All points in the disturbance that constitutes a standing wave oscillate in phase with one another, with the same frequency but different amplitudes. The nodes of a standing wave remain fixed.

To understand this in more detail, it is helpful to split the wave function into its real and imaginary parts. This is done by noting that $\mathrm{e}^{-\mathrm{i}E_nt/\hbar} = \cos(E_nt/\hbar) - \mathrm{i}\sin(E_nt/\hbar)$, giving

$$\mathrm{Re}(\Psi_n(x,t)) = \sqrt{\frac{2}{L}}\sin\left(\frac{n\pi x}{L}\right)\cos\left(\frac{E_nt}{\hbar}\right) \quad \text{for } 0 \le x \le L \tag{3.26}$$

$$\mathrm{Im}(\Psi_n(x,t)) = -\sqrt{\frac{2}{L}}\sin\left(\frac{n\pi x}{L}\right)\sin\left(\frac{E_nt}{\hbar}\right) \quad \text{for } 0 \le x \le L. \tag{3.27}$$

In each case the right-hand side represents a sinusoidal function of position, $\sin(n\pi x/L)$, scaled at each point by a time-dependent factor. The time-dependent factor is itself sinusoidal and oscillates with a period $T_n = 2\pi\hbar/E_n$. Various 'snapshots' of the real part of the wave function $\Psi_1(x,t)$, taken at different stages during its period T_1, are shown in Figure 3.9. If you imagine these snapshots forming an animated sequence you will get a good idea of what a standing wave is like. Different stationary-state wave functions, with different values of n, have different numbers of nodes and different periods. As n increases, so does the energy of the state and the number of nodes, while the period of oscillation, T_n, decreases.

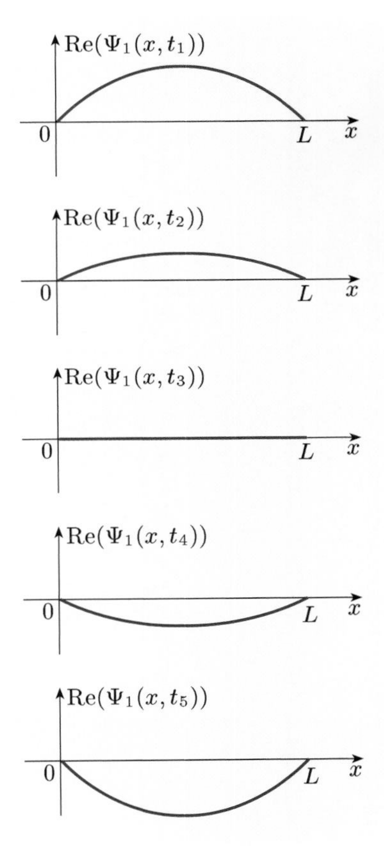

Figure 3.9 Snapshots of the real part of the ground state wave function $\Psi_1(x,t)$ from $t_1 = 0$ to $t_5 = T_1/2$. This is a standing wave with no nodes.

Figure 3.9 emphasizes the time-dependence of the wave function. However, it is equally important to appreciate that $\Psi_1(x,t)$ describes a stationary state in which the probability density $|\Psi_1(x,t)|^2$ is independent of time. Remember, that the diagrams in Figure 3.9 only show the real part of the wave function. No matter how the real part of the wave function $\Psi_1(x,t)$ changes at any particular value of x there will always be compensating changes in the imaginary part to guarantee that $|\Psi_1(x,t)|^2$ is independent of time, and the same is true for any stationary-state wave function, $\Psi_n(x,t)$.

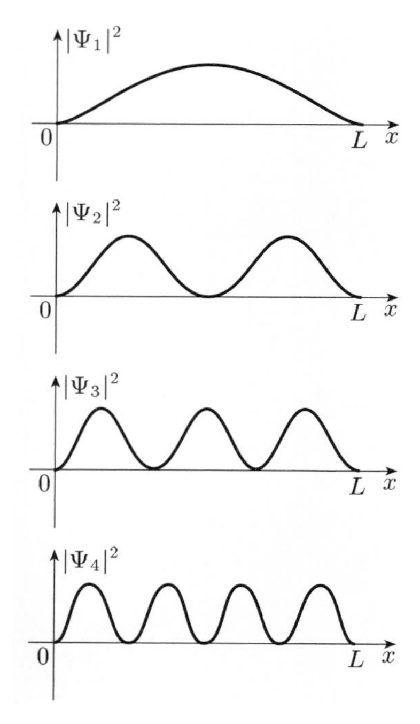

Figure 3.10 Plots of the probability density $|\Psi_n(x,t)|^2$ for $n = 1, 2, 3$ and 4. Note that in each case there are points at which the probability density is zero. The points at which $|\Psi_n(x,t)|^2 = 0$ correspond to the ends of the box and the nodes of $\Psi_n(x,t)$.

Exercise 3.6 Verify that the quantity $|\Psi_n(x,t)|^2$ is independent of time by using its equivalence to $\left[\text{Re}(\Psi_n(x,t))\right]^2 + \left[\text{Im}(\Psi_n(x,t))\right]^2$. ■

Figure 3.10 gives plots of the probability density, $|\Psi_n(x,t)|^2$, for $n = 1, 2, 3$ and 4. The probability of finding the particle in a small region of length δx, centred on x, is given by $|\Psi_n(x,t)|^2\,\delta x$. This remains independent of time.

It is worth giving special consideration to the points of zero probability density in Figure 3.10. In classical physics, a particle with positive kinetic energy trapped in a one-dimensional infinite square well would bounce back and forth between the impenetrable walls. In quantum physics, it is entirely inappropriate to think of a particle in a stationary state as bouncing back and forth at all. If you were to forget this fact, you might wonder how a particle can move from one side of the well to the other when there is a point of zero probability density in the way. How can a particle that is on the left side of the well move to the right side of the well without travelling through the point where the wave function vanishes and the corresponding probability density is zero? Even asking such a question reveals a failure to recognize just how radical quantum mechanics is.

In quantum mechanics, a particle has no position until it is measured, so it makes no sense to speak of the particle as 'moving' from place to place. There is no problem associated with the particle having to 'travel through' the nodal point, since it does not have a velocity any more than it has a position. The particle's position or velocity might be measured in an appropriate experiment, but any value obtained in this way only tells us about the properties of the particle immediately *after* the measurement is taken. Before the measurement, the particle is in a state described by a wave function $\Psi_n(x,t)$. This wave function gives the most complete description that it is possible to have of the state of the system, but it does not assign definite values to either the position or the velocity of the particle.

Exercise 3.7 As the above discussion indicates, a stationary-state wave function such as $\Psi_n(x,t)$ does not in any sense allow us, moment by moment, to 'track' a moving particle. Consequently we *cannot* say that the coordinate x that appears in the argument of $\Psi_n(x,t)$ represents the position of the particle at time t. What, then, does x represent? ■

3.1.8 Alternative descriptions of the well

Our description of a one-dimensional infinite square well of width L involved some special choices:

1. We took the potential energy function to be zero inside the well.
2. We placed the walls of the well at $x = 0$ and at $x = L$.

Neither of these assumptions is essential. The potential energy inside the well could be given any constant value, since this would still correspond to zero force inside the well, and the walls of the well could be placed at $x = x_0$ and $x = x_0 + L$, where x_0 is any coordinate along the x-axis. In this section we shall look at the effects of changing our description of the well. This will change our description of the energy eigenvalues, energy eigenfunctions and stationary-state wave functions, but it will not change the physical behaviour of the particle in the well, which cannot depend on our arbitrary choices.

Changing the zero of potential energy

In classical physics we have an arbitrary choice of where to place the zero of potential energy. Redefining the zero of potential energy has no influence on the behaviour of a system because only *differences* in potential energy are physically significant. The same is true in quantum mechanics.

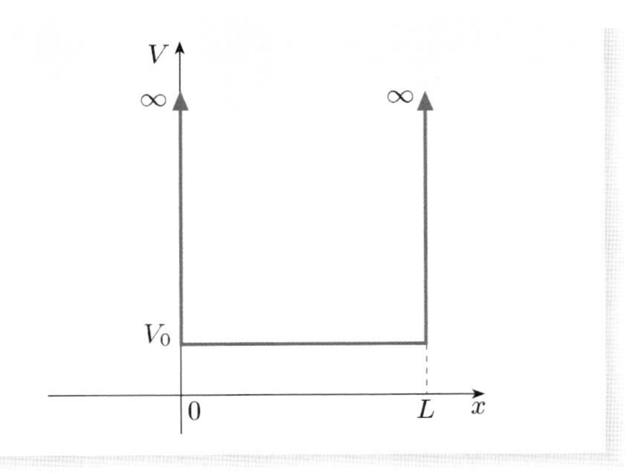

Figure 3.11 The potential energy function for a one-dimensional infinite square well when the potential energy associated with the bottom of the well has the value V_0.

Suppose that we take the potential energy at the bottom of the well to have the constant value $V_0 > 0$ (Figure 3.11). What difference does this make to our descriptions? In this case, the time-independent Schrödinger equation inside the well takes the form

$$-\frac{\hbar^2}{2m}\frac{\mathrm{d}^2\psi}{\mathrm{d}x^2} + V_0\psi = E\psi(x) \quad \text{for } 0 \le x \le L,$$

which can be rearranged to give

$$-\frac{\hbar^2}{2m}\frac{\mathrm{d}^2\psi}{\mathrm{d}x^2} = (E - V_0)\psi(x) \quad \text{for } 0 \le x \le L.$$

Comparing with Equation 3.7, we see that everything will be the same as before *except that* the energy E in our previous description must now be replaced by $E - V_0$. It follows that the energy levels of the well, E_n, obey the equation

$$E_n - V_0 = \frac{n^2\pi^2\hbar^2}{2mL^2} \quad \text{for } n = 1, 2, 3, \ldots$$

In other words all the energy levels are raised by the same constant value, V_0. This makes no difference to the observed behaviour of the particle. For example, the frequency of light emitted when the particle makes a transition from one energy level to another depends only on the *difference* between the energies involved, and this is independent of V_0.

The stationary-state wave functions still vanish outside the well. Inside the well, we adapt Equation 3.24 by replacing E_n by $E_n - V_0$ to obtain

$$\Psi_n(x,t) = \sqrt{\frac{2}{L}}\sin\left(\frac{n\pi x}{L}\right)\mathrm{e}^{-\mathrm{i}(E_n - V_0)t/\hbar} \quad \text{for } 0 \le x \le L.$$

Exercise 3.8 Why is it possible to say that these modified stationary-state wave functions describe the same behaviour as the original stationary-state wave function in Equation 3.24? ■

Since the location of the zero of energy makes no difference to the physical behaviour of a particle in an infinite well, we generally adopt the simplest choice, placing the zero of energy at the bottom of the well. However, the decision about where to place the walls of the well is less clear-cut. There is an alternative choice which has some advantages since it emphasizes the symmetry of the situation.

Choosing a symmetric well

Instead of placing the walls of the well at $x = 0$ and $x = L$, we can choose them to be $x = -L/2$ and $x = +L/2$ giving a potential energy function of the form

$$V(x) = 0 \quad \text{for } -L/2 \leq x \leq L/2,$$

$$V(x) = \infty \quad \text{elsewhere.}$$

Such a potential energy function is symmetrical either side of the origin, i.e. $V(-x) = V(x)$, so we will refer to it as a **symmetric well**.

We could obviously set about solving the time-independent Schrödinger equation for a symmetric infinite square well, following much the same procedure as used above for an unsymmetrical well. However, there is no need to go back to scratch in this way. Instead, we can take our previous solutions and adapt them by making a suitable change of variable.

If we let $x' = x + L/2$, a symmetric well that extends from $x = -L/2$ to $x = L/2$ will extend from $x' = 0$ to $x' = L$. Expressed in terms of the *primed* coordinate, x', this well is the same as that considered earlier, so we know that its energy eigenfunctions take the form

$$\sqrt{\frac{2}{L}} \sin\left(\frac{n\pi x'}{L}\right).$$

Returning to the unprimed coordinate, it follows that the energy eigenfunctions in the *symmetric* well are

$$\psi_n(x) = \sqrt{\frac{2}{L}} \sin\left(\frac{n\pi(x + L/2)}{L}\right) = \sqrt{\frac{2}{L}} \sin\left(\frac{n\pi x}{L} + \frac{n\pi}{2}\right).$$

This sine function on the right-hand side can be expanded to give

$$\sin\left(\frac{n\pi x}{L} + \frac{n\pi}{2}\right) = \sin\left(\frac{n\pi x}{L}\right)\cos\left(\frac{n\pi}{2}\right) + \cos\left(\frac{n\pi x}{L}\right)\sin\left(\frac{n\pi}{2}\right).$$

Further simplification is possible because n is an integer.

If n is even: $$\sin\left(\frac{n\pi x}{L} + \frac{n\pi}{2}\right) = \sin\left(\frac{n\pi x}{L}\right)\cos\left(\frac{n\pi}{2}\right),$$

If n is odd: $$\sin\left(\frac{n\pi x}{L} + \frac{n\pi}{2}\right) = \cos\left(\frac{n\pi x}{L}\right)\sin\left(\frac{n\pi}{2}\right).$$

The factors $\cos(n\pi/2)$ and $\sin(n\pi/2)$ that appear in these expressions are equal to $+1$ or -1. This means that they are phase factors (that is, factors of unit modulus). Such phase factors have no physical significance when they multiply eigenfunctions, and we are always free to ignore them. We therefore conclude that the energy eigenfunctions inside a symmetric infinite square well of length L can be written as

$$\psi_n(x) = \sqrt{\frac{2}{L}}\sin\left(\frac{n\pi x}{L}\right) \quad \text{for even values of } n, \tag{3.28}$$

$$\psi_n(x) = \sqrt{\frac{2}{L}}\cos\left(\frac{n\pi x}{L}\right) \quad \text{for odd values of } n. \tag{3.29}$$

The energy eigenvalues are the same as before and the stationary-state wave functions inside the well are

$$\Psi_n(x,t) = \sqrt{\frac{2}{L}}\sin\left(\frac{n\pi x}{L}\right)\mathrm{e}^{-\mathrm{i}E_nt/\hbar} \quad \text{for even values of } n, \tag{3.30}$$

$$\Psi_n(x,t) = \sqrt{\frac{2}{L}}\cos\left(\frac{n\pi x}{L}\right)\mathrm{e}^{-\mathrm{i}E_nt/\hbar} \quad \text{for odd values of } n. \tag{3.31}$$

Outside the well, the eigenfunctions and wave functions are equal to zero.

Worked Example 3.1

Show that the functions $\psi_n(x)$ in Equations 3.28 and 3.29 satisfy the time-independent Schrödinger equation and find the corresponding energy eigenvalues. State the continuity boundary conditions for a symmetric infinite square well of width L and verify that these functions satisfy them.

Essential skill

Verifying that a given function satisfies the time-independent Schrödinger equation together with appropriate boundary conditions

Solution

Differentiating either of the functions in Equations 3.28 and 3.29 twice with respect to x gives

$$\frac{\mathrm{d}^2\psi_n}{\mathrm{d}x^2} = -\left(\frac{n\pi}{L}\right)^2\psi_n(x).$$

Substituting either of the functions into the time-independent Schrödinger equation then gives

$$-\frac{\hbar^2}{2m}\left(-\frac{n^2\pi^2}{L^2}\right)\psi_n(x) = E\psi_n(x),$$

which is satisfied provided that $E = n^2\pi^2\hbar^2/2mL^2$. This is the energy eigenvalue corresponding to the energy eigenfunction $\psi_n(x)$.

For a symmetric infinite square well with walls at $x = -L/2$ and $x = L/2$, the continuity boundary conditions are

$$\psi(-L/2) = \psi(L/2) = 0.$$

The eigenfunctions in Equations 3.28 and 3.29 satisfy these boundary conditions because

$$\sin\left(-\frac{n\pi}{2}\right) = 0, \quad \sin\left(\frac{n\pi}{2}\right) = 0 \quad \text{for } n \text{ even,}$$

and

$$\cos\left(-\frac{n\pi}{2}\right) = 0, \quad \cos\left(\frac{n\pi}{2}\right) = 0 \quad \text{for } n \text{ odd.}$$

The energy eigenfunctions of a symmetric infinite square well have some special features that turn out to be very useful in calculations. These features relate to their *evenness* or *oddness*.

Odd and even functions

$f(x)$ is said to be an **even function** of x if $f(-x) = f(x)$ for all x.

$f(x)$ is said to be an **odd function** of x if $f(-x) = -f(x)$ for all x.

Most functions are neither even nor odd. However, $\cos x$ is an even function, and $\sin x$ is an odd function so, in a symmetric infinite square well, *every* energy eigenfunction is either even or odd. If n is odd, the eigenfunction $\psi_n(x)$ is an even function; if n is even, the eigenfunction $\psi_n(x)$ is an odd function. In particular, the ground state (with $n = 1$) is an *even* function, the first excited state (with $n = 2$) is an odd function, and so on. Moreover, the evenness and oddness of the eigenfunctions alternates as we climb upwards through the energy levels.

It turns out (although we shall not prove it here) that all one-dimensional *symmetric* wells (i.e. wells with $V(-x) = V(x)$) share these properties: all their energy eigenfunctions are either even or odd, the ground state is always even, and the evenness and oddness alternate as n increases.

3.2 Two- and three-dimensional infinite square wells

In this section we generalize the results obtained for the one-dimensional infinite square well, first to two dimensions and then to three dimensions. The two-dimensional case can be thought of as a sort of stepping stone to quantum dots, which confine particles in all three dimensions.

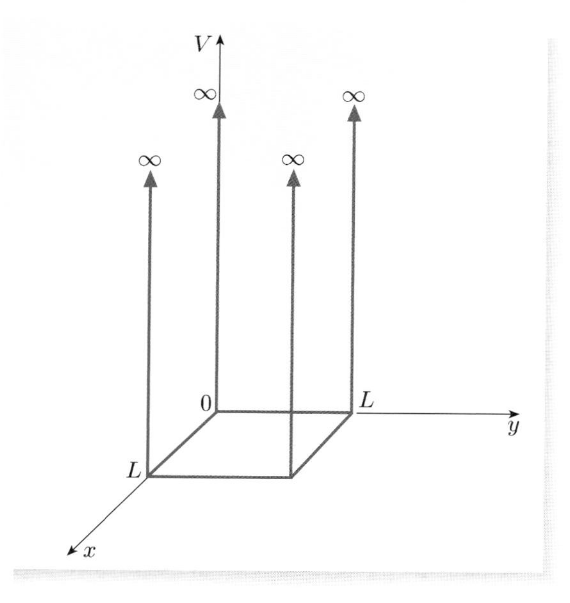

Figure 3.12 The potential energy function for the two-dimensional infinite square well.

3.2.1 The two-dimensional infinite square well

A two-dimensional infinite square well is one that restricts two of the coordinates of a particle. It is the kind of box that confines the electrons in a quantum wire in the dimensions perpendicular to the axis of the wire. For simplicity, we will assume that the box occupies a square region in two dimensions. Nevertheless, it is worth noting that the term 'square well' refers to the abrupt rise of the potential energy function, not to the region in which the potential energy is zero — it would be perfectly possible to have a *square* well that occupies a *circular* region in two dimensions, although we will not discuss such a case here.

The potential energy function we shall investigate takes the form

$$V(x,y) = 0 \quad \text{for } 0 \le x \le L \text{ and } 0 \le y \le L, \qquad (3.32)$$

$$V(x,y) = \infty \quad \text{elsewhere.} \qquad (3.33)$$

For convenience, we have aligned the walls of the well with the x- and y-coordinate axes, chosen one corner of the well to be at the origin, and taken the potential energy to be zero inside the well. This potential energy function is illustrated in Figure 3.12.

Starting from a classical system consisting of a particle of mass m, moving under the influence of the two-dimensional infinite square well $V(x,y)$, the usual procedure leads to the Schrödinger equation

$$\mathrm{i}\hbar\frac{\partial\Psi(x,y,t)}{\partial t} = -\frac{\hbar^2}{2m}\left(\frac{\partial^2\Psi}{\partial x^2} + \frac{\partial^2\Psi}{\partial y^2}\right) + V(x,y)\Psi(x,y,t). \tag{3.34}$$

Because we are interested in finding the energy levels of the system, we will again look for stationary-state solutions. In this case we start by seeking solutions of the form

$$\Psi(x,y,t) = \psi(x,y)T(t) \quad \text{for } 0 \le x \le L \text{ and } 0 \le y \le L, \tag{3.35}$$

$$\Psi(x,y,t) = 0 \quad \text{elsewhere.} \tag{3.36}$$

We shall now concentrate on the region *inside* the well. Using the potential energy function of Equation 3.32, and substituting the product wave function of Equation 3.35 into Schrödinger's equation, the method of separation of variables leads to the following equations for the region inside the well:

$$\mathrm{i}\hbar\frac{\mathrm{d}T}{\mathrm{d}t} = ET(t), \tag{3.37}$$

$$-\frac{\hbar^2}{2m}\left(\frac{\partial^2\psi}{\partial x^2} + \frac{\partial^2\psi}{\partial y^2}\right) = E\psi(x,y). \tag{3.38}$$

As always, Equation 3.37 is satisfied by a function of the form

$$T(t) = \mathrm{e}^{-\mathrm{i}Et/\hbar},$$

while Equation 3.38 is the *time-independent Schrödinger equation* for this problem. This is the eigenvalue equation for energy; its solutions $\psi(x,y)$ are the energy eigenfunctions and the corresponding values of E are the energy eigenvalues.

Equation 3.38 can be further separated by looking for solutions of the form $\psi(x,y) = X(x)Y(y)$. Substituting this expression into Equation 3.38 and dividing both sides by $X(x)Y(y)$ leads to the equation

$$-\frac{\hbar^2}{2m}\left(\frac{1}{X(x)}\frac{\mathrm{d}^2X(x)}{\mathrm{d}x^2} + \frac{1}{Y(y)}\frac{\mathrm{d}^2Y(y)}{\mathrm{d}y^2}\right) = E.$$

The left-hand side of this equation can be written as the sum of a function of x and a function of y, with the property that the value of this sum is constant. Given our freedom to choose the values of x and y independently, how can the sum of a function of x and a function of y remain constant? The only way is if the functions are themselves constants (not necessarily the same). This allows us to write

$$-\frac{\hbar^2}{2m}\frac{1}{X(x)}\frac{\mathrm{d}^2X(x)}{\mathrm{d}x^2} = E_X, \tag{3.39}$$

$$-\frac{\hbar^2}{2m}\frac{1}{Y(y)}\frac{\mathrm{d}^2Y(y)}{\mathrm{d}y^2} = E_Y, \tag{3.40}$$

where E_X and E_Y are constants that satisfy $E_X + E_Y = E$.

Now, Equations 3.39 and 3.40 have the same form as the one-dimensional time-independent Schrödinger equation that was solved in Section 3.1. Even the boundary conditions are the same, since $X(x)$ must vanish at $x = 0$ and $x = L$ and $Y(y)$ must vanish at $y = 0$ and $y = L$. So, inside the well,

$$X(x) = A_{n_x}\sin\left(\frac{n_x\pi x}{L}\right) \quad \text{and} \quad Y(y) = A_{n_y}\sin\left(\frac{n_y\pi y}{L}\right), \tag{3.41}$$

where A_{n_x} and A_{n_y} are constants and n_x and n_y are positive integers with

$$\frac{n_x^2\pi^2\hbar^2}{2mL^2} = E_X \quad \text{and} \quad \frac{n_y^2\pi^2\hbar^2}{2mL^2} = E_Y. \tag{3.42}$$

The energy eigenfunction inside the well can therefore be written as

$$\psi_{n_x,n_y}(x,y) = X(x)Y(y) = A_{n_x,n_y}\sin\left(\frac{n_x\pi x}{L}\right)\sin\left(\frac{n_y\pi y}{L}\right),$$

where $A_{n_x,n_y} = A_{n_x}A_{n_y}$ is a constant that can be found by normalizing the eigenfunction. In the present context, this implies that

$$\int_{-\infty}^{\infty}\int_{-\infty}^{\infty}|\psi_{n_x,n_y}(x,y)|^2\,\mathrm{d}x\,\mathrm{d}y = 1,$$

that is,

$$\left|A_{n_x,n_y}\right|^2\int_{-\infty}^{\infty}\sin^2\left(\frac{n_x\pi x}{L}\right)\mathrm{d}x\int_{-\infty}^{\infty}\sin^2\left(\frac{n_y\pi y}{L}\right)\mathrm{d}y = 1. \tag{3.43}$$

Each of the definite integrals on the left-hand side of Equation 3.43 is of a form already met in Section 3.1.4, and is equal to $\sqrt{L/2}$. Taking the normalization constant to be real and positive, we conclude that the normalized energy eigenfunctions are

$$\psi_{n_x,n_y}(x,y) = \frac{2}{L}\sin\left(\frac{n_x\pi x}{L}\right)\sin\left(\frac{n_y\pi y}{L}\right), \tag{3.44}$$

inside the well, with $\psi_{n_x,n_y}(x,y) = 0$ outside. Using $E = E_X + E_Y$, together with Equation 3.42, the corresponding energy eigenvalues are

Note that in two dimensions, the energy eigenfunctions, energy eigenvalues and stationary-state wave functions are labelled by *two* quantum numbers, n_x and n_y.

$$E_{n_x,n_y} = \frac{(n_x^2+n_y^2)\pi^2\hbar^2}{2mL^2}. \tag{3.45}$$

Finally, putting everything together, the normalized stationary-state wave functions are

$$\Psi_{n_x,n_y}(x,y,t) = \frac{2}{L}\sin\left(\frac{n_x\pi x}{L}\right)\sin\left(\frac{n_y\pi y}{L}\right)\mathrm{e}^{-\mathrm{i}E_{n_x,n_y}t/\hbar}, \tag{3.46}$$

inside the well, with $\Psi_{n_x,n_y}(x,y,t) = 0$ outside.

Figure 3.13 provides snapshots of the real part of one of these wave functions, for the case $n_x = 1$ and $n_y = 2$. The snapshots are taken at five instants t_1 to t_5, starting with t_1. Note that the real part of the wave function again takes the form of a standing wave, though this time it is two-dimensional, like the standing wave that might be observed on the head of a rectangular drum or some other stretched membrane with fixed edges. Notice also that $\Psi_{1,2}(x, y, t)$ contains a *nodal line* at $y = L/2$, along which the wave function is permanently zero. Such lines correspond to the nodal points (nodes) that were present in the one-dimensional case.

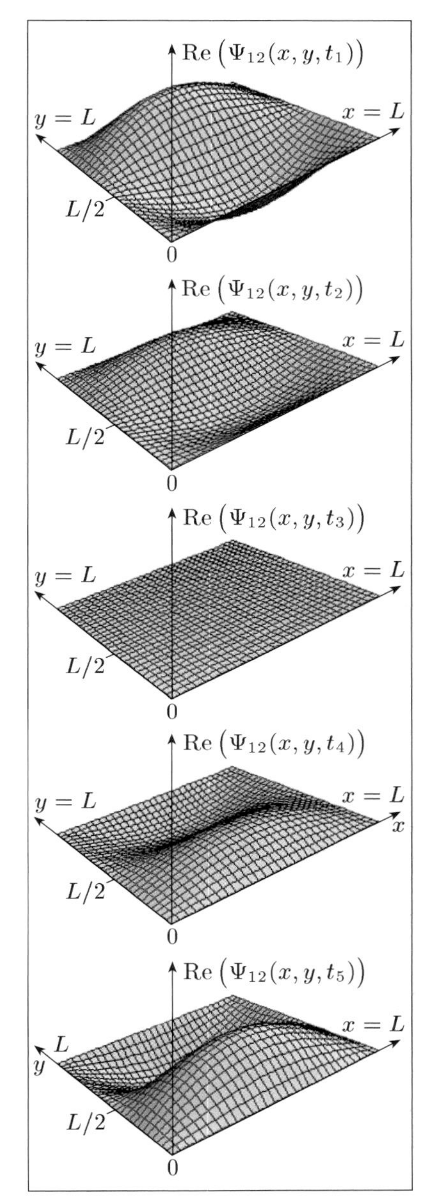

Figure 3.13 Snapshots of the real part of $\Psi_{1,2}(x, y, t)$ at five instants, starting with t_1.

The probability of finding the particle in a small rectangular region of area $\delta x \times \delta y$ centred on the point (x, y) is $|\Psi_{1,2}(x, y, t)|^2 \, \delta x \, \delta y$. The corresponding probability density $|\Psi_{1,2}(x, y, t)|^2$ is independent of time, since the time-dependent phase factor has unit modulus, just as in the one-dimensional case. This is what you would expect for a stationary state.

Exercise 3.9 Show that for a stationary state described by the wave function $\Psi_{n_x,n_y}(x, y, t)$, the corresponding probability density is independent of time. ■

3.2.2 Degeneracy

The first few energy eigenvalues of a two-dimensional infinite square well are shown in the energy-level diagram of Figure 3.14 (overleaf). Here, each state is specified by an ordered pair of integer quantum numbers (n_x, n_y), and the corresponding energy level is indicated by the position of a horizontal line on a vertical scale. Each state is distinct, for example the wave function $\Psi_{1,2}(x, y, t)$ is not the same function as $\Psi_{2,1}(x, y, t)$. Yet some of the states have the same energy. This is the case for the states described by $\Psi_{1,2}(x, y, t)$ and $\Psi_{2,1}(x, y, t)$ for example, and for the states corresponding to $\Psi_{1,3}(x, y, t)$ and $\Psi_{3,1}(x, y, t)$.

The occurrence of different quantum states (i.e. states described by different wave functions) with the same energy is referred to as **degeneracy**. An energy level that corresponds to two or more quantum states is said to be **degenerate**. Moreover, states that share the same energy are said to be *degenerate with one another*.

The number of states that correspond to an energy level is called the *degree of degeneracy* of that level. As Figure 3.14 shows, several of the energy levels are *doubly degenerate* or *two-fold degenerate*; that is to say there are two different states that correspond to that particular energy eigenvalue. This degree of degeneracy is not altogether surprising when the two-dimensional well occupies a square region of space. It is a consequence of the square's symmetry, and could be removed by distorting the square region of space into a rectangular one. However, other degeneracies are also possible. For instance, in the case of the square well occupying a square region of space, the level of energy $50\pi^2\hbar^2/2mL^2$ is three-fold degenerate since 50 may be written as $7^2 + 1^2$, $1^2 + 7^2$ and $5^2 + 5^2$.

It is interesting to note that degeneracy is a phenomenon that first reveals itself in two dimensions. There are no degeneracies in one-dimensional wells.

Exercise 3.10 Which other states of the particle in a two-dimensional infinite square well are degenerate with the $n_x = 7$, $n_y = 4$ state? What is the degree of degeneracy of the corresponding energy level? ■

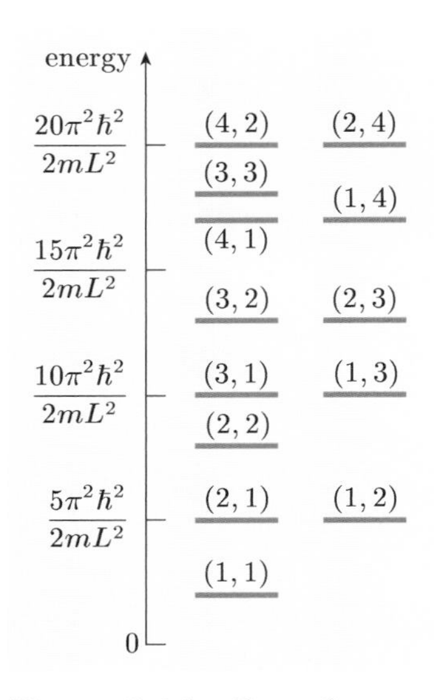

Figure 3.14 Part of an energy-level diagram for a particle in a two-dimensional infinite square well.

3.2.3 The three-dimensional infinite square well

A three-dimensional infinite square well is one that traps the particle in all three dimensions. It is the kind of box that confines the electrons in a quantum dot. Given our discussion of one- and two-dimensional infinite square wells, the quantum mechanics of a particle of mass m in a three-dimensional infinite square well should hold few surprises, so we proceed by stating the conclusions.

For simplicity, we assume that the box occupies a cubic region, with sides of length L. Then the states are described by three quantum numbers n_x, n_y and n_z, each of which may take any of the values $1, 2, 3, 4, \ldots$ etc. The energy level corresponding to any specific choice of the positive integers n_x, n_y and n_z is

$$E_{n_x,n_y,n_z} = \frac{\left(n_x^2 + n_y^2 + n_z^2\right)\pi^2\hbar^2}{2mL^2}. \tag{3.47}$$

Inside the well, the normalized stationary-state wave function describing the state characterized by n_x, n_y and n_z is

$$\Psi_{n_x,n_y,n_z}(x, y, z, t) = \psi_{n_x,n_y,n_z}(x, y, z)\,\mathrm{e}^{-\mathrm{i}E_{n_x,n_y,n_z}t/\hbar}, \tag{3.48}$$

where $\psi_{n_x,n_y,n_z}(x, y, z)$ is a normalized energy eigenfunction that satisfies the time-independent Schrödinger equation for the problem. The form of that equation, and of $\psi_{n_x,n_y,n_z}(x, y, z)$, depends on the precise location of the square well. If the well occupies the region in which $0 \le x \le L$ and $0 \le y \le L$ and $0 \le z \le L$ then the normalized energy eigenfunction inside the well is

In three dimensions, the energy eigenfunctions, energy eigenvalues and stationary-state wave functions are labelled by *three* quantum numbers n_x, n_y and n_z.

$$\psi_{n_x,n_y,n_z}(x, y, z) = \left(\frac{2}{L}\right)^{3/2} \sin\left(\frac{n_x\pi x}{L}\right) \sin\left(\frac{n_y\pi y}{L}\right) \sin\left(\frac{n_z\pi z}{L}\right). \tag{3.49}$$

Everywhere outside the well, the eigenfunction is equal to zero.

If any of these results (including the normalization constant $(2/L)^{3/2}$) comes as a surprise, you should review the treatment of the two-dimensional infinite square well, or, if necessary, work through the three-dimensional problem in detail for yourself.

Exercise 3.11 (a) Obtain an expression for the difference in energy between the two lowest energy levels of a particle of mass m confined to a cube with sides of length L by a three-dimensional infinite square well potential energy function.

(b) What is the degree of degeneracy of each of these two energy levels?

Exercise 3.12 What are acceptable SI units for the stationary-state wave functions in infinite square wells in one dimension, two dimensions and three dimensions? Explain why your answers are consistent with Born's rule. ■

3.2.4 F-centres and quantum dots

The quantum-mechanical system of a particle trapped in a three-dimensional infinite square well provides a reasonable model for some real physical systems.

We shall briefly consider two of these systems, F-centres and quantum dots, including experimental results that support some of our theoretical analysis.

First consider a crystal — that is, a regular arrangement of atoms. A simple example is common salt, sodium chloride (NaCl), in which the chlorine atoms (Cl) form the arrangement shown in Figure 3.15, while equal numbers of smaller sodium atoms (Na) fill the spaces between. In fact, each sodium atom loses one electron, and each chlorine atom gains one, so it is more accurate to say that the crystal is composed of *ions* (denoted by Na^+ and Cl^-). The electrical attraction between oppositely-charged ions is responsible for keeping the sodium chloride crystal bound together: it is said to be an **ionic crystal**.

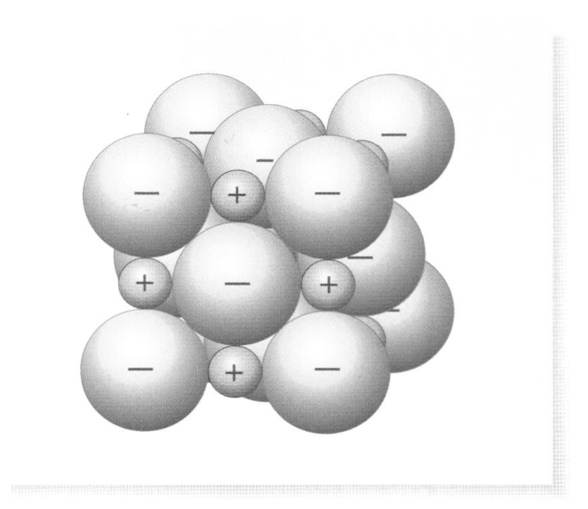

Figure 3.15 The arrangement of sodium (+) and chlorine (−) ions in crystalline salt (NaCl).

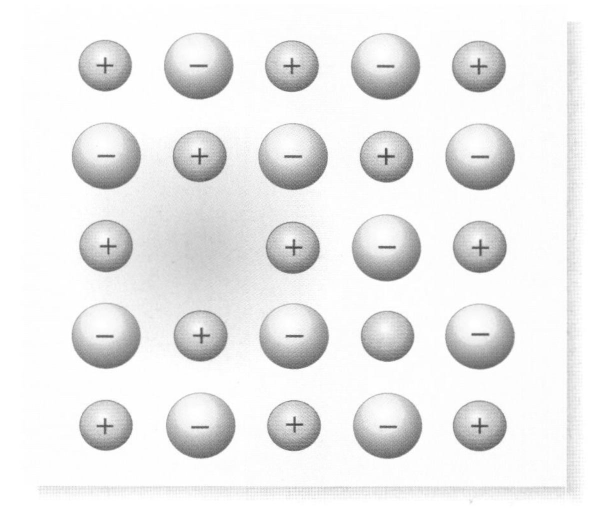

Figure 3.16 An F-centre created by the ejection of a negative ion and its replacement by an electron. The shading schematically indicates the probability density for the electron in one particular state of the F-centre.

A perfect crystal of sodium chloride would be transparent, but perfect crystals are rare. Table salt looks white because each grain contains many defects in its otherwise regular arrangement of ions. One special type of defect, called an **F-centre**, is especially interesting for us. It is possible for a negative chlorine ion to be ejected from the crystal and have its place taken by an electron, as indicated in Figure 3.16. The electron, whose closest neighbours are positive ions (Na^+), behaves as a particle trapped in a three-dimensional well. The exact shape of the well depends on the details of the crystal, but the crucial point is that the electron has its own characteristic set of energy levels, and transitions between these energy levels allow the F-centre to absorb electromagnetic radiation. In the case of NaCl, the absorbed radiation is visible light, in the violet part of the spectrum. Consequently, a sufficient concentration of F-centres, produced by irradiating an NaCl crystal with gamma rays for example, has the effect of turning the salt crystals an orange–brown colour, since that is the complement of violet (i.e. the result of subtracting violet from white light).

The term 'F-centre' comes from the German word 'Farbe' meaning colour.

We can roughly model an F-centre by treating it as a cubical three-dimensional infinite square well containing a trapped particle. The length L of a side of the cube is taken to be proportional to the separation of two similar ions in the crystal (the so-called *lattice spacing*). In sodium chloride and similar crystals, any visible light absorbed by an F-centre corresponds to transitions between neighbouring energy levels. One obvious prediction, emerging from Equation 3.47, is that the energy difference between neighbouring levels is proportional to $1/L^2$. So, if we take a range of different ionic crystals, all with F-centres, we might expect the characteristic frequency of absorption to be proportional to the inverse square of the lattice spacing. Figure 3.17 shows that this prediction is quite accurate. There are some minor discrepancies, but this is only to be expected bearing in mind that an electron in an F-centre is in a well of finite depth, and experiences a force when it is inside the well.

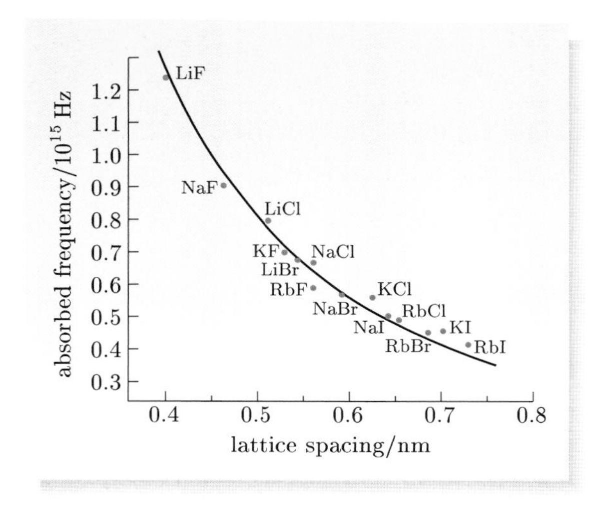

Figure 3.17 A plot of the frequency of absorbed light against lattice spacing for a range of ionic crystals containing F-centres. These data are broadly consistent with the relationship: frequency $\propto 1/(\text{lattice spacing})^2$.

A similar pattern is observed in quantum dots. The variation of colour shown in Figure 3.2 arises because small quantum dots have energy levels that are far apart, leading to high frequencies of absorbed light, while larger quantum dots have energy levels that are closer together, leading to smaller frequencies of absorbed light. Quantum dots are more complicated than F-centres because the trapped electrons are now in a semiconducting medium, rather than in empty space. This complicates the picture, as you will see later in the course, but in many cases the frequency of emitted light is a linear function of $1/L^2$, which can be regarded as being a consequence of Equation 3.47.

3.3 The finite square well

The wells considered so far have all had infinite depth, ensuring that the particle remains completely trapped inside their walls. In this final section, we consider a one-dimensional well of finite depth. You will see that the particle then stands some chance of being found outside the well.

As before, we start by giving a classical description of the system; we then write down Schrödinger's equation, solve it for stationary states subject to appropriate boundary and normalization conditions, and interpret the resulting wave functions.

3.3.1 The system and its Schrödinger equation

Figure 3.18 shows the potential energy function of a one-dimensional **finite square well** of width L and depth V_0. In this case we have chosen to consider a *symmetric* well with boundaries at $x = \pm L/2$. In another change of convention we have chosen to associate zero potential energy with the region *outside* the well. This is a reasonable convention in the present case, because it corresponds to saying that the particle has zero potential energy when it is far from the influence of the well (and therefore free).

According to classical physics, a particle trapped in the well will have a negative potential energy $-V_0$, a positive kinetic energy in the range $0 \leq E_{\text{kin}} < V_0$, and a negative total energy in the range $-V_0 \leq E < 0$. Energy considerations then imply that the regions outside the box, where the particle's total energy would have to be positive, are classically forbidden to the negative-energy particle bound inside the well.

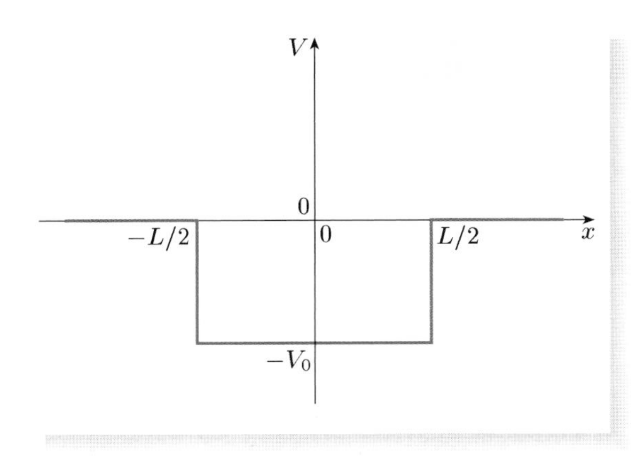

Figure 3.18 The potential energy function of a symmetric one-dimensional finite square well of width L and depth V_0.

The potential energy function may be written as

$$V(x) = -V_0 \quad \text{for } -L/2 \leq x \leq L/2, \tag{3.50}$$

$$V(x) = 0 \qquad \text{elsewhere.} \tag{3.51}$$

Apart from the potential energy function, the only other ingredient of our classical system is the particle itself which is characterized by a mass m. So the parameters describing the system are m, V_0 and L, which are all positive quantities.

The quantum-mechanical description of the system is based on Schrödinger's equation

$$\mathrm{i}\hbar\frac{\partial\Psi(x,t)}{\partial t} = -\frac{\hbar^2}{2m}\frac{\partial^2\Psi(x,t)}{\partial x^2} + V(x)\Psi(x,t). \tag{3.52}$$

We are looking for stationary-state solutions of the form $\Psi(x,t) = \psi(x)\mathrm{e}^{-\mathrm{i}Et/\hbar}$, where $E < 0$ and the function $\psi(x)$ satisfies the time-independent Schrödinger equation,

$$-\frac{\hbar^2}{2m}\frac{\mathrm{d}^2\psi(x)}{\mathrm{d}x^2} + V(x)\psi(x) = E\psi(x). \tag{3.53}$$

Substituting the finite square well potential energy function (Equations 3.50 and 3.51) into this equation gives

$$-\frac{\hbar^2}{2m}\frac{\mathrm{d}^2\psi(x)}{\mathrm{d}x^2} - V_0\psi(x) = E\psi(x) \quad \text{for } -L/2 \leq x \leq L/2, \tag{3.54}$$

$$-\frac{\hbar^2}{2m}\frac{\mathrm{d}^2\psi(x)}{\mathrm{d}x^2} = E\psi(x) \quad \text{elsewhere.} \tag{3.55}$$

We now have two differential equations to solve, one inside the well and the other outside. The allowed solutions are subject to boundary conditions. Firstly, we must ensure that $\psi(x)$ does not diverge as $x \to \pm\infty$. Secondly, we must make sure that the solutions to Equation 3.54 join on smoothly to the solutions to Equation 3.55. Although the potential energy function has abrupt steps, it does not become infinite, so the continuity boundary conditions require that both the eigenfunction $\psi(x)$ and its derivative $\mathrm{d}\psi/\mathrm{d}x$ are continuous. In particular,

$$\psi(x) \quad \text{is continuous at } x = \pm L/2, \tag{3.56}$$

$$\frac{\mathrm{d}\psi(x)}{\mathrm{d}x} \quad \text{is continuous at } x = \pm L/2. \tag{3.57}$$

In the next section we will solve Equations 3.54 and 3.55 *subject to these boundary conditions.*

3.3.2 Solving the time-independent Schrödinger equation

The region outside the well

Let us start by investigating the part of the eigenfunction outside the well, which must satisfy Equation 3.55. At first sight this seems to be similar to the problem we solved inside the infinite square well, so we might expect to simply modify the solution given in Equations 3.10 and 3.11. This approach would lead to a solution of the form

$$\psi(x) = A\sin(kx) + B\cos(kx) \quad \text{for } x < -L/2 \text{ or } x > L/2,$$

where A and B are arbitrary constants and $k = \sqrt{2mE}/\hbar$. However, in the present case the energy eigenvalues E are negative, so k is imaginary. Under these circumstances it is better to take a different approach; to consider each part of the outside region separately and write the general solution to Equation 3.55 in the form

$$\psi(x) = A\mathrm{e}^{\alpha x} + B\mathrm{e}^{-\alpha x} \quad \text{for } x < -L/2, \tag{3.58}$$

$$\psi(x) = F\mathrm{e}^{\alpha x} + G\mathrm{e}^{-\alpha x} \quad \text{for } x > L/2, \tag{3.59}$$

where A, B, F, and G, are arbitrary constants, and α is the positive constant

$$\alpha = \frac{\sqrt{2m(-E)}}{\hbar}. \tag{3.60}$$

Remember, we are considering the case $E < 0$.

Exercise 3.13 Show that $\psi(x) = A\mathrm{e}^{\alpha x} + B\mathrm{e}^{-\alpha x}$ is the general solution to the time-independent Schrödinger equation for the one-dimensional finite square well in the region $x < -L/2$, provided that $\alpha = \sqrt{2m(-E)}/\hbar$. ■

Section 8.2 explains what is meant by the general solution of a differential equation.

One reason for writing the solutions outside the well in the form of Equations 3.58 and 3.59 is that we can easily identify how these functions behave far from the well. In Equation 3.58 (where x is negative) the term $B\mathrm{e}^{-\alpha x}$ diverges as $x \to -\infty$, so we conclude that $B = 0$. Similarly, in Equation 3.59 (where x is positive) the term $F\mathrm{e}^{\alpha x}$ diverges as $x \to \infty$, so we conclude that $F = 0$. Imposing these conditions we can now say that outside the well

$$\psi(x) = A\mathrm{e}^{\alpha x} \quad \text{for } x < -L/2, \tag{3.61}$$

$$\psi(x) = G\mathrm{e}^{-\alpha x} \quad \text{for } x > L/2. \tag{3.62}$$

In addition, we are dealing with a symmetric well. In any symmetric well, the eigenfunctions are always either even or odd functions, so we conclude that:

for even eigenfunctions $\quad A = G$,

for odd eigenfunctions $\quad A = -G$.

When dealing with one-dimensional wells, it is always possible to choose the energy eigenfunctions to be real functions. This allows us to draw meaningful graphs of $\psi(x)$ for the one-dimensional finite square well, just as we did in the earlier case of the infinite square well. Figure 3.19 shows the relevant graphs for the parts of the eigenfunction that lie outside the well in cases of even eigenfunctions and odd eigenfunctions, as described by Equations 3.61 and 3.62.

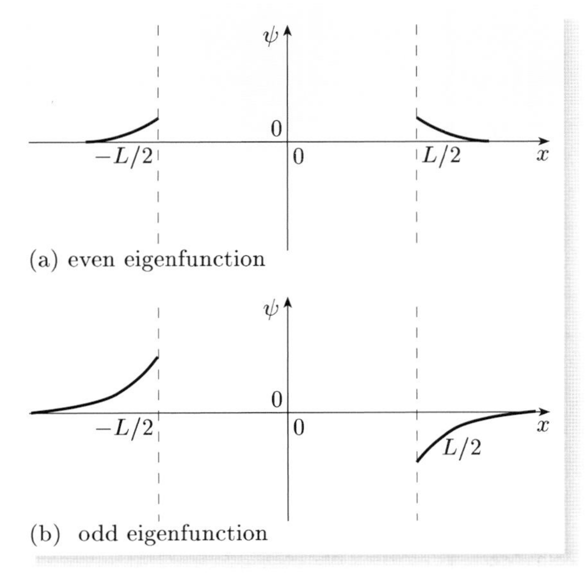

Figure 3.19 The behaviour of: (a) an even eigenfunction and (b) an odd eigenfunction outside a one-dimensional finite square well. In each case, the eigenfunction decreases exponentially with distance from the walls of the well. For the purposes of graph plotting, A has been taken to be positive.

These diagrams assume that A and G are non-zero and you will see shortly that this is always the case. They reveal that the energy eigenfunction for a state with negative energy extends outside the well. This implies that, in quantum physics, there is a possibility of detecting the particle in the classically forbidden region. This purely quantum phenomenon, often referred to as **barrier penetration**, has no classical analogue. It is another indication of the vast gulf that separates classical from quantum physics.

Outside the well, the energy eigenfunctions approach zero *exponentially* with increasing distance from the walls of the well. Whatever the value of the eigenfunction at the edge of the well, its value will have decreased by a factor of $1/\mathrm{e} = 0.368$ at a distance $1/\alpha = \hbar/\sqrt{2m(-E)}$ beyond the wall. This implies that the eigenfunctions corresponding to the lowest energy eigenvalues (i.e. those with the largest values of the positive quantity $-E$) will decrease most rapidly with distance from the walls of the well.

The region inside the well

To find the energy eigenfunctions *inside* the well, we must solve Equation 3.54. In this case we can expect the solutions to be similar to those we met in the case of the *symmetric* infinite square well, so we shall immediately state that:

$$\text{for even eigenfunctions} \quad \psi(x) = D\cos(kx), \tag{3.63}$$

$$\text{for odd eigenfunctions} \quad \psi(x) = C\sin(kx), \tag{3.64}$$

where C and D are arbitrary constants and x lies in the range from $-L/2$ to $L/2$.

Substituting Equation 3.63 or 3.64 into Equation 3.54 shows that the positive quantity k must be related to the energy eigenvalue E and the well depth V_0 as follows:

$$k = \frac{\sqrt{2m(E+V_0)}}{\hbar}. \tag{3.65}$$

Although E is negative, the quantity $E + V_0$ is positive; it is the energy above the bottom of the well, and may be thought of as the analogue of the classical kinetic energy.

Figure 3.20 shows two of these eigenfunctions in the region inside the well. There is no reason to suppose that the eigenfunctions vanish at the edges of the well; what matters is that they should join on smoothly to the eigenfunctions outside the well plotted in Figure 3.19.

Matching solutions at boundaries

Following our experience with the infinite square well, we can expect the allowed energy eigenfunctions and eigenvalues to arise from the continuity boundary conditions imposed at the edges of the well. The requirements that $\psi(x)$ and $\mathrm{d}\psi/\mathrm{d}x$ should each be continuous at $x = -L/2$ lead to the following conditions for the even eigenfunctions:

Remember $\cos(-x) = \cos x$.

$$A\mathrm{e}^{-\alpha L/2} = D\cos(kL/2), \tag{3.66}$$

$$\alpha A\mathrm{e}^{-\alpha L/2} = kD\sin(kL/2), \tag{3.67}$$

and to the following conditions for the odd eigenfunctions:

$$Ae^{-\alpha L/2} = -C\sin(kL/2), \tag{3.68}$$

Remember $\sin(-x) = -\sin x$.

$$\alpha Ae^{-\alpha L/2} = kC\cos(kL/2). \tag{3.69}$$

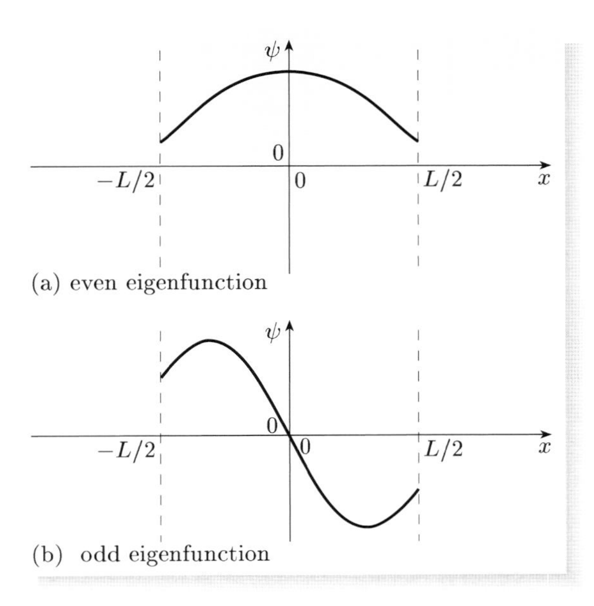

Figure 3.20 The behaviour of: (a) an even eigenfunction and (b) an odd eigenfunction inside a one-dimensional finite square well. In each case, the eigenfunction has an appreciable non-zero value at the walls.

Dividing each side of Equation 3.67 by the corresponding side of Equation 3.66, and rearranging, gives

$$\text{for even eigenfunctions:} \quad k\tan(kL/2) = \alpha. \tag{3.70}$$

Similarly, dividing each side of Equation 3.69 by the corresponding side of Equation 3.68, and rearranging, gives

$$\text{for odd eigenfunctions:} \quad k\cot(kL/2) = -\alpha. \tag{3.71}$$

Recalling that $\alpha = \sqrt{2m(-E)}/\hbar$ and $k = \sqrt{2m(E + V_0)}/\hbar$, we see that Equations 3.70 and 3.71 implicitly determine the energy eigenvalues, E in terms of the parameters m, V_0 and L that characterize the system. There is no way of solving these equations algebraically to get an explicit expression for E. However, it is relatively easy to solve them *numerically*, using a computer. When this is done, a finite number of negative energy eigenvalues is obtained: E_1, $E_2, \ldots$. For a one-dimensional finite square well, there is always at least one eigenvalue, but the precise number depends on the width and depth of the well and the mass of the particle.

For each energy eigenvalue, E_n, a corresponding eigenfunction $\psi_n(x)$ can be found from Equations 3.61, 3.62, 3.63 and 3.64. Given the value of E_n, we can determine α and k and go back to Equations 3.66 and 3.68 to find the ratios D/A and C/A. Finally, we impose the normalization condition by requiring that

$$\int_{-\infty}^{\infty} |\psi_n(x)|^2 \,dx = 1. \tag{3.72}$$

Note that, the integral runs from $-\infty$ to ∞ rather than from $-L/2$ to $L/2$. This is because we can no longer presume that a particle with negative energy is restricted to the well, even though it would be so restricted in classical physics. The normalization condition (Equation 3.72) leads to a value for the constant A and hence to an explicit expression for the normalized energy eigenfunction, $\psi_n(x)$.

Figure 3.21 shows the energy eigenvalues and the corresponding eigenfunctions that result from one choice of m, V_0 and L. For the particular parameters chosen, there are three energy levels below the top of the well, but this number will vary from well to well. Many of the features of the eigenfunctions are similar to those of an infinite square well. In the lowest energy level, E_1, the eigenfunction $\psi_1(x)$ is even and has no nodes. In the next energy level, E_2, the eigenfunction $\psi_2(x)$ is odd and has one node. The alternating pattern of evenness and oddness continues, with each successive eigenfunction gaining one extra node. In general, the nth eigenfunction has $(n-1)$ nodes and is an even/odd function according to whether n is odd/even.

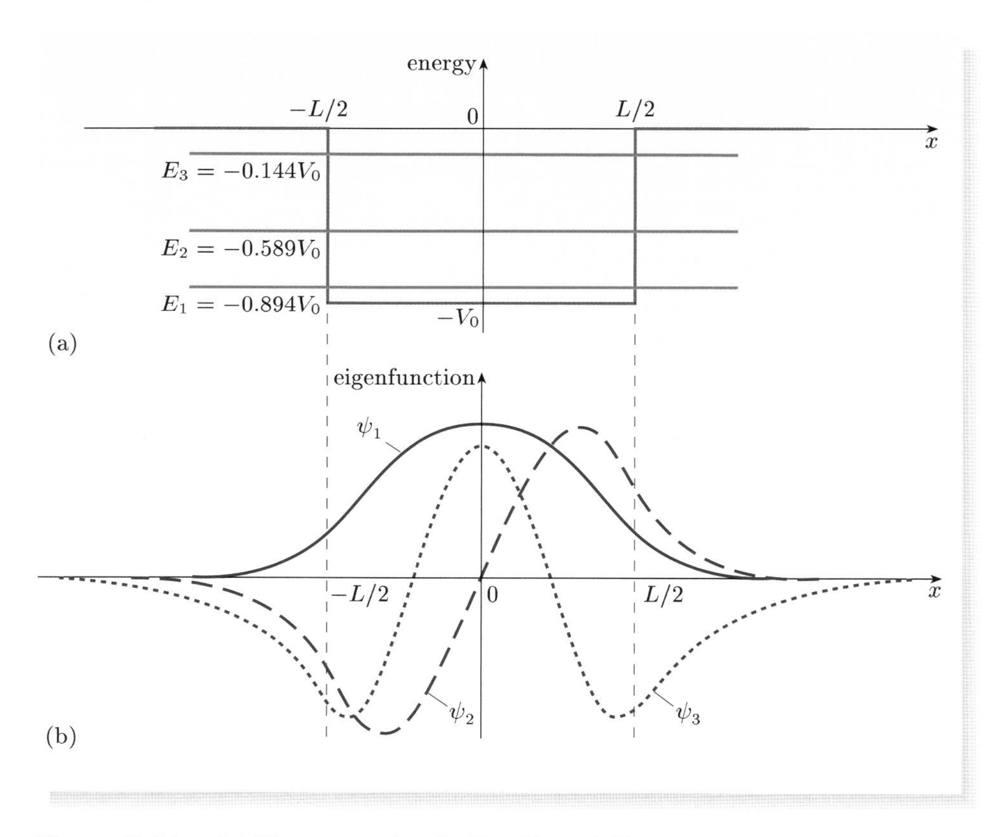

Figure 3.21 (a) The energy levels E_1, E_2 and E_3 of a one-dimensional finite square well; (b) the eigenfunctions ψ_1, ψ_2 and ψ_3 plotted on a single set of axes.

The time-independent probability densities $|\psi_n(x)|^2$ corresponding to these three energy eigenfunctions are shown in Figure 3.22. The probability density drops to zero at each node of the eigenfunction, indicating that there is little chance of finding the particle near these points.

The probability densities for a finite well extend beyond the boundaries of the well, showing that a particle of negative energy may be found outside the well (the phenomenon of barrier penetration). The particle penetrates further into the classically forbidden region as its energy increases. For energy levels very near the top of the well, the particle may even be more likely to be found outside the well than inside it.

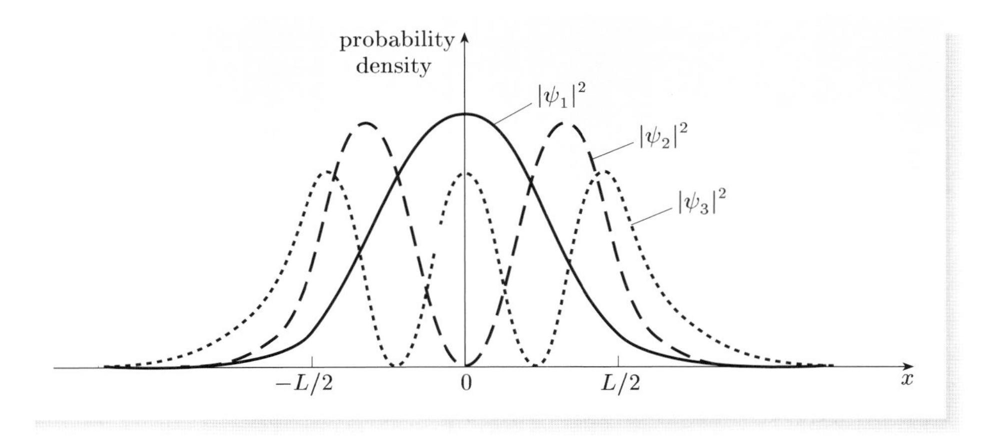

Figure 3.22 The probability densities $|\psi_1(x)|^2$, $|\psi_2(x)|^2$ and $|\psi_3(x)|^2$ for the case shown in Figure 3.21.

Something like this happens in the **deuteron** — a proton and a neutron bound together by attractive nuclear forces. Of course, this is a three-dimensional system, which leads to some differences in detail. Nevertheless, it is possible to devise a model for the deuteron involving a finite square well of depth $38.5\,\mathrm{MeV}$ occupying a spherical region of radius $1.63 \times 10^{-15}\,\mathrm{m}$. Using this model, only one energy level is found below the top of the well. This is the ground state of the deuteron; there are no excited states in which the proton and neutron remain bound together. In the ground state, the eigenfunction describing the separation of the proton and neutron extends far beyond the confines of the well. In fact, the root-mean-square separation of the proton and neutron is $4 \times 10^{-15}\,\mathrm{m}$, and there is about a 50% chance of finding the particles in a classically forbidden region.

So far, we have restricted attention to **bound states**. These are states in which the particle has a vanishingly small probability of being found very far from the well. They are described by normalized wave functions and have negative energies, below the top of the well. You might wonder what happens for positive energies, above the top of the well. It is certainly possible to find energy eigenvalues and eigenfunctions in this region. The eigenfunctions oscillate sinusoidally both inside and outside the well. However, these eigenfunctions carry on oscillating infinitely far from the well, and cannot be normalized. It follows that a single energy eigenfunction cannot describe an unbound particle. Later in this book, you will see that unbound particles are best described by linear combinations of energy eigenfunctions (wave packets).

3.3.3 Computations and generalizations

It should be clear by now that a particle in a one-dimensional finite square well is a more difficult system to analyse than a particle in an infinite square well. This is why we dealt with the infinite square well first, despite the fact that, from a mathematical point of view, the infinite square well is best regarded as a limiting case of the finite square well. To make further investigations of particles in boxes, you should now use the computer package *One-dimensional wells*. The package should give you insight into the way that parameters, such as m, L and V_0, influence the energy eigenvalues and eigenfunctions. The package will also allow you to investigate the solutions that arise for more general one-dimensional wells, such as those shown in Figure 3.23.

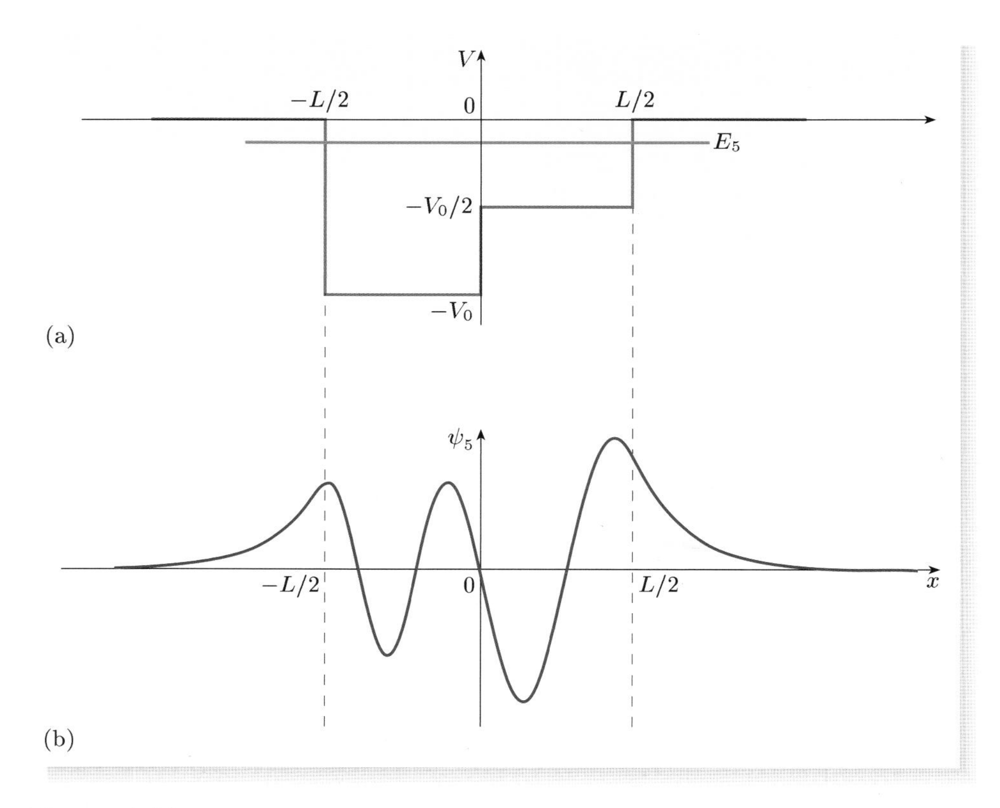

Figure 3.23 A more general type of one-dimensional well: (a) an energy eigenvalue E_5 and (b) the corresponding energy eigenfunction ψ_5.

When using the package, you should bear in mind the following points, which apply to all wells in one-dimension.

- Bound states have discrete energy eigenvalues.
- In one dimension, there is no degeneracy: each energy level corresponds to one eigenfunction.
- In one dimension, energy eigenfunctions may be taken to be real.
- The ground state eigenfunction has no nodes. When arranged in order of increasing energy, each successive eigenfunction has one more node than its predecessor.

- For a symmetric well, each eigenfunction is either an even or an odd function. When arranged in order of increasing energy, this property of evenness or oddness alternates from one eigenfunction to the next.

Install and run the computer package *One-dimensional wells*. Work through the whole of the package following the on-screen guidance notes.

Summary of Chapter 3

Section 3.1 A particle of mass m in a one-dimensional infinite square well, with walls at $x = 0$ and $x = L$, has an infinite set of discrete energy levels

$$E_n = \frac{n^2\pi^2\hbar^2}{2mL^2} \quad \text{for } n = 1, 2, 3, \ldots,$$

with corresponding normalized energy eigenfunctions

$$\psi_n(x) = \sqrt{\frac{2}{L}}\sin\left(\frac{n\pi x}{L}\right) \quad \text{for } 0 \leq x \leq L,$$

$$\psi_n(x) = 0 \quad \text{outside the well.}$$

The stationary-state wave functions

$$\Psi_n(x,t) = \sqrt{\frac{2}{L}}\sin\left(\frac{n\pi x}{L}\right)\mathrm{e}^{-\mathrm{i}E_n t/\hbar} \quad \text{for } 0 \leq x \leq L,$$

are complex standing waves describing states of definite energy, E_n.

These solutions are obtained by separating variables in the Schrödinger equation and solving the time-independent Schrödinger equation, subject to continuity boundary conditions. These conditions require that the energy eigenfunction $\psi(x)$ is continuous everywhere, and that $\mathrm{d}\psi/\mathrm{d}x$ is continuous at points where the potential energy function is finite. At the walls of an infinite square well we only require the continuity of $\psi(x)$, and this leads to energy quantization.

Shifting the walls of the well without altering its width, or assigning a different constant potential energy to the bottom of the well, will change the way the system is described, but not the way it behaves. Each energy eigenfunction in a one-dimensional *symmetric* well is either even or odd.

Section 3.2 The infinite square well can be generalized to two and three dimensions. For a three-dimensional infinite square well that occupies a cubic region of width L, the energy eigenvalues are given by

$$E_{n_x,n_y,n_z} = \frac{(n_x^2 + n_y^2 + n_z^2)\pi^2\hbar^2}{2mL^2}.$$

If the well occupies the region $0 \leq x \leq L$, $0 \leq y \leq L$ and $0 \leq z \leq L$, the corresponding normalized energy eigenfunctions are

$$\psi_{n_x,n_y,n_z}(x,y,z) = \left(\frac{2}{L}\right)^{3/2}\sin\left(\frac{n_x\pi x}{L}\right)\sin\left(\frac{n_y\pi y}{L}\right)\sin\left(\frac{n_z\pi z}{L}\right).$$

A three-dimensional infinite square well can be used to model the energy levels in F-centres and quantum dots. In two and three dimensions, energy levels may be degenerate.

Section 3.3 The time-independent Schrödinger equation for a particle in a finite one-dimensional square well can be solved numerically. For negative energies (below the top of the well) a finite number of energy eigenvalues and eigenfunctions are obtained. Energy quantization arises from the continuity boundary conditions, which now require both $\psi(x)$ and $\mathrm{d}\psi/\mathrm{d}x$ to be continuous at the walls of the well. The eigenfunctions, and the corresponding stationary-state wave functions, extend beyond the walls of the well. This means that there is a non-zero probability of finding the particle outside the well — a characteristic quantum phenomenon called *barrier penetration*. This phenomenon is observed in the case of a deuteron, whose constituent particles have a root-mean-square separation that is about twice the range of the force that binds them together.

Achievements from Chapter 3

After studying this chapter you should be able to:

3.1 Explain the meanings of the newly defined (emboldened) terms and symbols, and use them appropriately.

3.2 Write down the Schrödinger equation and the time-independent Schrödinger equation for the infinite and finite square well in one, two or three dimensions.

3.3 Write down and apply the continuity boundary conditions for energy eigenfunctions.

3.4 Verify that given functions are solutions of the Schrödinger equation and the time-independent Schrödinger equation (subject to certain boundary conditions); in the case of the infinite well, determine the corresponding energy eigenvalues.

3.5 Describe some characteristic features of energy eigenfunctions and eigenvalues, including energy quantization, degeneracy, and barrier penetration.

After studying the computer package One-dimensional wells, *you should also be able to:*

3.6 Describe in qualitative terms the generalization of the ideas you have met in the context of square wells to wells of a more general shape.

Chapter 4 The Heisenberg uncertainty principle

Introduction

Figure 4.1 Heisenberg grafitti.

The Heisenberg uncertainty principle is one of the most celebrated results of quantum mechanics. Many people who have not studied science in any detail have heard of this principle, and think of it as expressing the shifty evasiveness of the microscopic world, cloaking atoms and subatomic particles in a veil of mystery that we are unable to penetrate (Figure 4.1). This is not the scientific view. The final section of this chapter will discuss the uncertainty principle, and you will see that it makes a clear statement about the results of experiments, with consequences that can be tested and confirmed. In order to reach this understanding, we must first examine the extent to which predictions can be made in an unpredictable quantum world.

Chapter 1 emphasized the fact that quantum mechanics is indeterministic. For example, identical uranium nuclei can be in identical states and yet decay at very different times. What is more, identical nuclei in identical states can decay in different ways, with different products (Figure 4.2). This is not because of any differences between the nuclei, or their histories, or their surroundings; rather, it appears to be a fundamental fact about the world we live in. A given cause does not always produce the same effect. Yet there is a pattern behind the randomness. Each uranium nucleus has a definite *probability* to decay, and each *type* of decay in Figure 4.2 is also characterized by a probability. So physics is not brought to a halt by indeterminacy, but it is forced to deal directly with chance and probability. If we are to understand the quantum world, we have no option but to use probabilities, since this is what Nature appears to do.

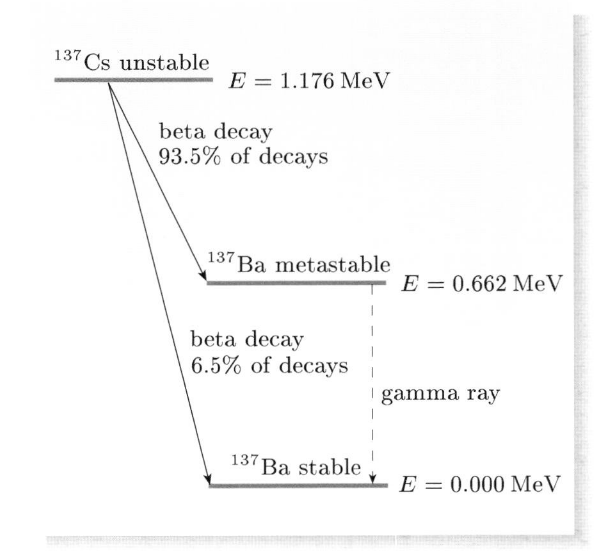

Figure 4.2 An example of the unpredictable nature of quantum mechanics. A nucleus in a definite state can decay in different ways, with different probabilities.

The first half of this chapter will show how quantum mechanics is used to establish the possible outcomes of a measurement, *and their relative probabilities*. This is a crucial issue — as important, in its way, as Schrödinger's equation. Any probability distribution can be characterized by two quantities: its *expectation value* and its *uncertainty*. The expectation value is the mean value you would expect to get in the long run, while the uncertainty gives the spread of values around this mean value. The second half of the chapter will show how expectation values and uncertainties are calculated in quantum mechanics, before explaining how uncertainties appear in the uncertainty principle. At the end of the chapter you will see how the uncertainty principle allows us to provide rough estimates for the size and ground state energy of a hydrogen atom.

Section 8.4 of the *Mathematical toolkit* reviews the concept of probability. You may like to read it as part of your study of this chapter.

4.1 Indeterminacy in quantum mechanics

The great French mathematician Pierre Simon Laplace told Napoleon that he would be able to predict the future of the Universe if he knew the present positions and velocities of all its particles. Napoleon asked what role God had in this Universe, and received the confident reply: 'I have no need of *that* hypothesis'.

Although Laplace greatly exaggerated his powers of calculation, the story does emphasize an important point about Newtonian mechanics — it is deterministic. In Newtonian mechanics it is possible, in principle, to use exact knowledge about the present state of a system to make exact predictions about its future. By contrast, quantum mechanics is indeterministic; we can know as much as it is possible to know about the state of a system and still be unable to predict what will happen if we carry out a measurement on the system. This section will trace the origins of quantum indeterminism. You have met many of the ideas in previous chapters, so the task is partly one of revision.

4.1.1 The wave function and Schrödinger's equation

In wave mechanics, the state of a system is specified by giving its wave function. If we know the wave function of a system at a given time, we know as much as it is possible to know about the state of the system at that time. All predictions about the behaviour of the system are based on knowledge of its wave function.

The wave function depends on the spatial coordinates of all the particles in the system, and also on time. We shall concentrate on the simplest possible case, where a single particle moves along the x-axis. In this case, the wave function $\Psi(x,t)$ is a function of the position coordinate x and the time t. In general, this function is complex, with both real and imaginary parts.

The time-evolution of the wave function is given by Schrödinger's equation,

$$\mathrm{i}\hbar\,\frac{\partial\Psi}{\partial t} = \widehat{\mathrm{H}}\Psi, \tag{4.1}$$

where $\widehat{\mathrm{H}}$ is the Hamiltonian operator of the system. For a single particle moving along the x-axis, the Hamiltonian operator is

$$\widehat{\mathrm{H}} = -\frac{\hbar^2}{2m}\frac{\partial^2}{\partial x^2} + V(x),$$

where $V(x)$ is the potential energy function, and Schrödinger's equation takes the form

$$\mathrm{i}\hbar\,\frac{\partial\Psi}{\partial t} = -\frac{\hbar^2}{2m}\frac{\partial^2\Psi}{\partial x^2} + V(x)\,\Psi(x,t). \tag{4.2}$$

This partial differential equation is of first order with respect to time. Consequently, if we know the wave function at all points *now*, Schrödinger's equation tells us what the wave function will be at all points in the future (and what it was in the past). This time-evolution is entirely predictable, and does not shed any light on the indeterministic nature of quantum mechanics.

4.1.2 Relating the wave function to measurements

Indeterminacy arises from the interpretation of the wave function itself. As an example, consider position measurements. Imagine placing a bank of small Geiger counters all along the x-axis, as in Figure 4.3. We will assume that each Geiger counter is 100% efficient, so that it clicks if it contains a charged particle. Now suppose that a charged particle is in a state described by the wave function $\Psi(x,t)$, which extends over many Geiger counters. Then, if all the counters are switched on simultaneously, one or other of them will click, *but we cannot predict which one.* Instead, we have Born's rule, which states that the probability of detecting the particle at time t in a small interval of length δx, centred on the position x, is given by

$$\text{probability} = |\Psi(x,t)|^2\,\delta x = \Psi^*(x,t)\,\Psi(x,t)\,\delta x.$$

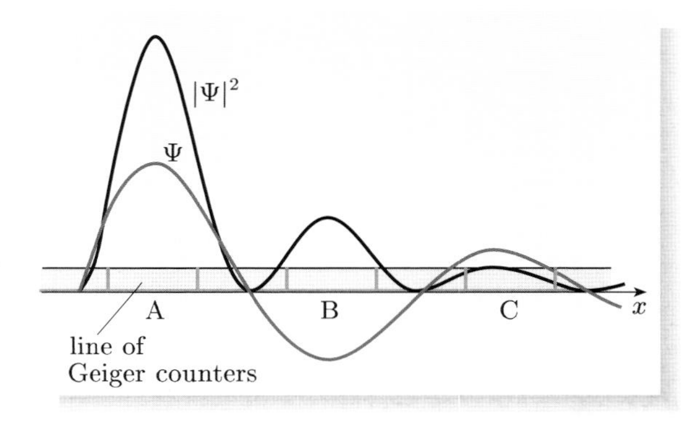

Figure 4.3 Detecting the position of a particle with a bank of Geiger counters.

Figure 4.3 shows both the wave function Ψ and the square of its modulus, $|\Psi|^2$. Using Born's rule, and taking appropriate areas under the graph of $|\Psi|^2$, we can say that, at time t, Geiger counter A is about ten times more likely to click than Geiger counter B, but we cannot say which Geiger counter will actually click. In other words, the wave function allows us to predict the possible experimental outcomes and their relative probabilities, but it does not allow us to predict exactly which outcome will occur in any given instance.

It is also worth considering the situation immediately after the particle has been detected. If Geiger counter C has clicked, we can be sure that the particle is somewhere in the vicinity of this Geiger counter — the particle cannot have strayed too far because it travels at a finite speed (certainly less than the speed of light). Since the square of the modulus of the wave function tells us the probability of detecting the particle in various regions, the wave function describing the particle immediately after the click must be strongly localized around the Geiger counter that happened to click (counter C in this case). So the wave function immediately after the click is drastically different from the wave function before the click. This phenomenon is called the **collapse of the wave function**. We cannot predict exactly how the wave function will collapse. If Geiger counter A had clicked, the wave function immediately after the click would have been strongly localized around counter A. However, it is clear that the act of measurement drastically affects the wave function; *the state of the system after the measurement is not the same as the state before the measurement.*

This has an important consequence for experiments designed to test the predictions of quantum physics. If we take a single system and perform a sequence of measurements on it, the wave function will change unpredictably with each measurement, making comparison between theory and experiment difficult.

It is generally better to start with a large number of *identical* systems all in the *same* state — for example, hydrogen atoms all in the ground state — and perform the *same* measurement of each system. Although each system is in exactly the same state, a spread of results is typically obtained. For example, we could take measurements of position at time t in the state described by the wave function $\Psi(x,t)$ and plot the results as a histogram. In the long run of many measurements, this histogram is expected to follow the shape of the graph of $|\Psi(x,t)|^2$.

It is straightforward to use a wave function to find the probability distribution for position. However, quantum mechanics assigns a much greater role to wave functions than this. In wave mechanics, the wave function gives a *complete* description of the state of a system, and therefore contains information about the probability distributions of other quantities, such as energy and momentum. This information is contained in the shape of the wave function. One of the main aims of this chapter is to explain how this information can be extracted.

4.1.3 Stationary states and energy values

Schrödinger's equation has a set of special solutions called *stationary states* in which the wave function takes the form

$$\Psi_n(x,t) = \psi_n(x)\,\mathrm{e}^{-\mathrm{i}E_nt/\hbar}, \tag{4.3}$$

where $\psi_n(x)$ satisfies the *time-independent Schrödinger equation*

$$\widehat{\mathrm{H}}\,\psi_n(x) = E_n\,\psi_n(x). \tag{4.4}$$

We generally find that this equation gives acceptable solutions only for certain values of E_n, or when E_n lies within certain ranges. The special values of E_n that allow a solution to be found are called *energy eigenvalues*, and the corresponding functions $\psi_n(x)$ are called *energy eigenfunctions*. Figure 4.4 shows some of the energy eigenvalues and eigenfunctions for a particle in a one-dimensional infinite square well.

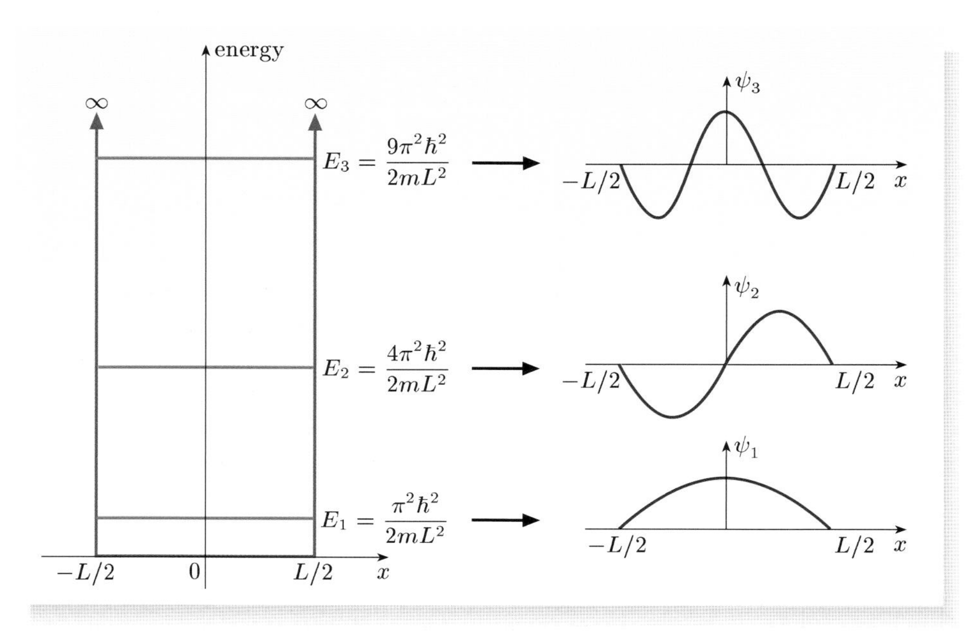

Figure 4.4 Energy eigenvalues and eigenfunctions for a particle in a one-dimensional infinite square well.

The energy eigenvalues are the allowed energies of the system. No matter what state the system might be in, if we measure its energy, we always get one or other of the energy eigenvalues. The stationary states are special because they are states of definite energy. If the system is described by a stationary-state wave function $\Psi_n(x,t)$, and we measure its energy, we get the corresponding energy eigenvalue E_n *with certainty*. So not everything in quantum mechanics is unpredictable; sometimes it is possible to predict the result of a measurement with absolute certainty.

Exercise 4.1 An isolated system is described by the stationary-state wave function $\Psi_n(x,t) = \psi_n(x)\,\mathrm{e}^{-\mathrm{i}E_nt/\hbar}$. Show that, at any fixed time t, $\Psi_n(x,t)$ is an energy eigenfunction with energy eigenvalue E_n. What value of energy will be found in this stationary state? Does the time of the measurement make any difference to your answer?

Exercise 4.2 A particle of mass m is in a symmetrical infinite one-dimensional square well of width L. What are the chances of measuring an energy $10\pi^2\hbar^2/2mL^2$? ■

4.1.4 Linear combinations of stationary states

Simple treatments of quantum mechanics sometimes leave the impression that a system is always in one or other of its allowed energy levels, and that it passes instantaneously from one level to another. For example, in describing the emission of light by an atom, we often talk of the atom as being in an excited energy level, prior to emitting a photon and jumping down to a lower energy level. There is nothing wrong with this description, but it is important to realize that other types of state can exist.

It is perfectly possible for a system to be in a linear combination of stationary states, described by a wave function

$$\Psi(x,t) = a_1\Psi_1(x,t) + a_2\Psi_2(x,t) + \cdots, \tag{4.5}$$

where the $\Psi_i(x,t)$ are normalized stationary-state wave functions and the coefficients a_i are complex constants, scaled suitably to ensure that the wave function $\Psi(x,t)$ is normalized.

Exercise 4.3 Show that the wave function in Equation 4.5 satisfies Schrödinger's equation. ■

The linear combination specified by Equation 4.5 is not a product of separate functions of x and t, so it is not a stationary state. However, this function *does* satisfy Schrödinger's equation, so there is no reason to reject it as a possible wave function of the system. It turns out that states like this are not just theoretical curiosities; it is perfectly possible to prepare them in the laboratory. For example, an atom in its ground state can be excited into a linear combination of two closely-spaced excited states by exposing it to coherent laser light with a range of photon energies that spans the energy differences between the ground state and both of the excited states. Such linear combinations may eventually become of great practical and economic interest, as they are at the heart of the rapidly-developing subject of quantum computing. For both theoretical and practical reasons, we are therefore obliged to take the linear combination wave function of Equation 4.5 seriously. But how should we interpret it?

It is natural to ask what the energy of the system is in such a state. You might imagine that it would be some weighted sum of the energy eigenvalues. For example, in the state

$$\Psi(x,t) = a_1\Psi_1(x,t) + a_2\Psi_2(x,t), \tag{4.6}$$

you might guess that the energy would lie somewhere between E_1 and E_2 — presumably closer to E_1 when $|a_1| \gg |a_2|$, and closer to E_2 when $|a_2| \gg |a_1|$, with a smooth transition in between. *But this would be totally wrong.* We know that every energy measurement yields one or other of the energy eigenvalues, so it makes no sense to say that the system has an energy *between* two neighbouring eigenvalues E_1 and E_2.

In fact, we cannot say that the state in Equation 4.6 has any definite energy at all. In this state, every measurement of energy yields *either* E_1 *or* E_2, but these results appear randomly, and it is impossible to predict which one of them will occur in any single measurement. Nevertheless, a definite pattern emerges. Each result has a definite probability, and it is these probabilities that can be predicted by quantum mechanics.

Given the indeterministic nature of the quantum world, the best we can do is to ask two questions:

1. In a given system, what are the *possible* values of energy?
2. In a given state, what are the *probabilities* of the various possible energies?

The first question has already been answered. The possible energy values are the energy eigenvalues, obtained by solving the time-independent Schrödinger equation for the system under consideration. We now turn to the second question. The answer to this question is of major importance, as it is here that the predictions of quantum mechanics make direct contact with experiment.

4.2 Probability distributions

Quantum mechanics is indeterministic; it tells us what *can* happen rather than what *will* happen. Fortunately, it also tells us the probabilities of the various possible outcomes. This section will introduce the basic law that determines these probabilities.

To take a definite case, we consider measurements of energy in a system consisting of a single particle in a one-dimensional *infinite* potential energy well. This could be a square well (as in Figure 4.4) or an infinite well with walls of some other shape. In such a well, the energy eigenvalues are all *discrete*, which means that they occupy a set of isolated values rather than a continuum. The energy eigenvalues are also *non-degenerate*, which means that each energy eigenvalue corresponds to a unique energy eigenfunction (ignoring an arbitrary choice of phase factor, which has no physical significance). It follows that each energy eigenvalue, and its corresponding eigenfunction, can be labelled by an integer $n = 1, 2, 3, \ldots$.

4.2.1 The overlap rule

If we measure the energy of the system, we are bound to get one of its energy eigenvalues $E_1, E_2, \ldots$ — but how likely is each value? The answer is given by the following rule:

The overlap rule

If the energy of a system is measured at time t, the probability of obtaining the ith energy eigenvalue E_i is given by

$$p_i = \left| \int_{-\infty}^{\infty} \psi_i^*(x)\, \Psi(x,t)\, \mathrm{d}x \right|^2, \tag{4.7}$$

where $\Psi(x,t)$ is the wave function of the system and $\psi_i(x)$ is the energy eigenfunction corresponding to eigenvalue E_i. Both the wave function and the eigenfunction are assumed to be normalized.

The integral on the right-hand side of Equation 4.7 will be called the **overlap integral** of $\psi_i(x)$ and $\Psi(x,t)$.

Let us assume, for the moment, that $\psi_1(x)$ and $\Psi(x,t)$ are both real functions. This will help us to give a graphical interpretation of the overlap rule. Figure 4.5a shows a case in which the energy eigenfunction $\psi_1(x)$, corresponding to the eigenvalue E_1, does not overlap with the wave function $\Psi(x,t)$ at all. In this case, the overlap integral is equal to zero, so there is no chance of obtaining the value E_1 in an energy measurement. In Figure 4.5b, there is a small overlap between the energy eigenfunction and the wave function, leading to a small probability of obtaining the value E_1, while in Figure 4.5c, there is a much larger overlap, leading to a much larger probability of obtaining the value E_1 in an energy measurement. The probability of obtaining a particular energy eigenvalue is given by the square of the modulus of the corresponding overlap integral. In the example shown in Figure 4.5d, the overlap integral is negative, but this does not matter; the probability of getting the eigenvalue E_1 is just as high as in Figure 4.5c.

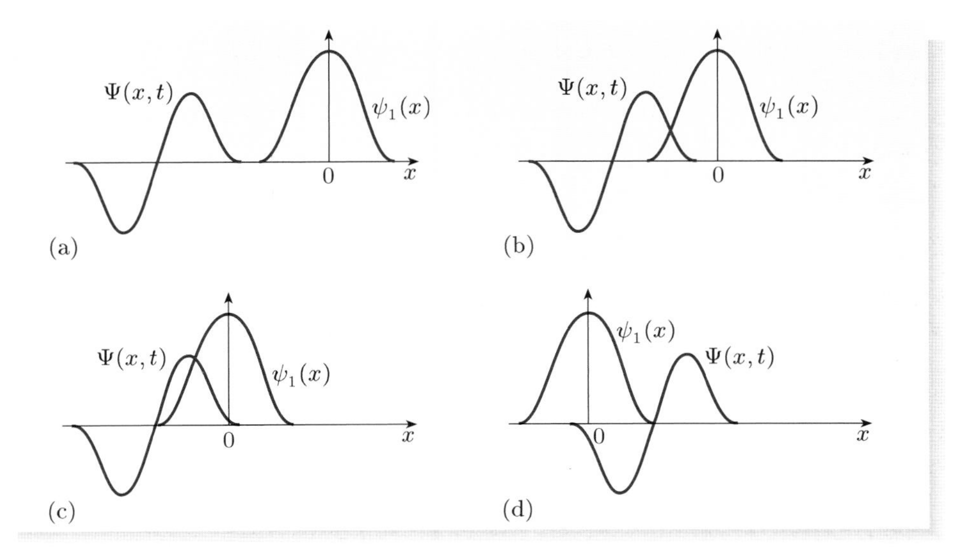

Figure 4.5 The overlap between an energy eigenfunction $\psi_1(x)$ and a wave function $\Psi(x,t)$: (a) no overlap; (b) a small amount of overlap; (c) and (d) large amounts of overlap.

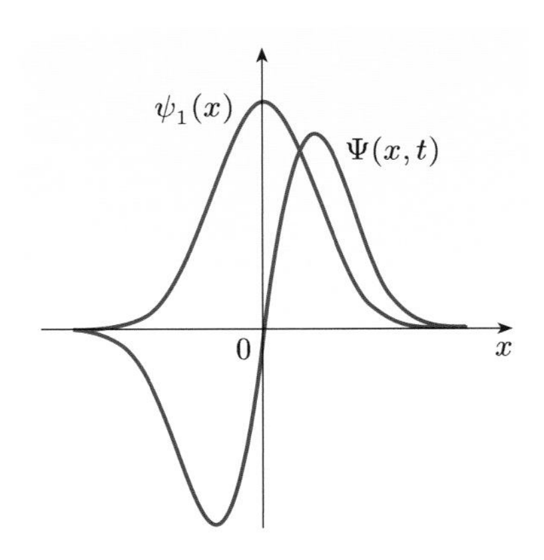

Figure 4.6 Cancellations can occur in an overlap integral.

It is important to note that cancellations can occur in an overlap integral. Figure 4.6 shows a case where a positive contribution to the overlap integral from $x > 0$ is exactly cancelled by a negative contribution from $x < 0$. The overlap integral in this case is equal to zero, so the probability of obtaining the eigenvalue E_1 is equal to zero. In general, two functions with a vanishing overlap integral are said to be **orthogonal** to one another. The word 'orthogonal' is borrowed from geometry, where it means 'perpendicular to'. In the present context, we can think of the square of the modulus of the overlap integral of two normalized functions as being a measure of their similarity; this quantity is equal to 1 for two identical normalized functions, and it is equal to zero for two orthogonal functions. So orthogonal functions are thought of as being *completely dissimilar*.

In order to draw graphs, we have restricted attention to functions with real values. In general, both the wave function and the energy eigenfunctions may have complex values, so the graphs in Figures 4.5 and 4.6 should be regarded as emblems of something more general.

Apart from the difficulties of drawing graphs, there are no special difficulties in handling complex wave functions or eigenfunctions. We can still calculate the appropriate overlap integral, and then take the square of its modulus to find the probability of a given energy value. It is important to note that Equation 4.7 involves $\psi_i^*(x)$, the *complex conjugate* of the ith energy eigenfunction. In many cases, the energy eigenfunctions can be chosen to be real, so complex conjugation may seem unnecessary. However, you will eventually meet energy eigenfunctions that are complex, so it is best to include complex conjugation from the outset, since it is needed in the general case.

You may recall from Chapter 1 that a *probability amplitude* is something whose modulus squared is equal to a probability. The overlap rule can therefore be interpreted as telling us that the **probability amplitude** for measuring the energy eigenvalue E_i is given by

$$\text{probability amplitude} = \int_{-\infty}^{\infty} \psi_i^*(x)\,\Psi(x,t)\,\mathrm{d}x. \tag{4.8}$$

Hence, in wave mechanics, a probability amplitude can be found by evaluating an overlap integral. The probability is found by taking the square of the modulus of the probability amplitude. Do not confuse probabilities and probability amplitudes. While the probability amplitude may be a negative or complex number, the probability is always real and positive.

The overlap rule applied to a stationary state

The overlap rule is a fundamental principle of quantum mechanics, as important in its own way as Schrödinger's equation. We will now explore some of its consequences. We begin with the simplest case — a stationary state. Suppose that a system is described by the stationary-state wave function

$$\Psi_i(x,t) = \psi_i(x)\,\mathrm{e}^{-\mathrm{i}E_i t/\hbar},$$

corresponding to the energy eigenfunction $\psi_i(x)$ and energy eigenvalue E_i. We know that any measurement of the energy in this state is bound to give the value E_i. Let's see how this fact emerges naturally from the overlap rule.

For simplicity, let us suppose that the energy measurement is made at time $t = 0$, when the wave function is $\Psi(x,0) = \psi_i(x)\,\mathrm{e}^0 = \psi_i(x)$. Then the overlap rule tells us that the probability of obtaining the value E_i is

$$p_i = \left|\int_{-\infty}^{\infty} \psi_i^*(x)\,\psi_i(x)\,\mathrm{d}x\right|^2 = 1,$$

where the last step follows because the eigenfunctions are assumed to be normalized. The fact that this probability is equal to 1 confirms that the eigenvalue E_i is certain, as expected.

If we accept the overlap rule as a general principle of quantum mechanics, another far-reaching conclusion emerges. It is a simple matter of logic to note that if the value E_i is certain, then any other value $E_j \neq E_i$ is impossible. The overlap rule therefore requires that

$$p_j = \left|\int_{-\infty}^{\infty} \psi_j^*(x)\,\psi_i(x)\,\mathrm{d}x\right|^2 = 0 \quad \text{for } E_j \neq E_i.$$

Bearing in mind that the energy eigenvalues are non-degenerate, the condition $E_j \neq E_i$ is equivalent to saying that $j \neq i$. Moreover, the equation $|z|^2 = 0$ can be satisfied only if $z = 0$. We therefore conclude that

$$\int_{-\infty}^{\infty} \psi_j^*(x)\,\psi_i(x)\,\mathrm{d}x = 0 \quad \text{for } j \neq i. \tag{4.9}$$

Using the definition of orthogonal functions given earlier, this means that:

Any two energy eigenfunctions with different energy eigenvalues are orthogonal to one another.

Combining the two words *orthogonal* and *normalized*, the set of energy eigenfunctions is said to be **orthonormal**, which means that

$$\int_{-\infty}^{\infty} \psi_j^*(x)\,\psi_i(x)\,\mathrm{d}x = \begin{cases} 1 & \text{for } j = i, \\ 0 & \text{for } j \neq i. \end{cases} \tag{4.10}$$

It is helpful to introduce a special symbol for the right-hand side of this equation. The **Kronecker delta symbol** δ_{ij} is defined by

$$\delta_{ji} = \begin{cases} 1 & \text{for } j = i, \\ 0 & \text{for } j \neq i. \end{cases} \tag{4.11}$$

The orthonormality of energy eigenfunctions is then expressed compactly as

$$\int_{-\infty}^{\infty} \psi_j^*(x)\,\psi_i(x)\,\mathrm{d}x = \delta_{ji}. \tag{4.12}$$

We have been led to the orthonormality of energy eigenfunctions by considering implications of the overlap rule, but the energy eigenfunctions are solutions of the time-independent Schrödinger equation, so there must something special about the time-independent Schrödinger equation, guaranteeing that its normalized solutions obey Equation 4.12. Later in the course you will see that this is indeed the case. For the moment, we shall simply accept that Equation 4.12 is valid; the following exercise asks you to verify it in one special case.

Use the list of standard integrals inside the back cover.

Exercise 4.4 A one-dimensional infinite square well has its walls at $x = 0$ and $x = L$. The normalized energy eigenfunctions for this well are given by

$$\psi_n(x) = \sqrt{\frac{2}{L}} \sin\left(\frac{n\pi x}{L}\right) \quad \text{for } 0 \leq x \leq L \qquad \text{(Eqn 3.22)}$$

$$\psi_n(x) = 0 \quad \text{for } x < 0 \text{ and } x > L. \qquad \text{(Eqn 3.23)}$$

where $n = 1, 2, 3, \ldots$. Verify that these eigenfunctions are orthonormal. ■

The overlap rule for a linear combination of stationary states

The overlap rule is rather dull for stationary states because we already know that a measurement of energy in a stationary state is certain to give the corresponding energy eigenvalue. We now turn to the more interesting case of a linear combination of stationary states. Let us suppose that the system is described by the wave function

$$\Psi(x,t) = a_1\Psi_1(x,t) + a_2\Psi_2(x,t) + \cdots, \qquad (4.13)$$

where the $\Psi_i(x,t)$ are stationary-state wave functions and the coefficients a_i are constants. The stationary-state wave functions are of the form

$$\Psi_i(x,t) = \psi_i(x)\,\mathrm{e}^{-\mathrm{i}E_i t/\hbar},$$

where $\psi_i(x)$ is an energy eigenfunction. So, if we introduce the notation

$$c_i(t) = a_i\,\mathrm{e}^{-\mathrm{i}E_i t/\hbar}, \qquad (4.14)$$

the wave function in Equation 4.13 can be expressed as a linear combination of energy eigenfunctions:

$$\Psi(x,t) = c_1(t)\,\psi_1(x) + c_2(t)\,\psi_2(x) + \cdots. \qquad (4.15)$$

Wave functions of this type describe states of *indefinite* energy. When the energy of the system is measured, a variety of different results may be obtained, each with its own probability. The overlap rule allows us to calculate these probabilities.

To see how this works, let's form the overlap integral of $\psi_1(x)$ with both sides of Equation 4.15. This gives

$$\int_{-\infty}^{\infty} \psi_1^*(x)\,\Psi(x,t)\,\mathrm{d}x = c_1(t)\int_{-\infty}^{\infty} \psi_1^*(x)\,\psi_1(x)\,\mathrm{d}x + c_2(t)\int_{-\infty}^{\infty} \psi_1^*(x)\,\psi_2(x)\,\mathrm{d}x + \cdots,$$

where factors that are independent of x have been taken outside the integrals. By orthogonality, only the first of the integrals on the right-hand side is non-zero, and this is equal to 1 because the energy eigenfunctions are normalized. We therefore have

$$\int_{-\infty}^{\infty} \psi_1^*(x)\,\Psi(x,t)\,\mathrm{d}x = c_1(t).$$

The overlap integral on the left is the probability amplitude for getting the value E_1 in an energy measurement carried out at time t. Taking the square of the modulus of this probability amplitude, the corresponding probability is

$$p_1 = \left| \int_{-\infty}^{\infty} \psi_1^*(x)\, \Psi(x,t)\, \mathrm{d}x \right|^2 = |c_1(t)|^2.$$

A similar argument applies to any other energy eigenvalue, so we are led to the following rule.

The coefficient rule

If the wave function $\Psi(x,t)$ of a system is expressed as a linear combination of energy eigenfunctions, the probability of obtaining the ith eigenvalue E_i is

$$p_i = |c_i(t)|^2 \,, \tag{4.16}$$

where $c_i(t)$ is the coefficient of the ith energy eigenfunction in the wave function at the instant of measurement. The coefficient $c_i(t)$ can therefore be interpreted as a *probability amplitude* whose modulus squared gives the probability of measuring the value E_i.

We assume that the wave function and the eigenfunctions are normalized.

For example, suppose that the wave function at the time of measurement is

$$\Psi(x,t_0) = 0.6\mathrm{i}\, \psi_1(x) - 0.8\, \psi_2(x),$$

where $\psi_1(x)$ and $\psi_2(x)$ are normalized energy eigenfunctions with eigenvalues E_1 and E_2. Then the probabilities of getting energies E_1 and E_2 are

$$p_1 = |0.6\mathrm{i}|^2 = 0.36 \quad \text{and} \quad p_2 = |-0.8|^2 = 0.64,$$

while the probability of getting any other eigenvalue (such as E_3 or E_4) is zero because the corresponding eigenfunctions are absent from the wave function, and therefore have zero coefficients.

It might appear that the probability given by Equation 4.16 is time-dependent, but this is not the case. Because the coefficients $c_i(t)$ take the special form given in Equation 4.14, we have

$$p_i = |c_i(t)|^2 = \left|a_i \mathrm{e}^{-\mathrm{i}E_i t/\hbar}\right|^2 = |a_i|^2 \left|\mathrm{e}^{-\mathrm{i}E_i t/\hbar}\right|^2 = |a_i|^2, \tag{4.17}$$

which is clearly independent of time. In classical mechanics, the energy of any isolated system remains constant in time; this is the law of **conservation of energy**. In quantum mechanics, the system may have an indefinite energy, so it is generally meaningless to say that *the* energy of a system remains constant in time. However, we can say that an isolated system has a *probability distribution* of energy that remains constant in time; this is the quantum-mechanical version of energy conservation.

The probability distribution of energy may change when a measurement is made, but any measurement involves interaction with a measuring instrument, so the system is no longer isolated.

The coefficient rule can also be derived in a more concise way, using the Kronecker delta symbol. It is worth going through this alternative derivation carefully because it introduces some useful skills. The starting point is to express Equation 4.15 more compactly, using a summation sign:

$$\Psi(x,t) = \sum_j c_j(t)\, \psi_j(x).$$

Taking the overlap integral of both sides with the ith energy eigenfunction then gives

$$\int_{-\infty}^{\infty} \psi_i^*(x)\,\Psi(x,t)\,\mathrm{d}x = \sum_j c_j(t) \int_{-\infty}^{\infty} \psi_i^*(x)\,\psi_j(x)\,\mathrm{d}x$$
$$= \sum_j c_j(t)\,\delta_{ij},$$

because the eigenfunctions are orthonormal (Equation 4.12). Although the sum on the right-hand side contains many terms, nearly all of these terms are equal to zero because the Kronecker delta symbol vanishes whenever $j \neq i$. The only surviving term is that for which $j = i$, and in this case the Kronecker delta symbol is equal to 1. We therefore conclude that

$$\int_{-\infty}^{\infty} \psi_i^*(x)\,\Psi(x,t)\,\mathrm{d}x = c_i(t).$$

Finally, taking the square of the modulus of both sides, and using the overlap rule, we obtain $p_i = |c_i(t)|^2$, as before.

When using the coefficient rule, it is important to remember that both the wave function and the energy eigenfunctions must be normalized. This imposes a constraint on the coefficients, which can be established using the Kronecker delta technique. We write

$$\Psi^*(x,t)\,\Psi(x,t) = \Big(\sum_j c_j(t)\,\psi_j(x)\Big)^* \Big(\sum_i c_i(t)\,\psi_i(x)\Big)$$
$$= \sum_i \sum_j c_j^*(t)\,c_i(t)\,\psi_j^*(x)\,\psi_i(x). \qquad (4.18)$$

Note very carefully that we have chosen different indices, i and j, in the two sums in this product. This is essential. If we had used the same index in both sums, we would incorrectly omit terms such as $c_1^*(t)\,c_2(t)\,\psi_1^*(x)\,\psi_2(x)$. The index used in any single sum is arbitrary, but you should always avoid using the same index in two sums that are multiplied together.

Integrating Equation 4.18 over all x and using orthonormality of the energy eigenfunctions (Equation 4.12), we then obtain

$$\int_{-\infty}^{\infty} \Psi^*(x,t)\,\Psi(x,t)\,\mathrm{d}x = \sum_i \sum_j c_j^*(t)\,c_i(t)\,\delta_{ji}.$$

Most of the terms in the sum over j on the right-hand side are equal to zero. The only non-zero terms are those with $j = i$, for which the Kronecker delta symbol is equal to 1, so the double sum reduces to the single sum $\sum_i c_i^*(t)\,c_i(t)$. Normalization of the wave function therefore requires that

Remember, $z^*z = |z|^2$ for any complex number z.

$$\sum_i |c_i(t)|^2 = 1.$$

This makes good sense because the probability of getting the ith energy eigenvalue is $p_i = |c_i(t)|^2$, and the sum of the p_i, taken over all the energy eigenvalues, is the probability of obtaining *one or other* of the eigenvalues. This must be equal to 1 because we are bound to get one or other of these allowed energies. Because $|c_i(t)|^2$ is independent of time, a wave function that is normalized at a given time remains normalized forever.

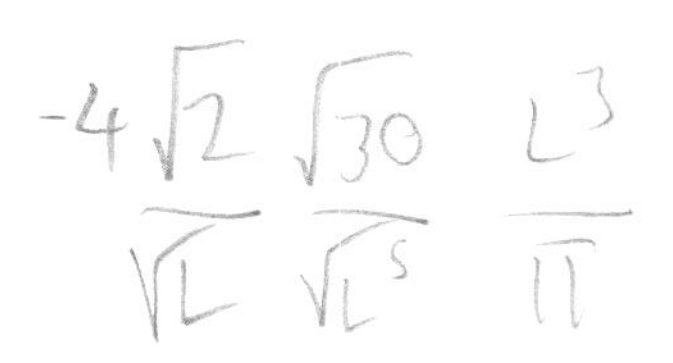

Exercise 4.5 A wave function takes the form

$$\Psi(x,t) = A\Big(2\,\psi_1(x)\,e^{-iE_1t/\hbar} - 3\,\psi_2(x)\,e^{-iE_2t/\hbar}\Big),$$

where A is a constant and $\psi_1(x)$ and $\psi_2(x)$ are normalized energy eigenfunctions with eigenvalues E_1 and E_2. Find a constant A that normalizes this wave function. What are the probabilities of finding the values E_1 and E_2 in an energy measurement taken on this state? ■

The overlap rule in the general case

If the wave function is presented as a sum of energy eigenfunctions, it is easy to calculate probabilities by reading off the coefficients and then taking their modulus squared. More commonly, however, the wave function is specified as a single function that has not been split into a sum of energy eigenfunctions. In such a case, we use the overlap rule directly. The following worked example shows how this is done. You will need to remember that the energy eigenfunctions in a *symmetric* one-dimensional infinite square well of width L take the form

$$\psi_n(x) = \sqrt{\frac{2}{L}}\cos\left(\frac{n\pi x}{L}\right) \quad \text{for } n = 1, 3, 5, \ldots, \qquad \text{(Eqn 3.29)}$$

$$\psi_n(x) = \sqrt{\frac{2}{L}}\sin\left(\frac{n\pi x}{L}\right) \quad \text{for } n = 2, 4, 6, \ldots, \qquad \text{(Eqn 3.28)}$$

inside the well, with $\psi_n(x) = 0$ outside.

Essential skill

Using the overlap rule

Worked Example 4.1

A particle of mass m is in a symmetric one-dimensional infinite square well from $x = -L/2$ to $x = +L/2$. At time $t = 0$, the state of the particle is given by the normalized wave function

$$\Psi(x,0) = \sqrt{\frac{30}{L^5}}\left(x^2 - \frac{L^2}{4}\right) \quad \text{for } -L/2 \le x \le L/2.$$

If the energy is measured at this time, what is the probability of obtaining the ground-state energy $E_1 = \pi^2\hbar^2/2mL^2$?

Solution

Using the overlap rule, the probability amplitude for energy E_1 is

$$\begin{aligned} c_1(0) &= \int_{-\infty}^{\infty} \psi_1^*(x)\,\Psi(x,0)\,dx \\ &= \sqrt{\frac{2}{L}}\sqrt{\frac{30}{L^5}}\int_{-L/2}^{L/2}\cos\left(\frac{\pi x}{L}\right)\left(x^2 - \frac{L^2}{4}\right)dx. \end{aligned}$$

The integral can be evaluated by changing the variable of integration from x to $y = \pi x/L$. The limits of integration change from $x = \pm L/2$ to $y = \pm\pi/2$, and $dx = (L/\pi)\,dy$, so we obtain

$$c_1(0) = \frac{\sqrt{60}}{L^3}\left[\left(\frac{L}{\pi}\right)^3\int_{-\pi/2}^{\pi/2} y^2\cos y\,dy - \frac{L^2}{4}\times\frac{L}{\pi}\int_{-\pi/2}^{\pi/2}\cos y\,dy\right].$$

Then, using a standard integral given inside the back cover of the book,

$$c_1(0) = \frac{\sqrt{60}}{L^3}\left[\left(\frac{L}{\pi}\right)^3 \frac{\pi^2 - 8}{2} - \frac{L^3}{4\pi} \times 2\right] = -\frac{4\sqrt{60}}{\pi^3},$$

so the probability of energy E_1 is

$$p_1 = |c_1(0)|^2 = 960/\pi^6 = 0.9986.$$

Comment: This result shows that any measurement of the energy is very likely to give the value E_1, with just over one case in 1000 giving some other value. The answer makes good sense because the wave function is quite similar in shape to the ground-state energy eigenfunction, so there is a large overlap between them.

Exercise 4.6

This useful result can save a lot of unnecessary work.

(a) Show that the definite integral of any *odd* function $f(x)$, taken over a range that is centred on $x = 0$, is equal to zero.

(b) Use the result of part (a) to show that the probability of getting the energy eigenvalue E_2 in a state described by the wave function of Worked Example 4.1 is equal to zero.

Exercise 4.7 What is the probability of getting the energy eigenvalue E_3 in a state described by the wave function of Worked Example 4.1? ■

4.2.2 Extension to other observables

So far, we have concentrated on energy measurements in a one-dimensional infinite well, but the ideas introduced above can be extended to other quantities and to other systems. The purpose of this subsection is to sketch how this is done. Subsequent sections of the chapter do not rely on this material, so you should read it quickly, without lingering on the details; the main point to grasp is that the overlap rule is a general principle of quantum mechanics, and is not confined to energy measurements.

Let us consider an observable quantity A, which need not be energy. In quantum mechanics, A is represented by a linear operator $\widehat{\mathrm{A}}$, and we can construct the eigenvalue equation

$$\widehat{\mathrm{A}}\,\phi_i(x) = \alpha_i\,\phi_i(x).$$

We shall assume that the eigenvalues α_i are discrete, and that each corresponds to a single eigenfunction $\phi_i(x)$. Then the eigenvalues are the possible values of the observable A, and the probability of obtaining the eigenvalue α_i is given by

$$p_i = \left|\int_{-\infty}^{\infty} \phi_i^*(x)\,\Psi(x,t)\,\mathrm{d}x\right|^2, \tag{4.19}$$

where $\Psi(x,t)$ is the wave function of the system at the time of measurement, and $\phi_i(x)$ is the eigenfunction of $\widehat{\mathrm{A}}$ with eigenvalue α_i. This extends the *overlap rule* to observables other than energy. The important point to note is that it involves the *eigenfunctions of the observable* A *that is being measured.*

It is also possible to extend the *coefficient rule*. Suppose that the wave function of the system is expressed as a linear combination of the eigenfunctions $\phi_i(x)$ of $\widehat{\text{A}}$:

$$\Psi(x,t) = \sum_i c_i(t)\,\phi_i(x).$$

Then a measurement of A, carried out at time t, will yield the eigenvalue α_i with probability

$$p_i = |c_i(t)|^2. \tag{4.20}$$

In general, the coefficients $c_i(t)$ will be complicated functions of time, and the probabilities p_i will depend on time. However, the important point to note is that the coefficient rule can be used for observables other than energy, provided that the wave function is expressed as a linear combination of the *eigenfunctions of the observable* A *that is being measured.*

The above descriptions assume that the eigenvalues of $\widehat{\text{A}}$ are discrete. However, it must be admitted that many observables do not have purely discrete eigenvalues. For example, position and momentum are not quantized. If we measure the position of a particle, we might obtain any value in a continuous range. A different type of description is needed for observables with continuous values. Precise measurements of continuous variables make no sense from an experimental point of view. In practice, one cannot measure the exact value of a continuous variable, but only its value *to within some finite resolution*. When a Geiger counter clicks, for example, it tells us only that the particle is somewhere inside the counter; it does not reveal the exact point occupied by the particle.

It is actually rather pointless to talk about the probability of detecting a particle at a single point, P. An infinite number of points lie arbitrarily close to P, so the chances of finding the particle *exactly* at P are completely negligible. It is much more sensible to talk about the probability of finding the particle in a small *region* around P. This is what Born's rule does. In one dimension this rule states that the probability of finding the particle in a small interval of length δx, centred on the point x, is $|\Psi(x,t)|^2\,\delta x$. In other words, the probability density for position (the probability per unit length) is $|\Psi(x,t)|^2$. At first sight, this looks very different from the overlap rule, but this is not really so.

To express Born's rule in terms of a (modified) overlap rule, we introduce the 'top-hat' function $T_{x_0}(x)$ shown in Figure 4.7. This function has a constant value $1/\sqrt{w}$ throughout a small range of width w, centred on the point x_0, and is equal to zero outside this range. The top-hat function is normalized because

$$\int_{-\infty}^{\infty} |T_{x_0}(x)|^2\,\mathrm{d}x = \int_{x_0-\frac{w}{2}}^{x_0+\frac{w}{2}} \left|\frac{1}{\sqrt{w}}\right|^2 \mathrm{d}x = \frac{1}{w} \times w = 1.$$

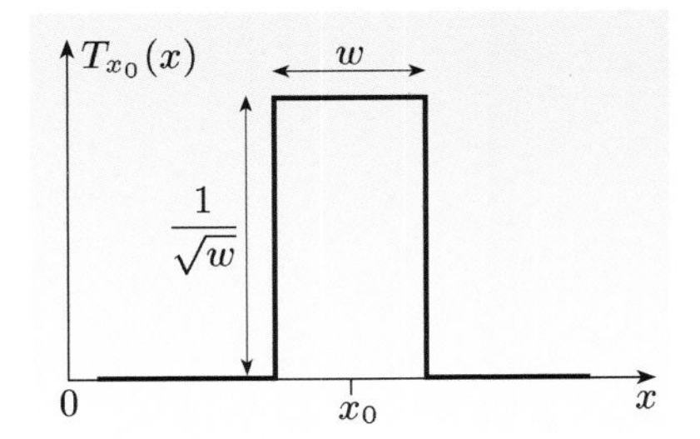

Figure 4.7 A top-hat function of width w, centred on the point x_0. The (informal) name of this function comes from its shape!

If we take the overlap integral of a *very narrow* top-hat function with the wave function $\Psi(x,t)$, we can make the approximation

$$\int_{-\infty}^{\infty} T^*_{x_0}(x)\,\Psi(x,t)\,\mathrm{d}x \simeq \frac{1}{\sqrt{w}}\,\Psi(x_0,t) \times w = \Psi(x_0,t)\sqrt{w},$$

and this approximation can be made as accurate as we like by taking a top-hat function that is narrow enough. Setting $w = \delta x$, and taking the square of the modulus of both sides, we therefore have

$$\left|\int_{-\infty}^{\infty} T^*_{x_0}(x)\,\Psi(x,t)\,\mathrm{d}x\right|^2 = |\Psi(x_0,t)|^2\,\delta x. \tag{4.21}$$

According to Born's rule, this is the probability of finding the particle within a small interval of width δx, centred on the point x. Born's rule can therefore be regarded as a modified version of the overlap rule, applied to a high-resolution measurement of position. (The only modification is that the left-hand side of Equation 4.21 involves a narrow top-hat function, rather than an eigenfunction arising explicitly from an eigenvalue equation, but this can be regarded as a technicality.)

Equation 4.21 is not useful for calculations; if you want to find the probability of detecting a particle in a small region, you should use Born's rule directly. However, the focus of this subsection has been mainly conceptual. Its purpose is accomplished if you remember one thing: the overlap rule is a universal principle, which applies to all measurements in wave mechanics, not just those involving energies.

4.3 Expectation values in quantum mechanics

The probability distribution for a given observable can be represented graphically — either as a bar chart of probabilities for a discrete variable, or as a graph of probability density for a continuous variable (Figure 4.8).

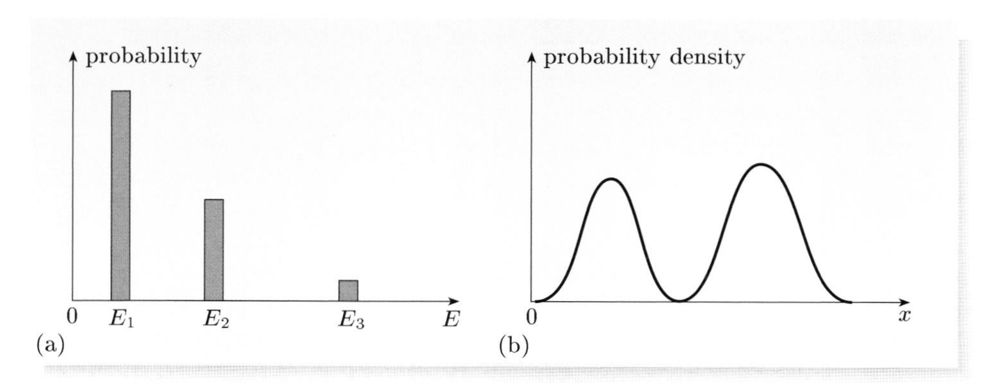

Figure 4.8 (a) A bar chart showing probabilities of discrete energy values. (b) A graph showing the probability density of position.

Probability distributions like these summarize the extent to which quantum mechanics can predict the likely results of measurements. However, for many purposes, it is convenient to give a less detailed description, characterizing the probability distribution by two measures — its *expectation value* (which represents the mean value of the distribution) and its *uncertainty* (which represents the spread in values around the mean). This section will discuss expectation values.

4.3.1 Mean values and expectation values

Suppose that an observable quantity A has a discrete set of possible values, $a_1, a_2, \ldots$. Then we can prepare N identical systems, all in the same state, and measure the value of A in each system. In general, we will get a spread of results. Let's suppose that result a_i is obtained in n_i of the measurements. Then we can form the quantity

$$\overline{A} = \frac{n_1 a_1 + n_2 a_2 + \cdots + n_N a_N}{N} = \sum_{i=1}^{N} \left(\frac{n_i}{N}\right) a_i. \tag{4.22}$$

This is the **mean value** of A taken over the set of N measurements. The quantity n_i/N is the relative frequency of result a_i; since $\sum_i n_i = N$, the sum of the relative frequencies is equal to 1.

The **expectation value** of an observable A is the quantum-mechanical prediction for the mean value. It is denoted by the symbol $\langle A \rangle$ and obtained by replacing the relative frequencies in Equation 4.22 by quantum-mechanical probabilities. Thus

$$\langle A \rangle = \sum_i p_i a_i, \tag{4.23}$$

where p_i is the probability of result a_i. It is in the nature of chance that the mean value $\overline{A}$, taken over a finite number of measurements, may deviate slightly from the expectation value $\langle A \rangle$. However, in the long run, as the number of measurements tends to infinity, the mean value is expected to converge to the expectation value. So a quantum-mechanical prediction of the expectation value can be compared with experiment.

The expectation value of an observable can be readily calculated if we know its probability distribution. For example, suppose that a system is in the state described by the normalized wave function

$$\Psi(x,t) = \frac{1}{\sqrt{2}} \left(\psi_1(x)\,\mathrm{e}^{-\mathrm{i}E_1 t/\hbar} + \psi_2(x)\,\mathrm{e}^{-\mathrm{i}E_2 t/\hbar}\right),$$

where $\psi_1(x)$ and $\psi_2(x)$ are energy eigenfunctions with eigenvalues E_1 and E_2. In this case, the coefficient rule shows that the probabilities of obtaining the values E_1 and E_2 are $p_1 = 1/2$ and $p_2 = 1/2$. All other values have zero probability, so the expectation value of the energy in this state is

$$\langle E \rangle = p_1 E_1 + p_2 E_2 = \frac{E_1 + E_2}{2},$$

which is midway between the two allowed energy values. Note that this expectation value is not one of the allowed energy values. A *single* energy measurement gives either E_1 or E_2, but never $(E_1 + E_2)/2$. This is not surprising because the mean value of a set of numbers need not be equal to any of the numbers in the set. One is reminded of the fact that an average family in the UK contains 1.7 children, although no real family has such a composition.

Exercise 4.8 Find the expectation value of energy in the state described in Exercise 4.5, expressing your answer in terms of the energy eigenvalues E_1 and E_2. ■

We can also define the expectation values of A^2, A^3, or any other power of A by using appropriate powers of the eigenvalues. For example,

$$\langle A^2 \rangle = \sum_i p_i a_i^2, \tag{4.24}$$

where the sum runs over all the possible values a_i.

Expectation values can also be defined for observables with a continuous set of possible values, such as position. The probability of obtaining a value of position in the small interval between $x - \delta x/2$ and $x + \delta x/2$ is given by

$$\text{probability} = |\Psi(x,t)|^2\, \delta x,$$

where $|\Psi(x,t)|^2$ is the probability density for position at the time of measurement. The expectation value of position is then defined by the integral

$$\langle x \rangle = \int_{-\infty}^{\infty} |\Psi(x,t)|^2\, x\, \mathrm{d}x, \tag{4.25}$$

and the expectation value of x^2 is given by

$$\langle x^2 \rangle = \int_{-\infty}^{\infty} |\Psi(x,t)|^2\, x^2\, \mathrm{d}x. \tag{4.26}$$

Comparing with the discrete case (Equations 4.23 and 4.24), we see that a sum over allowed values weighted by probabilities is replaced by an integral over allowed values weighted by a probability density function.

4.3.2 The sandwich integral rule

Our expression for the expectation value of an observable (Equation 4.23 in the discrete case) can be evaluated if we know all the possible values and their probabilities. In principle, the probabilities can be found using the methods of Section 4.2, but this could involve a great deal of work, because the number of possible values could be vast, or even infinite. Fortunately, there is another way of calculating expectation values, using the following rule.

The sandwich integral rule for expectation values

In a state described by the wave function $\Psi(x,t)$, the expectation value of an observable A at time t is given by

$$\langle A \rangle = \int_{-\infty}^{\infty} \Psi^*(x,t)\, \widehat{\mathrm{A}}\, \Psi(x,t)\, \mathrm{d}x, \tag{4.27}$$

The operator $\widehat{\mathrm{A}}$ acts on $\Psi(x,t)$. The resulting function is then multiplied by $\Psi^*(x,t)$ and this product is integrated over all x.

where $\widehat{\mathrm{A}}$ is the quantum-mechanical operator corresponding to A. The operator $\widehat{\mathrm{A}}$ is 'sandwiched' between the two functions $\Psi^*(x,t)$ and $\Psi(x,t)$; for this reason, we informally call the integral on the right-hand side a **sandwich integral**.

It is easy to see that the sandwich integral rule works for the expectation value of position. The position operator $\widehat{\mathrm{x}}$ simply tells us to multiply by x, so Equation 4.27 is just Equation 4.25 in disguise.

Equation 4.27 can be justified in other cases as well. As an illustration, we will calculate the sandwich integral

$$\int_{-\infty}^{\infty} \Psi^*(x,t)\,\widehat{\mathrm{H}}\,\Psi(x,t)\,\mathrm{d}x$$

for a system with discrete, non-degenerate energy eigenvalues.

Let's assume that the wave function $\Psi(x,t)$ is a linear combination of stationary-state wave functions, which means that

$$\Psi(x,t) = \sum_i a_i\,\psi_i(x)\,\mathrm{e}^{-\mathrm{i}E_i t/\hbar} = \sum_i c_i(t)\,\psi_i(x), \tag{4.28}$$

where $c_i(t) = a_i\,\mathrm{e}^{-\mathrm{i}E_i t/\hbar}$, and $\psi_i(x)$ is an energy eigenfunction with eigenvalue E_i. Then applying the Hamiltonian operator to this wave function gives

$$\widehat{\mathrm{H}}\,\Psi(x,t) = \sum_i c_i(t)\,\widehat{\mathrm{H}}\,\psi_i(x) = \sum_i c_i(t)\,E_i\,\psi_i(x), \tag{4.29}$$

where we have used the fact that $\widehat{\mathrm{H}}$ is a linear operator to move it inside the summation sign, and remembered that $\psi_i(x)$ is an eigenfunction of $\widehat{\mathrm{H}}$ with eigenvalue E_i.

Inserting Equations 4.28 and 4.29 into the sandwich integral gives

$$\int_{-\infty}^{\infty} \Psi^*(x,t)\,\widehat{\mathrm{H}}\,\Psi(x,t)\,\mathrm{d}x = \int_{-\infty}^{\infty} \Big(\sum_j c_j(t)\,\psi_j(x)\Big)^* \Big(\sum_i c_i(t)\,E_i\,\psi_i(x)\Big)\,\mathrm{d}x.$$

Note carefully that we have used different indices, j and i, in the two sums in the integrand. The reason for this was explained on page 106, in the context of normalizing a wave function; if we did not take this precaution, 'cross-product' terms would be incorrectly omitted.

The integral on the right-hand side may look complicated, but it can be simplified by interchanging the summation and integral signs, and taking terms that do not depend on x outside the integral. This gives

$$\int_{-\infty}^{\infty} \Psi^*(x,t)\,\widehat{\mathrm{H}}\,\Psi(x,t)\,\mathrm{d}x = \sum_{ij} c_j^*(t)\,c_i(t)\,E_i \int_{-\infty}^{\infty} \psi_j^*(x)\,\psi_i(x)\,\mathrm{d}x.$$

We can use the orthonormality of the energy eigenfunctions to replace the remaining integral by a Kronecker delta symbol δ_{ji}. This removes all terms with $i \neq j$, so the double sum over i and j becomes a single sum over i:

$$\int_{-\infty}^{\infty} \Psi^*(x,t)\,\widehat{\mathrm{H}}\,\Psi(x,t)\,\mathrm{d}x = \sum_{ij} c_j^*(t)\,c_i(t)\,E_i\,\delta_{ji} = \sum_i c_i^*(t)\,c_i(t)\,E_i.$$

Finally, the coefficient rule allows us to interpret $c_i^*(t)\,c_i(t) = |c_i(t)|^2$ as the probability p_i of obtaining the eigenvalue E_i, so we conclude that

$$\int_{-\infty}^{\infty} \Psi^*(x,t)\,\widehat{\mathrm{H}}\,\Psi(x,t)\,\mathrm{d}x = \sum_i p_i E_i = \langle E\rangle,$$

as required.

This justification of the sandwich integral rule was based on a special case, but the sandwich integral rule is valid for any observable in any state in wave mechanics. For example, the expectation value of momentum is given by

Remember:

$$\widehat{\mathrm{p}}_x = -\mathrm{i}\hbar\frac{\partial}{\partial x}.$$

$$\langle p_x \rangle = \int_{-\infty}^{\infty} \Psi^*(x,t)\left(-\mathrm{i}\hbar\frac{\partial}{\partial x}\right)\Psi(x,t)\,\mathrm{d}x, \tag{4.30}$$

and the expectation value of p_x^2 is

$$\begin{aligned}\langle p_x^2 \rangle &= \int_{-\infty}^{\infty} \Psi^*(x,t)\left(-\mathrm{i}\hbar\frac{\partial}{\partial x}\right)^2\Psi(x,t)\,\mathrm{d}x \\ &= -\hbar^2\int_{-\infty}^{\infty} \Psi^*(x,t)\,\frac{\partial^2}{\partial x^2}\Psi(x,t)\,\mathrm{d}x. \end{aligned} \tag{4.31}$$

Because observables are represented by linear operators, it is safe to assume that

$$\langle \alpha A + \beta B \rangle = \alpha\langle A \rangle + \beta\langle B \rangle \tag{4.32}$$

for any constants α and β. However, care is needed when dealing with powers of observables. In general,

$$\langle A^2 \rangle \neq \langle A \rangle^2, \tag{4.33}$$

as you will see in the following worked example.

Essential skill

Using the sandwich integral rule to calculate expectation values

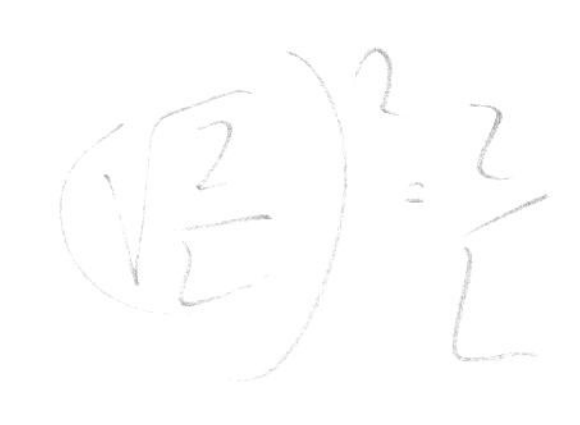

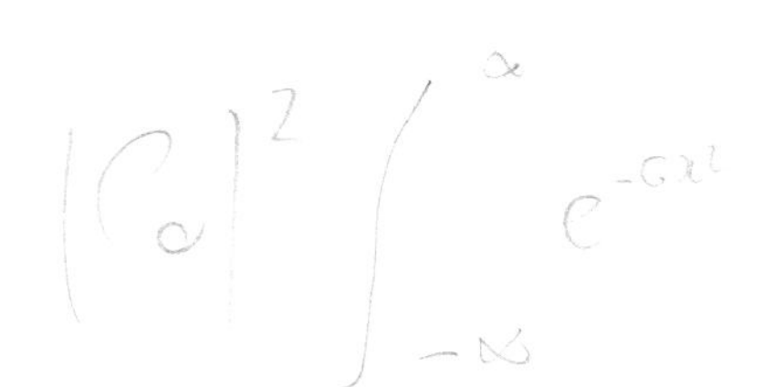

Worked Example 4.2

Find the expectation values of x and x^2 in the ground state of a symmetric one-dimensional infinite square well, with walls at $x = -L/2$ and $x = +L/2$.

Solution

The ground-state wave function vanishes outside the well. Inside the well (for $-L/2 \leq x \leq L/2$), it is

$$\Psi(x,t) = \sqrt{\frac{2}{L}}\cos\left(\frac{\pi x}{L}\right)\mathrm{e}^{-\mathrm{i}E_1 t/\hbar}.$$

Using the sandwich integral rule (Equation 4.27), the expectation value of x is given by

$$\langle x \rangle = \frac{2}{L}\int_{-L/2}^{L/2}\left[\cos\left(\frac{\pi x}{L}\right)\mathrm{e}^{\mathrm{i}E_1 t/\hbar}\right] x \left[\cos\left(\frac{\pi x}{L}\right)\mathrm{e}^{-\mathrm{i}E_1 t/\hbar}\right]\mathrm{d}x.$$

The time-dependent phase factors combine to give 1, and we are left with

$$\langle x \rangle = \frac{2}{L}\int_{-L/2}^{L/2} x\cos^2\left(\frac{\pi x}{L}\right)\mathrm{d}x.$$

This integral vanishes because the integrand is an odd function of x and the range of integration is centred on the origin. Hence

$$\langle x \rangle = 0.$$

Proceeding in a similar way, the expectation value of x^2 is

$$\langle x^2 \rangle = \frac{2}{L} \int_{-L/2}^{L/2} x^2 \cos^2\left(\frac{\pi x}{L}\right) dx.$$

Changing the variable of integration to $y = \pi x/L$, we obtain

$$\langle x^2 \rangle = \frac{2}{L} \times \left(\frac{L}{\pi}\right)^3 \int_{-\pi/2}^{\pi/2} y^2 \cos^2 y \, dy.$$

Finally, using a standard integral given inside the back cover of the book, we obtain

$$\langle x^2 \rangle = \frac{2L^2}{\pi^3} \left(\frac{\pi^3}{24} - \frac{\pi}{4}\right) = 0.0327L^2.$$

Clearly, $\langle x^2 \rangle \neq \langle x \rangle^2$.

Exercise 4.9 Find the expectation values of p_x and p_x^2 in the ground state of a symmetric one-dimensional infinite square well, with walls at $x = -L/2$ and $x = +L/2$.

Exercise 4.10 A particle of mass m is in a symmetric one-dimensional infinite square well of width L. Find the expectation value of the energy of this particle in a state described by the normalized wave function

$$\Psi(x, 0) = \sqrt{\frac{30}{L^5}} \left(x^2 - \frac{L^2}{4}\right) \quad \text{for } -L/2 \leq x \leq L/2,$$

with $\Psi(x, 0) = 0$ outside the well. ■

4.4 Uncertainties in quantum mechanics

The two sets of data values in Figure 4.9 below are very different, even though they have the same mean value; those in Figure 4.9b are much more spread out than those in Figure 4.9a. For many purposes, it is useful to have a measure of the amount by which a set of data values spreads out on either side of the mean value. The standard deviation $\sigma(A)$ gives us such a measure.

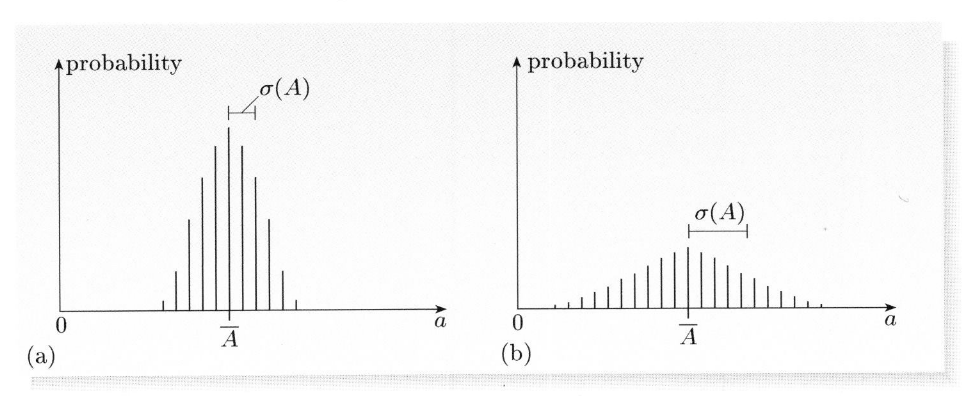

Figure 4.9 A set of data values with (a) a small spread and (b) a large spread.

For each value a_i, we can form its deviation from the mean value, $(a_i - \overline{A})$. We might try to measure the spread of values by taking the average value of these deviations. Unfortunately, this is useless because the average value of the deviations is equal to zero. (The negative deviations and the positive deviations exactly cancel out.) The **standard deviation**, $\sigma(A)$, is defined by taking the *squares* of the deviations before taking the average, and then taking the *square root* of the result. This produces a quantity with the same units as A. The squares of the deviations add cumulatively, without any cancellation, so $\sigma(A)$ gives a sensible measure of the spread of values of A.

Just as an expectation value is a theoretical prediction for a mean value, so an *uncertainty* is a theoretical prediction for a standard deviation. For a system in a given state Ψ, the **uncertainty** ΔA of an observable A is defined by

$$\Delta A = \left\langle (A - \langle A \rangle)^2 \right\rangle^{1/2}, \tag{4.34}$$

where the expectation values are calculated in the state Ψ. If many measurements of A are taken on identical systems all in the state Ψ, then the standard deviation $\sigma(A)$ of the results is expected to converge to the uncertainty ΔA as the number of measurements tends to infinity. So a quantum-mechanical prediction for the uncertainty can be compared with experiment.

The uncertainty can also be expressed in a more convenient form. Squaring both sides and expanding the round brackets in Equation 4.34 gives

$$(\Delta A)^2 = \left\langle A^2 - A\langle A \rangle - \langle A \rangle A + \langle A \rangle^2 \right\rangle.$$

Remembering that $\langle A \rangle$ is a constant, and using Equation 4.32, we obtain

$$(\Delta A)^2 = \langle A^2 \rangle - 2\langle A \rangle\langle A \rangle + \langle A \rangle^2 = \langle A^2 \rangle - \langle A \rangle^2,$$

and hence

$$\Delta A = \left(\langle A^2 \rangle - \langle A \rangle^2\right)^{1/2}. \tag{4.35}$$

This formula is an efficient way of calculating uncertainties, with the expectation values of A and A^2 found by using the sandwich integral rule.

Essential skill

Calculating uncertainties

Worked Example 4.3

Calculate the uncertainties of position and momentum in the ground state of a one-dimensional infinite square well of width L, centred on the origin.

Solution

It simplifies the working to consider the *square* of the uncertainties until the last step in the calculation. For position,

$$(\Delta x)^2 = \langle x^2 \rangle - \langle x \rangle^2.$$

From Worked Example 4.2, we have $\langle x \rangle = 0$ and $\langle x^2 \rangle = 0.0327L^2$, so

$$\Delta x = \sqrt{0.0327L^2} = 0.18L.$$

For momentum,

$$(\Delta p_x)^2 = \langle p_x^2 \rangle - \langle p_x \rangle^2.$$

From the solution to Exercise 4.9, we have $\langle p_x \rangle = 0$ and $\langle p_x^2 \rangle = \pi^2\hbar^2/L^2$, so

$$\Delta p_x = \sqrt{\frac{\pi^2\hbar^2}{L^2}} = \frac{\pi\hbar}{L}.$$

Exercise 4.11 Show that any stationary state has zero uncertainty in energy.

Exercise 4.12 A system is in a linear combination of stationary states described by the wave function

$$\Psi(x,t) = \frac{1}{\sqrt{2}}\big(\Psi_1(x,t) + \Psi_2(x,t)\big),$$

where $\Psi_1(x,t)$ and $\Psi_2(x,t)$ are stationary states with energies E_1 and E_2. What is the uncertainty in energy in this state? ■

4.5 The Heisenberg uncertainty principle

The basic principles of quantum mechanics were worked out in a remarkably short period, from 1925 to 1927. Towards the end of this revolution, Heisenberg (Figure 4.10) proposed his celebrated uncertainty principle. It is possible to prove the uncertainty principle from basic quantum-mechanical rules, and this will be done later in the course. Here, we shall concentrate on its meaning, interpretation and applications, and on how it is used to make order-of-magnitude estimates.

Figure 4.10 Werner Heisenberg (1901–1976) proposed the uncertainty principle in 1927.

The Heisenberg uncertainty principle can be stated as follows:

The Heisenberg uncertainty principle

In any system, in any state, the uncertainties of position and momentum components obey the inequality

$$\Delta x \, \Delta p_x \geq \frac{\hbar}{2}. \tag{4.36}$$

This inequality involves position and momentum components along the *same* axis. Similar results apply to the uncertainty products $\Delta y \, \Delta p_y$ and $\Delta z \, \Delta p_z$, but a product such as $\Delta x \, \Delta p_y$, which involves position and momentum components along different axes, does not obey an inequality and can, in principle, be equal to zero.

You have seen how the uncertainties Δx and Δp_x are defined and calculated. The important thing to remember is that these uncertainties refer to a particular quantum-mechanical state. Even though the state is known precisely, we still obtain a spread of values when we measure position, and another spread of values when we measure momentum. These spreads in values correspond to uncertainties that must obey the Heisenberg uncertainty principle. For example, using the results of Worked Example 4.3, we see that the uncertainties in the ground state of a one-dimensional infinite square well satisfy

$$\Delta x\,\Delta p_x = 0.18L \times \frac{\pi\hbar}{L} = 0.57\hbar > \frac{\hbar}{2},$$

which is consistent with the uncertainty principle.

The uncertainty principle tells us that it is impossible to find a state in which a particle has definite values of both position and momentum. It therefore demolishes the classical picture of a particle following a well-defined trajectory. This must have been very satisfying to Heisenberg, who was driven by the conviction that the orbits of Bohr's atomic model had no part to play in quantum mechanics. As early as 1925, Heisenberg wrote:

> 'my entire meagre efforts go towards killing off and suitably replacing the concept of orbital paths that one cannot observe.'

With the uncertainty principle, he effectively barred any attempt to revive classical orbits. Instead, wave mechanics describes the state of a system by a wave function, which implies a spread of position and momentum values. The uncertainty principle requires that a narrow spread in one of these quantities is offset by a wide spread in the other.

It is important to remember that the uncertainty principle is an inequality, not an equation. If Δx is small, it follows that Δp_x is large — at least as large as $\hbar/(2\,\Delta x)$. However, if Δx is large, the uncertainty principle does not guarantee that Δp_x is small. It only tells us that Δp_x is greater than or equal to $\hbar/(2\,\Delta x)$, which is a weak constraint when Δx is large. In fact, it is perfectly possible for both Δx and Δp_x to be large. This is consistent with the uncertainty principle and turns out to be quite common in states with high quantum numbers. Nevertheless, states with low quantum numbers usually have a product of the position and momentum uncertainties that is close to the minimum value of $\hbar/2$. For this reason, physicists sometimes suppose that $\Delta x\,\Delta p_x \sim \hbar/2$ for the lowest levels in a system. Unlike the uncertainty principle, this statement is not rigorous, but it can be used as a rough rule of thumb.

Interpretation

The uncertainty principle is sometimes interpreted in terms of the disturbance produced by a measurement. For example, we know that a measurement of position disturbs the system, and it is tempting to suppose that this 'explains' why position and momentum cannot both have definite values at the same time. However, such interpretations seem to miss the main point. It is true that measurements create disturbances, and that the state of a system is generally altered by a measurement. *But this is not the subject of the uncertainty principle.* The uncertainty principle tells us nothing about the disturbance caused by a measurement: instead, it tells us about the indeterminacy that is inherent in a given state before the measurement is taken. One way of demonstrating this indeterminacy is to prepare many identical systems in identical states. If we measure position in some of these systems, and momentum in others, we will get spreads of values of position and momentum that are consistent with the uncertainty principle. However, this reflects the nature of the state *before* any measurements are taken, not the disturbance that the measurement creates in individual systems.

4.5.1 The uncertainty principle in action

A striking illustration of the uncertainty principle is provided by a gas of atoms that has been cooled down to a very low temperature. In 1995, Eric Cornell and Carl Wieman cooled a sample of about 2000 rubidium-87 atoms to tens of nanokelvins. In order to keep the sample cool, it was trapped in a small volume of empty space by magnetic fields and beams of light; this kept it away from the relatively warm walls of the container. Figure 4.11 shows the density profile of the cloud as it was cooled below 400 nK. The peak that emerges below this temperature reveals a remarkable phenomenon, known as Bose–Einstein condensation, in which a large proportion of the atoms suddenly occupy a single quantum state, the ground state of the well in which they are trapped.

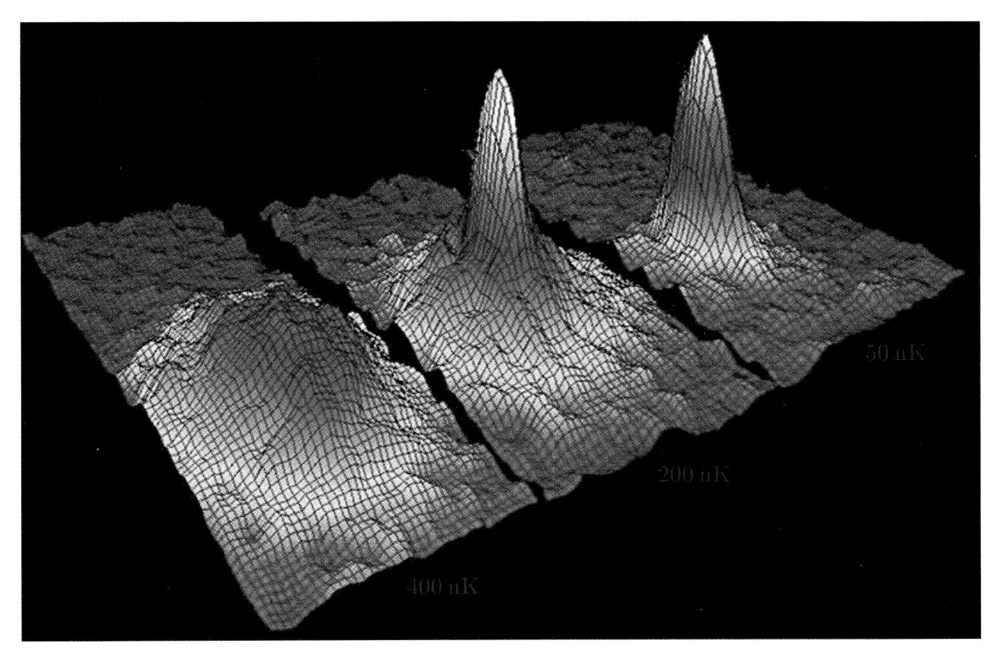

Figure 4.11 Density distributions of ^{87}Rb (rubidium-87) atoms below 400 nK reveal Bose–Einstein condensation. Cornell and Wieman received the 2001 Nobel prize for physics for this work.

By allowing the gas cloud to expand freely in a horizontal direction and taking snapshots of its shape, a measure could be gained of its momentum distribution. The spread of momenta in the peak was found to be close to the minimum allowed by the uncertainty principle. On closer inspection, the peak was observed to be anisotropic, with a greater spread of momentum in one direction than another. This was explained by the fact that the trapping region was itself anisotropic. The direction with the widest spread of momentum was that with the narrowest spatial confinement, just as one would expect from the uncertainty principle.

Another example of the uncertainty principle in action is provided by liquid helium. At atmospheric pressure, helium remains liquid no matter how much it is cooled, and this is largely due to the uncertainty principle. Each helium atom can be thought of as being trapped in a small cage formed by its nearest neighbours. It therefore has a small uncertainty in position, which must be accompanied by a large uncertainty in momentum. It follows that a typical helium atom in the liquid has a significant momentum magnitude p, and a significant kinetic energy $p^2/2m$, even at absolute zero. Because helium atoms have low mass m and bind together very weakly, this kinetic energy is enough to disrupt the formation of a solid. Any clumps of solid helium that do manage to assemble are rapidly shaken to pieces so the helium remains liquid.

4.5.2 Making estimates with the uncertainty principle

If we know how a particle is confined, we can generally make a rough estimate of its uncertainty in position. For example, we might guess that a particle trapped in a square well of width L has $\Delta x \simeq L/2$. Actually, this is not a very good estimate. The uncertainty of any quantity is defined by Equation 4.35, and you have seen that this gives $\Delta x = 0.18L$ in the ground state of an infinite square well. This is significantly less than $L/2$ because the ground-state eigenfunction tails off towards the edges of the well. In a deep finite square well, where the penetration into the classically forbidden region is small, the uncertainty of position in the ground state is still of order $0.18L$. Using this value as a guide, and recognizing that $\langle p_x \rangle = 0$, we have

$$\langle p_x^2 \rangle = (\Delta p_x)^2 \geq \left(\frac{\hbar}{2 \times 0.18L} \right)^2 \simeq \frac{7.7\hbar^2}{L^2}.$$

In a three-dimensional well with smallest linear dimension L, the average kinetic energy of a particle of mass m obeys

$$\langle E_{\text{kin}} \rangle = \frac{\langle p_x^2 \rangle}{2m} + \frac{\langle p_y^2 \rangle}{2m} + \frac{\langle p_z^2 \rangle}{2m} = \frac{3\langle p_x^2 \rangle}{2m} \geq \frac{23\hbar^2}{2mL^2}. \tag{4.37}$$

The lower bound of this inequality is comparable with the ground-state energy $3\pi^2\hbar^2/2mL^2$ of a particle in a three-dimensional cubic box, obtained by solving Schrödinger's equation.

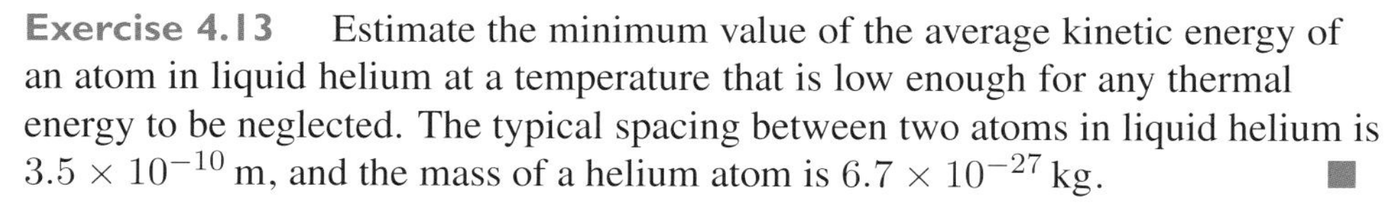
Exercise 4.13 Estimate the minimum value of the average kinetic energy of an atom in liquid helium at a temperature that is low enough for any thermal energy to be neglected. The typical spacing between two atoms in liquid helium is 3.5×10^{-10} m, and the mass of a helium atom is 6.7×10^{-27} kg. ■

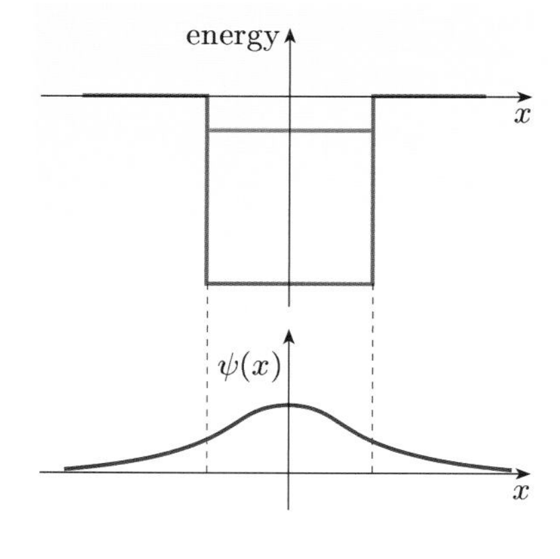

Figure 4.12 A weakly-bound eigenfunction near the top of a potential energy well can extend far beyond the attractive region.

Our estimate $\Delta x \simeq 0.18L$ is reasonable for states near the bottom of a square well, but becomes less reliable near the top of the well. Figure 4.12 shows an eigenfunction very near the top of a square well. This function extends far beyond the confines of the well because the particle penetrates deep into the classically forbidden region. Such an eigenfunction represents a bound state — a state in which the particle is trapped somewhere in the vicinity of the well rather than escaping to infinity. However, in spite of being trapped, the particle is quite likely to be found outside the well, and has $\Delta x \gg 0.18L$.

States like this really do exist. For example, a lithium-11 nucleus is five times larger than normal because two of its neutrons occupy a sort of halo around the central core. The wave function of these neutrons extends far beyond the range of the nuclear force that keeps them bound to the nucleus. In solid-state physics, two electrons in a superconductor can join together to form an entity called a Cooper pair. The wave function of a Cooper pair extends over a distance that is at least 100 times larger than the range of the force that binds the electrons together.

Because it is may not be obvious what value to take for Δx, an effective way of using the uncertainty principle is to treat Δx as an adjustable parameter and to vary it in order to find the minimum energy consistent with the uncertainty principle. This minimum energy is usually a good approximation to the ground-state energy of the system. The following worked example shows how this method works.

Worked Example 4.4

In classical physics, a particle undergoing simple harmonic motion has total energy

$$E = \frac{p_x^2}{2m} + \tfrac{1}{2}m\omega_0^2 x^2,$$

where m is the mass of the particle and ω_0 is the angular frequency of the oscillation. Use the uncertainty principle to estimate the energy of the ground state and the uncertainties in position and momentum in the ground state.

Essential skill

Using the uncertainty principle to estimate energies and uncertainties in the ground state

Solution

Let the uncertainty in position be $\Delta x = s$; then the uncertainty in momentum obeys $\Delta p_x \geq \hbar/2s$. The potential energy well is symmetric about the origin, so $\langle x \rangle = \langle p_x \rangle = 0$ and the uncertainties in position and momentum provide typical values of position and momentum. It follows that a typical value of the total energy obeys the inequality

$$E \geq \frac{1}{2m}\left(\frac{\hbar}{2s}\right)^2 + \tfrac{1}{2}m\omega_0^2 s^2 = \frac{\hbar^2}{8ms^2} + \tfrac{1}{2}m\omega_0^2 s^2.$$

The minimum value of the right-hand side is the smallest energy consistent with the uncertainty principle; we can take this minimum value as giving a good estimate of the ground-state energy. Denoting the function on the right-hand side by $f(s)$, we can minimize $f(s)$ by the usual methods of calculus. This involves taking the derivative of $f(s)$ with respect to s, and setting the result equal to zero. We obtain

$$\frac{\mathrm{d}f}{\mathrm{d}s} = \frac{\mathrm{d}}{\mathrm{d}s}\left(\frac{\hbar^2}{8ms^2} + \tfrac{1}{2}m\omega_0^2 s^2\right) = -\frac{\hbar^2}{4ms^3} + m\omega_0^2 s = 0,$$

so $s^2 = \hbar/2m\omega_0$, and the ground-state energy is estimated to be

$$E_{\text{min}} = \frac{\hbar^2}{8m}\frac{2m\omega_0}{\hbar} + \tfrac{1}{2}m\omega_0^2\frac{\hbar}{2m\omega_0} = \tfrac{1}{2}\hbar\omega_0.$$

The uncertainty in position in the ground state is $\Delta x = s = (\hbar/2m\omega_0)^{1/2}$. Finally, because we are dealing with a low-lying state, the uncertainty in momentum can be roughly estimated by taking the lowest value allowed by the uncertainty principle. This gives $\Delta p_x \simeq \hbar/2s = (\hbar m\omega_0/2)^{1/2}$.

These estimates turn out to be *exactly* correct (as you will see in the next chapter). This lucky coincidence does not extend to other systems, but the method remains useful in providing order-of-magnitude estimates when Schrödinger's equation is difficult to solve directly.

ε_0 is a fundamental constant that determines the strength of electrostatic forces; it is called the *permittivity of free space*. The values of some important physical constants are listed inside the back cover.

Exercise 4.14 The classical expression for the total energy of an electron interacting with a proton in a hydrogen atom is

$$E = \frac{p_x^2 + p_y^2 + p_z^2}{2m} - \frac{e^2}{4\pi\varepsilon_0 r},$$

where r is the distance between the electron (of charge $-e$) and the proton (of charge e). Taking the typical distance of the electron from the proton to be s, and assuming for simplicity that $\Delta x = \Delta y = \Delta z \simeq s$, estimate the ground-state energy and radius of a hydrogen atom. ■

Summary of Chapter 4

Section 4.1 In wave mechanics, the state of a system is completely specified by its wave function. The time-evolution of the wave function is fully predicted by Schrödinger's equation. Nevertheless, quantum mechanics is indeterministic because the result of measuring an observable in a state described by a given wave function is unpredictable. The act of measurement drastically alters the state of the system, so the wave function after the measurement is not the same as the wave function before the measurement. To test the predictions of quantum physics, we often take a large number of *identical* systems all in the *same* state, and perform the *same* measurement on each system.

Not everything in quantum mechanics is unpredictable. Every energy measurement yields one or other of the energy eigenvalues. In a stationary state corresponding to the energy eigenvalue E_i, we are certain to obtain the energy E_i.

Section 4.2 We consider a system with a discrete set of non-degenerate energy eigenvalues (a particle in a one-dimensional infinite well). If the energy of a system is measured at time t, then the probability of obtaining the ith energy eigenvalue E_i is given by the overlap rule

$$p_i = \left| \int_{-\infty}^{\infty} \psi_i^*(x)\, \Psi(x,t)\, \mathrm{d}x \right|^2,$$

where $\Psi(x,t)$ is the wave function of the system at the instant of measurement, and $\psi_i(x)$ is the energy eigenfunction with eigenvalue E_i. Both the wave function and the eigenfunction are assumed to be normalized. For this rule to be consistent, energy eigenfunctions with different eigenvalues must be orthogonal to one another.

If, at the time of measurement, the wave function is a linear combination of energy eigenfunctions, then the probability of obtaining the energy eigenvalue E_i is given by $|c_i(t)|^2$, where $c_i(t)$ is the coefficient of the energy eigenfunction $\psi_i(x)$ in the wave function.

These rules can be extended to other observables and systems.

Section 4.3 If an observable A has a discrete set of possible values, the expectation value of A in a given state Ψ is defined by

$$\langle A \rangle = \sum_i p_i a_i,$$

where p_i is the probability of obtaining value a_i in the state Ψ. Expectation values can be calculated using the sandwich integral rule

$$\langle A \rangle = \int_{-\infty}^{\infty} \Psi^*(x,t)\,\widehat{\mathrm{A}}\,\Psi(x,t)\,\mathrm{d}x.$$

Sections 4.4 and 4.5 The uncertainty of an observable A is defined by

$$\Delta A = \left\langle (A - \langle A \rangle)^2 \right\rangle^{1/2} = \left(\langle A^2 \rangle - \langle A \rangle^2 \right)^{1/2}.$$

With this definition, the Heisenberg uncertainty principle states that

$$\Delta x\,\Delta p_x \geq \frac{\hbar}{2}.$$

Achievements from Chapter 4

After studying this chapter you should be able to:

4.1 Explain the meanings of the newly defined (emboldened) terms and symbols, and use them appropriately.

4.2 Recall that the allowed values of a discrete observable are given by its eigenvalues.

4.3 Use the overlap rule and/or the coefficient rule to calculate the probability distribution of an observable in a given state.

4.4 Recall that the overlap rule can be extended to an observable with a continuum of possible values.

4.5 Define the expectation value of an observable in a given state. Calculate expectation values in simple cases.

4.6 Define the uncertainty of an observable in a given state. Calculate uncertainties in simple cases.

4.7 State and explain the meaning of the Heisenberg uncertainty principle. Use the uncertainty principle to make order-of-magnitude estimates.

Chapter 5 Simple harmonic oscillators

Introduction

Oscillations are a familiar type of motion. A guitar string, the suspension system of a car and a pendulum all oscillate in various ways. If the amplitude is small, it is generally a good approximation to suppose that the oscillation is **simple harmonic**, with a displacement that depends sinusoidally on time. Any system that performs motion of this sort is called a simple harmonic oscillator, or a **harmonic oscillator** for short.

Oscillations are also common in the microscopic world. For example, a nitrogen molecule consists of two nitrogen atoms separated by an equilibrium distance of 1.09×10^{-10} m. But molecules are not rigid. From a classical perspective, we can picture a nitrogen molecule as stretching and compressing, producing an oscillation similar to that of two balls joined by a spring. Atoms in solids also oscillate around equilibrium positions, and some atomic nuclei wobble about equilibrium shapes. Moreover, sinusoidal waves are closely related to sinusoidal oscillations, so the quantum physics of waves, including light, is generally based on the behaviour of harmonic oscillators. The founders of quantum mechanics knew that oscillations were important. When Heisenberg took the first tentative steps towards quantum mechanics in 1925, the first system he tackled was the harmonic oscillator.

The physics of harmonic oscillators has two aspects. Obviously, there is the oscillation itself — a time-dependent process in which something swings back and forth; in classical mechanics, we concentrate on this almost exclusively. In quantum mechanics, however, another issue must be tackled first. We need to investigate the stationary states and energy levels of the system. This requires us to solve the time-independent Schrödinger equation for a harmonic oscillator and interpret its solutions.

In a stationary state $\Psi_n(x,t) = \psi_n(x)\,\mathrm{e}^{-\mathrm{i}E_n t/\hbar}$, the probability density $|\Psi_n(x,t)|^2$ is independent of time. So a stationary state describes a sort of 'suspended animation' — it has a definite energy, but it does not provide a description of motion through space. This seems strange because we have no direct experience of stationary states in everyday life, but the physics of microscopic systems, such as atoms or molecules, is dominated by them. For example, when molecules collide with one another or absorb or emit electromagnetic radiation, their stationary-state wave functions determine the transitions that are likely to occur. The transitions between vibrational energy levels produce infrared spectra and influence the rates of chemical reactions; it has also been suggested that they affect our sense of smell.

It turns out that the to-and-fro motion observed in a simple harmonic oscillator is well described by *linear combinations* of stationary states. This is the main topic of the *next* chapter (Chapter 6); the present chapter deals only with single stationary states. It is organized as follows. Section 5.1 gives a brief description of classical simple harmonic motion. Section 5.2 gives a quantum-mechanical description; it introduces the time-independent Schrödinger equation and describes its solutions, the energy eigenfunctions, together with the energy eigenvalues. These eigenfunctions and eigenvalues are justified in Section 5.3, which also introduces some useful new tools — raising and lowering operators. Finally, in Section 5.4, we calculate expectation values and uncertainties for a harmonic oscillator, and derive a selection rule that helps explain the spectral lines emitted by vibrating molecules.

5.1 Classical harmonic oscillators

It is sensible to begin by reviewing the classical physics of simple harmonic motion. This will allow us to introduce some terminology and provide a reference against which the quantum-mechanical description can be compared.

Consider a particle of mass m that moves to and fro along the x-axis, with a position of stable equilibrium at the origin, $x = 0$. When the particle moves away from the origin, it feels a force pulling it back towards the origin. In simple harmonic motion, the force is proportional to the displacement:

$$F_x = -Cx. \tag{5.1}$$

This is **Hooke's law**. The constant C is called the **force constant**, and x is the displacement of the particle from the equilibrium position at $x = 0$. The minus sign ensures that F_x is a **restoring force** — one that pulls the particle back towards its equilibrium position.

In classical physics, the motion is governed by Newton's second law

$$m\,\frac{\mathrm{d}^2x}{\mathrm{d}t^2} = F_x = -Cx,$$

which can be rearranged to give

$$\frac{\mathrm{d}^2x}{\mathrm{d}t^2} + \omega_0^2 x = 0, \tag{5.2}$$

where

$$\omega_0 = \sqrt{\frac{C}{m}} \tag{5.3}$$

is a constant. The solution of Equation 5.2 can be written as

$$x(t) = A\cos(\omega_0 t + \phi), \tag{5.4}$$

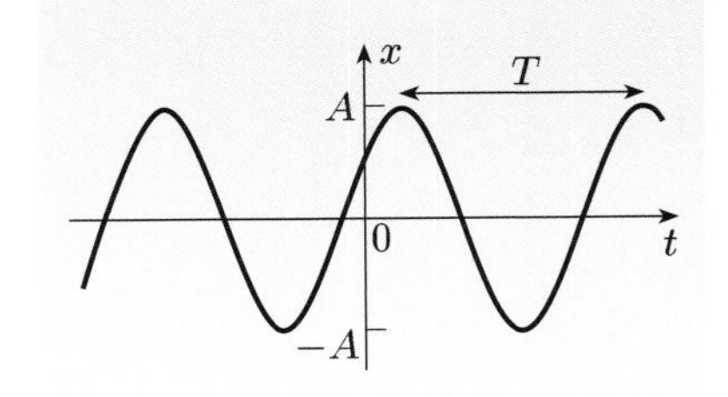

Figure 5.1 The displacement of a particle undergoing classical simple harmonic motion.

which describes a sinusoidal oscillation, as illustrated in Figure 5.1. The to-and-fro motion of the particle is shown by the fact that the graph oscillates above and below $x = 0$. There is no damping in our model, so the oscillation continues indefinitely, each cycle of period T an exact repetition of the last.

In addition to the parameter ω_0 that characterizes the oscillator, Equation 5.4 contains two *arbitrary* constants A and ϕ. This number of arbitrary constants is appropriate for the general solution of a second-order differential equation. The meaning of all these constants is as follows.

Note that ω_0 is *not* an arbitrary constant because it is determined by the system parameters m and C.

- A is the **amplitude** of the oscillation. This is the magnitude of the maximum displacement of the particle from its equilibrium position.

- ϕ is the **phase constant** of the oscillation. This depends on when the particle reaches a position of maximum displacement. If the particle reaches its maximum positive displacement at $t = 0$, then $\phi = 0$.
- ω_0 is the **angular frequency** of the oscillation. It is related to the frequency f and period T by

$$\omega_0 = 2\pi f = \frac{2\pi}{T}. \tag{5.5}$$

Since $\omega_0 = \sqrt{C/m}$, oscillations of high angular frequency are associated with large restoring forces acting on particles of low mass, just as you would expect. The angular frequency is independent of the amplitude and the phase constant, and so characterizes the oscillator under study, no matter how it has been set in motion. This is a special feature of simple harmonic motion; it relies on the restoring force being exactly proportional to the displacement, and is not true more generally.

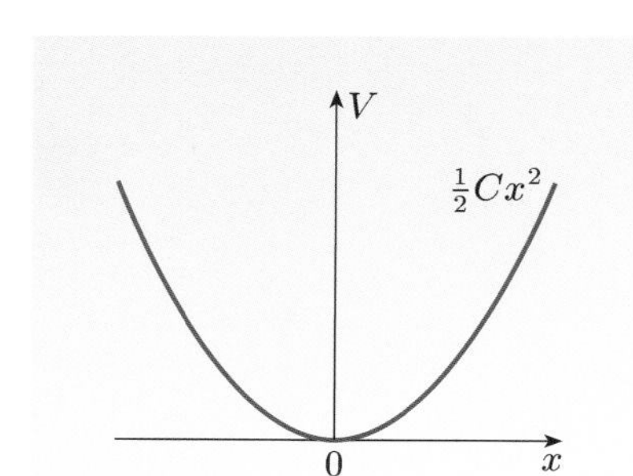

Figure 5.2 A simple harmonic potential energy well. The zero of potential energy is taken to be at the bottom of the well for a harmonic oscillator.

Simple harmonic motion can also be analyzed in terms of energy. The potential energy function of a simple harmonic oscillator takes the form

$$V(x) = \tfrac{1}{2}Cx^2,$$

where C is the force constant. This function is sketched in Figure 5.2. It has its minimum value $V = 0$ at the equilibrium position, $x = 0$, and increases either side of the equilibrium position as the displacement grows in magnitude.

The particle has momentum $p_x = mv_x$ and kinetic energy

$$E_{\text{kin}} = \tfrac{1}{2}mv_x^2 = \frac{p_x^2}{2m}.$$

Adding both contributions to the energy together, the total energy is

$$E = \frac{p_x^2}{2m} + \tfrac{1}{2}Cx^2. \tag{5.6}$$

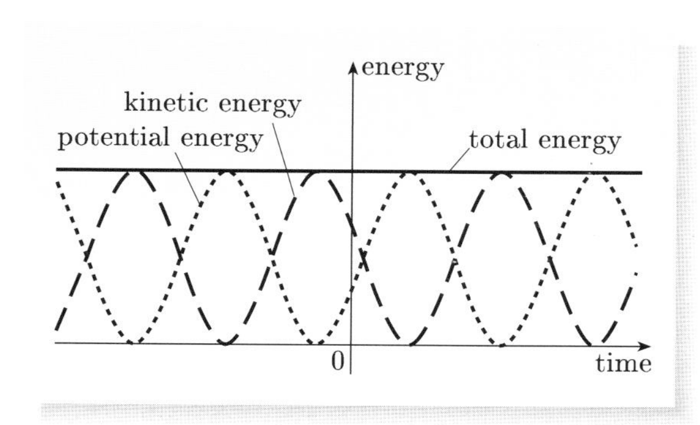

Figure 5.3 The kinetic, potential and total energies of a simple harmonic oscillator. The kinetic and potential energies interchange, but the total energy remains constant.

During the oscillation, energy is continually converted between kinetic and potential forms. When the particle is near its equilibrium position, most of the energy is in the form of kinetic energy. Near a point of maximum displacement, most of the energy is in the form of potential energy. However, the total energy always remains constant, in agreement with the law of conservation of energy. The kinetic, potential and total energies of a simple harmonic oscillator are illustrated in Figure 5.3. The graphs of kinetic and potential energy are similar in shape, but are out of phase with one another. It follows that the average kinetic energy is equal to the average potential energy, provided that the averages are taken over one or more complete periods of the oscillation.

The total energy E is related to the amplitude of the oscillation, as you will see in the next exercise.

Exercise 5.1 A particle performs simple harmonic motion of amplitude A in a harmonic oscillator with force constant C. Show that its total energy is

$$E = \tfrac{1}{2}CA^2.$$

Exercise 5.2 Use Equations 5.4 and 5.6 to verify that the total energy of a harmonic oscillator is independent of time. ■

5.1.1 Diatomic molecules in classical physics

Simple harmonic motion does not always involve the motion of a single particle about a fixed point. For example, molecules can vibrate internally. The simplest case is provided by a diatomic molecule — one consisting of two atoms. From a classical perspective, a diatomic molecule can be pictured as two balls joined by a spring (Figure 5.4). Several types of motion are possible: translational motion of the molecule as a whole, rotational motion around an axis, and vibrational motion in which the bond between the two atoms alternately stretches and compresses. We focus on the vibrational motion here, emphasizing its relationship to the basic description of simple harmonic motion given above.

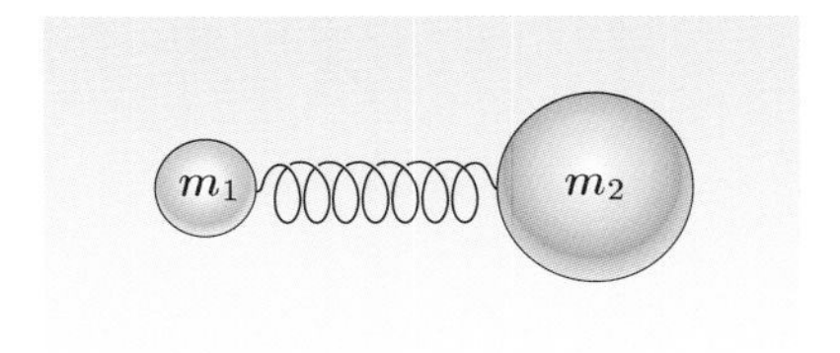

Figure 5.4 A classical model of a diatomic molecule.

In a diatomic molecule, such as CO, we have two particles (the carbon and oxygen atoms) which interact with one other, rather than a single particle responding to an external force. We neglect any external forces acting on the molecule. Then, ignoring rotational motion, and placing the x-axis along the line of separation of the atoms, the total energy of the molecule is

$$E = \frac{p_1^2}{2m_1} + \frac{p_2^2}{2m_2} + V(x_2 - x_1), \tag{5.7}$$

We have omitted the subscript x on p_1 and p_2 for simplicity.

where m_i, p_i and x_i are the mass, momentum and position of atom i. The first term in Equation 5.7 is the kinetic energy of atom 1, the second term is the kinetic energy of atom 2, and the last term is the potential energy of interaction between the two atoms, which depends only on the distance between them.

Now, a remarkable simplification can be achieved, merely by changing variables. We shall not spend time on the algebraic details, but just state the final result. It turns out that Equation 5.7 can be rewritten as

$$E = \frac{P^2}{2M} + \frac{p^2}{2\mu} + V(x), \tag{5.8}$$

where $M = m_1 + m_2$ and $P = p_1 + p_2$ are the total mass and total momentum of the molecule, and the other variables are defined by

$$x = x_2 - x_1, \quad \mu = \frac{m_1 m_2}{m_1 + m_2} \quad \text{and} \quad p = \mu\frac{\mathrm{d}x}{\mathrm{d}t}. \tag{5.9}$$

In Equation 5.8, the first term, $P^2/2M$, is the kinetic energy associated with translational motion of the molecule as a whole. This term has nothing to do with any internal vibration of the molecule, and we can set it equal to zero by agreeing to work in the **centre-of-mass frame** (the reference frame where the total momentum of the molecule is zero).

- Why is the centre-of-mass frame more suitable than a reference frame with its origin fixed on one of the two atoms?

○ With no external force acting on the molecule, the centre of mass moves uniformly, so it provides the origin of an **inertial frame**. Newton's laws apply in such a frame; the same cannot be said of a reference frame fixed on one of the vibrating atoms.

In the centre-of-mass frame, the energy becomes

$$E = \frac{p^2}{2\mu} + V(x), \tag{5.10}$$

which is just like the energy of a *single* particle with its potential energy provided by a fixed external agency. The mass

$$\mu = \frac{m_1 m_2}{m_1 + m_2} \tag{5.11}$$

is less than either m_1 or m_2, and is called the **reduced mass** of the molecule. So, here is the general point to remember:

> The *internal* motion of a two-particle system can be analyzed by considering the motion of a single particle with the *reduced* mass.

If $m_1 \ll m_2$, the reduced mass is only slightly less than m_1, leading to a small correction in the results that would be obtained by (wrongly) taking the more massive particle to be permanently at rest. In a molecule such as N_2, however, the reduced mass is only half the mass of either atom.

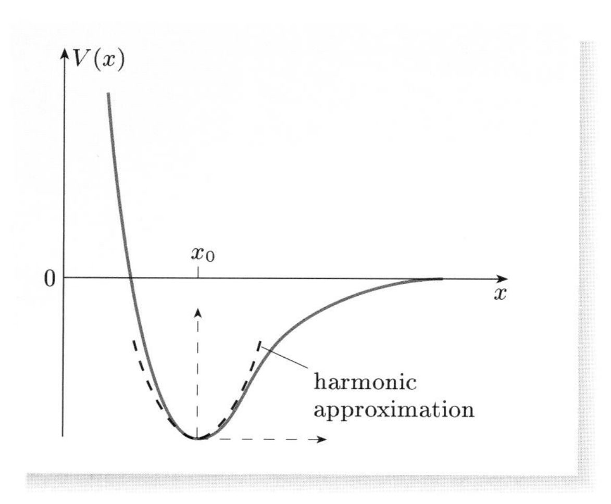

Figure 5.5 Solid curve: the potential energy of interaction for atoms in a typical diatomic molecule, plotted against their separation. Dashed curve: an approximation valid near the potential energy minimum; referred to the dashed axes, this is the potential energy function of a harmonic oscillator.

Finally, let us consider the form of the potential energy function. Figure 5.5 shows a graph of potential energy versus separation for two atoms in a typical diatomic molecule. The minimum in this graph corresponds to the equilibrium separation of the atoms. The potential energy rises either side of the equilibrium separation as the atoms resist being squashed too closely together or pulled too far apart.

Near the minimum at $x = x_0$, the potential energy function can be approximated by

$$V(x) \simeq V(x_0) + \tfrac{1}{2}C(x - x_0)^2. \tag{5.12}$$

With a suitable choice for the origin of coordinates and zero of potential energy (indicated by the dashed set of axes in Figure 5.5), this is the potential energy of a harmonic oscillator. The approximation is good for oscillations of relatively small amplitude; under these circumstances, we can say that a vibrating molecule undergoes simple harmonic oscillations, with a bond length that varies sinusoidally in time.

Exercise 5.3 A hydrogen molecule consists of two hydrogen atoms, each of mass 1.67×10^{-27} kg. The molecule has vibration frequency $f = 1.25 \times 10^{14}$ Hz. If the force between the two hydrogen atoms can be modelled by Hooke's law, what is the value of the force constant, C? ■

5.2 Quantum harmonic oscillators

5.2.1 Schrödinger's equation

We now turn to the quantum-mechanical description of a harmonic oscillator. It is easy to construct Schrödinger's equation for this system, following the usual three-step recipe of Chapter 2: write down the classical Hamiltonian function, replace variables by operators, and then assemble Schrödinger's equation.

The first step is already complete because Equation 5.6 is the total energy of the system expressed in terms of momentum, so it is the classical Hamiltonian function. Applying the conversion rule

$$\frac{\widehat{\mathrm{p}}_x^2}{2m} \longrightarrow -\frac{\hbar^2}{2m}\frac{\partial^2}{\partial x^2},$$

we obtain the Hamiltonian operator

$$\widehat{\mathrm{H}} = -\frac{\hbar^2}{2m}\frac{\partial^2}{\partial x^2} + \tfrac{1}{2}Cx^2.$$

It is convenient to express this in terms of the variable $\omega_0 = \sqrt{C/m}$ introduced in Equation 5.3. This gives

We shall call ω_0 the *classical* angular frequency, to show that we are not prejudging its quantum-mechanical significance.

$$\widehat{\mathrm{H}} = -\frac{\hbar^2}{2m}\frac{\partial^2}{\partial x^2} + \tfrac{1}{2}m\omega_0^2x^2.$$

Finally, inserting the Hamiltonian operator in the general form of Schrödinger's equation, $\mathrm{i}\hbar\,\partial\Psi/\partial t = \widehat{\mathrm{H}}\Psi$, we obtain

$$\mathrm{i}\hbar\frac{\partial\Psi}{\partial t} = -\frac{\hbar^2}{2m}\frac{\partial^2\Psi}{\partial x^2} + \tfrac{1}{2}m\omega_0^2x^2\,\Psi(x,t). \tag{5.13}$$

This is *Schrödinger's equation* for a harmonic oscillator. It tells us how any wave function describing the state of the oscillator evolves in time. However, we are interested in the energy levels of the oscillator, so we shall focus on states that have a definite energy — the stationary states. A stationary-state wave function takes the form

$$\Psi(x,t) = \psi(x)\,\mathrm{e}^{-\mathrm{i}Et/\hbar}, \tag{5.14}$$

where the energy eigenfunction $\psi(x)$ satisfies the *time-independent Schrödinger equation* for a harmonic oscillator:

$$-\frac{\hbar^2}{2m}\frac{\mathrm{d}^2\psi}{\mathrm{d}x^2} + \tfrac{1}{2}m\omega_0^2x^2\,\psi(x) = E\psi(x). \tag{5.15}$$

Exercise 5.4 By substituting Equation 5.14 into Equation 5.13, show that $\psi(x)$ satisfies Equation 5.15. ■

The time-independent Schrödinger equation is the eigenvalue equation for energy. It is accompanied by the usual requirement that its solutions should not diverge as $x \to \pm\infty$. If we pick an arbitrary value of E, and try to find a function $\psi(x)$ that satisfies Equation 5.15, and does not diverge at infinity, we will almost certainly fail. Only very special values of E allow this to happen. The special values of E that permit non-divergent solutions of Equation 5.15 are the *energy eigenvalues* of the harmonic oscillator. These energy eigenvalues give the possible energies of the oscillator. They form a discrete set, so the quantum harmonic oscillator has a discrete set of energy levels, which we shall label as $E_0, E_1, E_2, \ldots$.

Note that the numbering starts from 0 in this case.

Each energy eigenvalue E_i has a corresponding *energy eigenfunction*, $\psi_i(x)$, and a stationary-state wave function

$$\Psi_i(x,t) = \psi_i(x)\,\mathrm{e}^{-\mathrm{i}E_it/\hbar}.$$

This wave function describes a state in which the oscillator is *certain* to have energy E_i. The value of the stationary-state wave function at time $t = 0$ is equal to the energy eigenfunction, $\psi_i(x)$. Both the wave function and the eigenfunction are normalized:

$$\int_{-\infty}^{+\infty} \Psi_i^*(x,t)\,\Psi_i(x,t)\,\mathrm{d}x = \int_{-\infty}^{+\infty} \psi_i^*(x)\,\psi_i(x)\,\mathrm{d}x = 1,$$

expressing the fact that the particle is certain to be found *somewhere*.

5.2.2 Eigenvalues and eigenfunctions

To find the energy eigenfunctions and eigenvalues of a harmonic oscillator, we need to solve the time-independent Schrödinger equation for a harmonic oscillator (Equation 5.15). From the point of view of physical principles, this is similar to the task tackled in Chapter 3, where the time-independent Schrödinger equation was solved for an infinite square well. The only difference is one of mathematical difficulty: we have a tougher nut to crack in this case. We will therefore begin by *describing* the solutions that emerge. Later, in Section 5.3, you will see how these solutions can be obtained from Equation 5.15.

Energy eigenvalues

The energy eigenvalues of a harmonic oscillator turn out to be

$$E_n = (n + \tfrac{1}{2})\hbar\omega_0 \quad \text{for } n = 0, 1, 2, \ldots, \tag{5.16}$$

where $\omega_0 = \sqrt{C/m}$ is the classical angular frequency of the oscillator. These are the energy levels of the oscillator. As shown in Figure 5.6, the energy levels are equally spaced, starting with the lowest energy $\hbar\omega_0/2$, and with a constant spacing of $\hbar\omega_0$ between neighbouring levels.

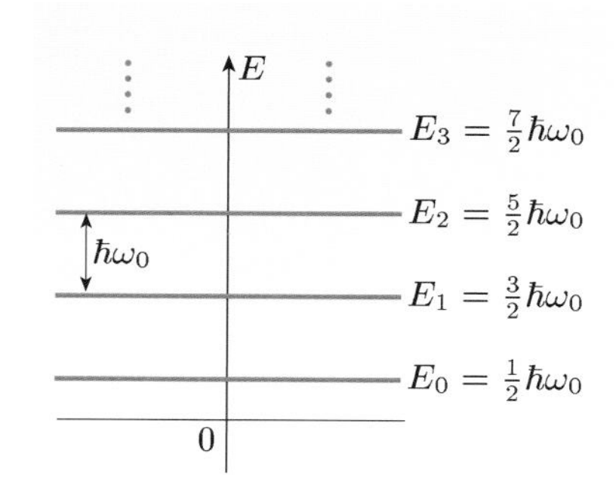

Figure 5.6 The first four energy levels of a harmonic oscillator, at the bottom of an infinite ladder of equally spaced energy levels.

In classical physics, the oscillating particle can come to rest at the bottom of the well, and can therefore have zero energy. By contrast, the lowest possible energy of a particle in a quantum harmonic oscillator is $E_0 = \hbar\omega_0/2$. This is called the **zero-point energy** of the oscillator because it persists even at the absolute zero of temperature, where one might expect the oscillator to be completely calm. Zero-point energy is not unexpected — you saw in Chapter 4 that all confined particles have some energy as a consequence of the uncertainty principle — but we now see the zero-point energy emerging directly from the energy eigenvalues. Note that the lowest energy level, E_0, is assigned quantum number $n = 0$. This is unlike the convention adopted for a square well, so you will need to remember this fact.

The pattern of equally-spaced energy levels shown in Figure 5.6 is a hallmark of a harmonic oscillator. In a one-dimensional infinite square well, the energy levels get further apart as the energy increases; in an un-ionized hydrogen atom, the energy levels get closer together as the energy increases, but the harmonic oscillator has an infinite ladder of equally-spaced energy levels; it is unique in this respect.

In practice, of course, a harmonic oscillator is an idealization — stretch any rubber band far enough, and it will break. However, we know that the atoms in a diatomic molecule interact via a potential energy function that is approximately harmonic near the bottom of the well. It is therefore interesting to ask whether real diatomic molecules have vibrational energy levels that are similar to those of a harmonic oscillator. Figure 5.7 shows the vibrational energy levels of a carbon monoxide molecule (CO) inferred from spectral lines observed in the infrared part of the spectrum. These energy levels are almost equally spaced, especially near the bottom of the well, but become slightly more closely spaced as the energy increases. In this case, a harmonic oscillator model provides a reasonable description of at least the first twenty-one vibrational energy levels.

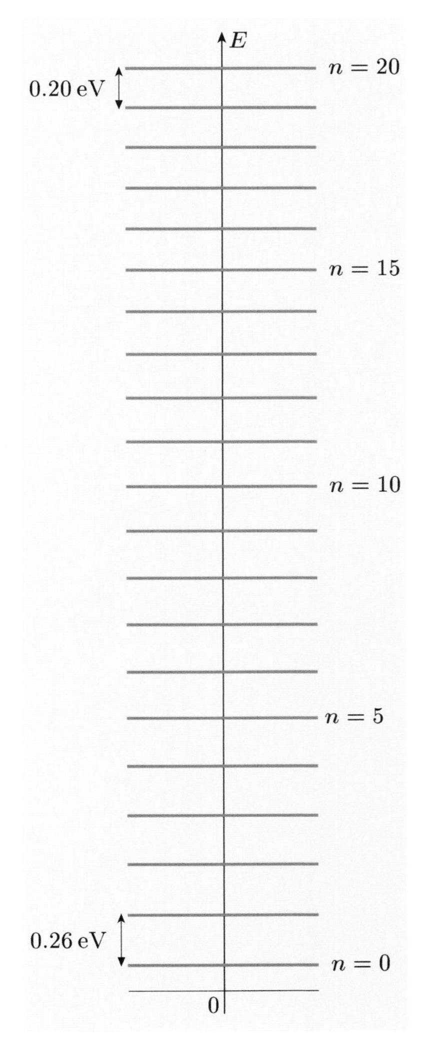

Figure 5.7 The first 21 vibrational energy levels of a CO molecule are almost, but not quite, equally spaced.

It is worth commenting on the relationship between the energy level diagram in Figure 5.7 and the spectral lines emitted by a CO molecule. At first sight, you might suppose that a molecule in a high energy level could jump to *any* lower energy level by emitting a photon of energy

$$\hbar\omega = E_i - E_j,$$

where E_i is the initial energy level, E_j is the final energy level and ω is the angular frequency of the emitted light.

However, the observed spectrum is much simpler than this. Ignoring very faint lines, the vibrational spectrum is produced by transitions between *neighbouring* energy levels — between E_{20} and E_{19}, for example, but not between E_{20} and E_{17}. A rule of this kind is called a **selection rule** because it *selects* the observed spectral lines from a wider set of possibilities. The selection rule we have just quoted is specific to vibrational energy levels. We shall explain why it is true at the end of this chapter.

The ground-state energy eigenfunction

Each energy eigenvalue E_i is associated with an energy eigenfunction $\psi_i(x)$. We begin by considering the eigenfunction $\psi_0(x)$, corresponding to the ground-state energy, $E_0 = \hbar\omega_0/2$. This eigenfunction turns out to have the form

$$\psi_0(x) = C_0 \mathrm{e}^{-x^2/2a^2}, \tag{5.17}$$

where a is a positive constant and C_0 is a constant of normalization. Functions of this type are called **Gaussian functions**, named after the great mathematician Carl Friedrich Gauss (1777–1855) who used them in his work on statistics. The function $\psi_0(x)$ is plotted in Figure 5.8 on page 133. This bell-shaped curve has no nodes, it decreases either side of a maximum value at $x = 0$, and it becomes very small for $|x| > 3a$.

We can verify that $\psi_0(x)$ is an energy eigenfunction of the harmonic oscillator by substituting it into the time-independent Schrödinger equation. The following worked example shows how this is done. As a bonus, we shall confirm that the corresponding eigenvalue is $\hbar\omega_0/2$ and find a suitable value for the constant a.

Worked Example 5.1

Show that the function $\psi_0(x) = C_0 \mathrm{e}^{-x^2/2a^2}$ is an energy eigenfunction for a harmonic oscillator, provided that the positive constant a is chosen appropriately. Find a suitable value for a, and determine the energy eigenvalue corresponding to this eigenfunction.

Essential skill

Verifying that a given function is a solution of the time-independent Schrödinger equation

Solution

We need to substitute the given function into the time-independent Schrödinger equation to see whether, under suitable circumstances, it satisfies this equation; these circumstances will give the value of a.

Schrödinger's equation involves the second derivative of the eigenfunction, so we must find the second derivative of $\psi_0(x)$. Differentiating once gives

$$\frac{d\psi_0}{dx} = \frac{d}{dx}\left(C_0 e^{-x^2/2a^2}\right) = -\frac{x}{a^2}\,C_0 e^{-x^2/2a^2}.$$

Differentiating again gives

$$\frac{d^2\psi_0}{dx^2} = \frac{d}{dx}\left(-\frac{x}{a^2}\,C_0 e^{-x^2/2a^2}\right) = \left(-\frac{1}{a^2} + \frac{x^2}{a^4}\right) C_0 e^{-x^2/2a^2}.$$

Substituting $\psi_0(x)$ into the time-independent Schrödinger equation (Equation 5.15) and cancelling the common factor $C_0 e^{-x^2/2a^2}$ on both sides, we obtain

$$-\frac{\hbar^2}{2m}\left(-\frac{1}{a^2} + \frac{x^2}{a^4}\right) + \tfrac{1}{2}m\omega_0^2 x^2 = E. \tag{5.18}$$

Because this equation is valid for all x, we can extract two separate conditions from it. First, setting $x = 0$ gives

$$E = \frac{\hbar^2}{2ma^2}.$$

Then, using this result for E, and setting $x \neq 0$ in Equation 5.18, we obtain

$$a^4 = \frac{\hbar^2}{m^2\omega_0^2},$$

so a suitable positive value of a is

$$a = \sqrt{\frac{\hbar}{m\omega_0}}.$$

With this choice of a, the above calculation confirms that $\psi_0(x) = C_0 e^{-x^2/2a^2}$ is an energy eigenfunction of the harmonic oscillator, and it also shows that the corresponding eigenvalue is

$$E = \frac{\hbar^2}{2ma^2} = \tfrac{1}{2}\hbar\omega_0.$$

The ground-state eigenfunction still needs to be normalized. This involves finding a suitable value of the normalization constant C_0.

Do not confuse the normalization constant C_0 with the force constant $C = m\omega_0^2$.

Exercise 5.5 Use a standard integral (given inside the back cover) to show that a suitable value of the normalization constant in Equation 5.17 is

$$C_0 = \left(\frac{1}{\sqrt{\pi}\,a}\right)^{1/2}.$$

■

To summarize, the normalized energy eigenfunction corresponding to the ground-state energy E_0 is

$$\psi_0(x) = \left(\frac{1}{\sqrt{\pi}a}\right)^{1/2} e^{-x^2/2a^2}, \tag{5.19}$$

where

$$a = \sqrt{\frac{\hbar}{m\omega_0}}. \tag{5.20}$$

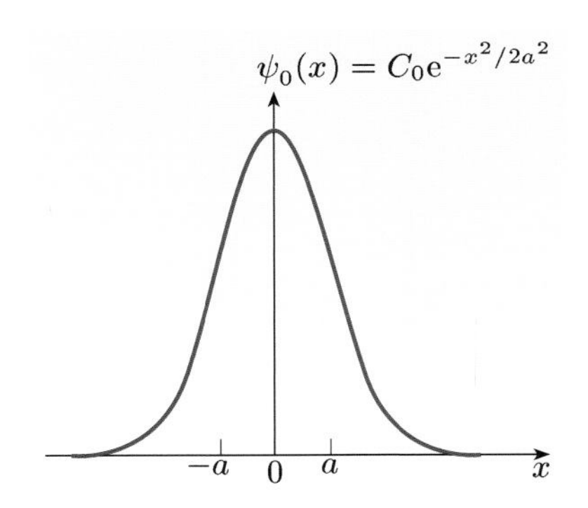

Figure 5.8 The ground-state energy eigenfunction of a harmonic oscillator is a Gaussian function.

The constant a is the only combination of $\hbar$, m and ω_0 (devoid of numerical factors) that has the dimensions of length. It can therefore be regarded as a characteristic length for the oscillator, setting a length scale for its quantum properties. We shall call a the **length parameter** of the oscillator.

Exercise 5.6

(a) Show that a has the dimensions of length.

(b) Show that a *classical* harmonic oscillator, with an energy equal to the ground-state energy $E_0 = \hbar\omega_0/2$, has an amplitude equal to a. ■

Exercise 5.6b prompts a comparison between classical and quantum physics. If classical physics were valid, a harmonic oscillator with energy E_0 would have amplitude a, and the oscillating particle would be confined to the region $-a \leq x \leq a$. The particle would not have enough energy to stray any further from the origin. But this is not what is found in quantum physics. Figure 5.8 shows that the energy eigenfunction $\psi_0(x)$ extends beyond the region $-a \leq x \leq a$, so there is some chance of finding the particle in the classically forbidden region $|x| > a$. This is an example of **barrier penetration**, a phenomenon you first met for a finite square well. You should appreciate that barrier penetration does not always involve the motion of a single particle. In the case of a diatomic molecule, it allows the two atoms to be further apart, or closer together, than would be possible for classical particles with the same energy.

Exercise 5.7 A particle is in the ground state of a harmonic oscillator. What is the probability of finding it in the classically forbidden region $|x| > a$?

(You may find the definite integral $\int_1^\infty e^{-x^2}\,dx = 0.139$ useful.) ■

Higher energy eigenfunctions

Finally, let us briefly survey the other energy eigenfunctions. This is mainly for reference purposes: you will not need to remember the details, but you should be aware of the general pattern that emerges for the eigenfunctions $\psi_0(x), \psi_1(x), \psi_2(x), \ldots$, corresponding to the eigenvalues $E_0, E_1, E_2, \ldots$.

It turns out that all the energy eigenfunctions in a harmonic oscillator are of the form

$$\text{constant} \times \left[\text{polynomial in } \frac{x}{a}\right] \times \left[\text{Gaussian function } e^{-x^2/2a^2}\right].$$

In the ground state, the polynomial is equal to 1, but higher energy eigenfunctions have increasingly complicated polynomials. The Gaussian function, however, always remains the same.

The polynomials that appear in the harmonic oscillator energy eigenfunctions are of a kind investigated by mathematicians in the nineteenth century. They are called **Hermite polynomials**, in honour of the French mathematician Charles Hermite (1822–1901). For this reason, the energy eigenfunctions are usually written as

Do not confuse the Hermite polynomial H_n with the Hamiltonian operator $\widehat{\mathrm{H}}$.

$$\psi_n(x) = C_n H_n(x/a)\,\mathrm{e}^{-x^2/2a^2}, \tag{5.21}$$

where $H_n(x/a)$ is the nth Hermite polynomial in x/a, and C_n is a normalization constant. The first few Hermite polynomials and normalization constants are listed in Table 5.1. In general, $H_n(x/a)$ is a polynomial of order n, so the highest power of x/a that appears in it is $(x/a)^n$. Only even powers of x/a are present when n is even, and only odd powers when n is odd.

Table 5.1 The first few Hermite polynomials, $H_n(x/a)$, and normalization constants, C_n.

n	$H_n(x/a)$	C_n
0	1	$\left(\dfrac{1}{\sqrt{\pi}a}\right)^{1/2}$
1	$2\left(\dfrac{x}{a}\right)$	$\left(\dfrac{1}{2\sqrt{\pi}a}\right)^{1/2}$
2	$4\left(\dfrac{x}{a}\right)^2 - 2$	$\left(\dfrac{1}{8\sqrt{\pi}a}\right)^{1/2}$
3	$8\left(\dfrac{x}{a}\right)^3 - 12\left(\dfrac{x}{a}\right)$	$\left(\dfrac{1}{48\sqrt{\pi}a}\right)^{1/2}$

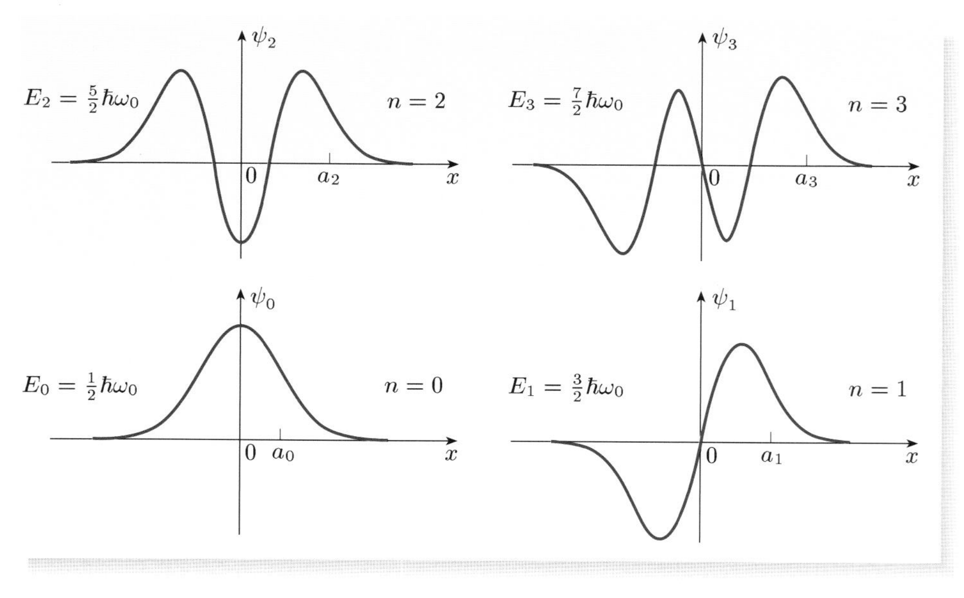

Figure 5.9 The first four energy eigenfunctions of a harmonic oscillator.

Using Equation 5.21 and Table 5.1, the first few energy eigenfunctions are plotted in Figure 5.9. Notice the general pattern: the eigenfunctions are even or odd, depending on whether the quantum number n is even or odd. The ground state $\psi_0(x)$ has no nodes, and each successive eigenfunction gains one additional node, so $\psi_n(x)$ has n nodes. Although not obvious from the figure, the energy eigenfunctions form an orthonormal set, so

$$\int_{-\infty}^{\infty} \psi_i^*(x)\,\psi_j(x)\,\mathrm{d}x = \delta_{ij} = \begin{cases} 1 & \text{if } i = j, \\ 0 & \text{if } i \neq j. \end{cases} \tag{5.22}$$

Each of the energy eigenfunctions exhibits the phenomenon of barrier penetration mentioned earlier. For an eigenfunction with quantum number n, the energy is $E_n = (n + \frac{1}{2})\hbar\omega_0$ and the classical amplitude of oscillation is

$$a_n = \sqrt{\frac{2E_n}{C}} = \sqrt{2n+1}\,a. \tag{5.23}$$

Refer to Exercise 5.1 and recall that $E_n = (n + \frac{1}{2})\hbar\omega_0$ and $C = m\omega_0^2$.

These amplitudes are marked on Figure 5.9, and you can see that each energy eigenfunction extends into the classically forbidden region, $|x| > a_n$.

In classical physics, an oscillating particle moves quickly through its equilibrium position, and more slowly near the points of maximum displacement, where it comes momentarily to rest. It follows that a classical oscillator spends more of its time near points of maximum displacement than near the equilibrium position. It is interesting to ask if a similar feature is found in quantum physics. Figure 5.10 plots the probability density functions, $|\psi_n(x)|^2$, for the ground state ($n = 0$) and a highly excited state ($n = 20$). The ground state defies our classical expectations, but the highly excited state is in broad agreement with the classical picture. It is often the case that quantum-mechanical predictions approach classical predictions in the limit of high quantum numbers. This is not altogether surprising, because familiar objects, which obey classical physics, do have very high quantum numbers. The inventors of quantum mechanics, especially Bohr and Heisenberg, were strongly guided by this general idea, and dignified it with a title — the **correspondence principle**.

Figure 5.10 The probability density $|\psi_n|^2$ plotted for two energy eigenfunctions of a harmonic oscillator: (a) $n = 0$; (b) $n = 20$. The dashed line shows the proportion of time spent by a classical harmonic oscillator in each region, averaged over a complete cycle.

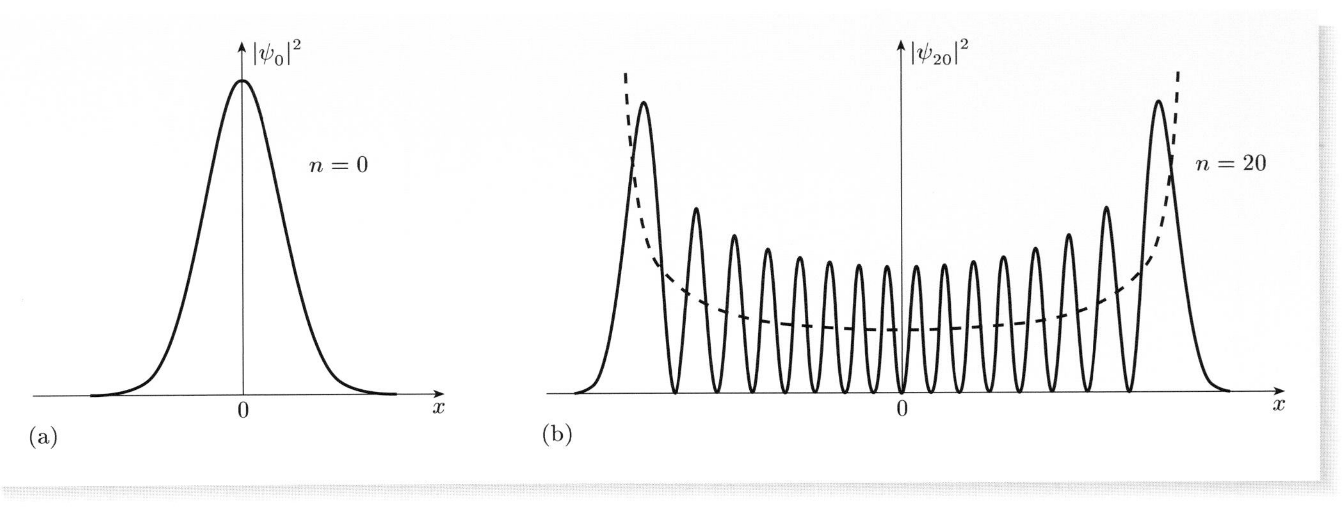

5.3 Solving the time-independent Schrödinger equation

This section will justify the statements made in the preceding section about the energy eigenvalues and eigenfunctions of a harmonic oscillator. In the course of doing so, it will introduce two new operators, called the *raising operator* and the *lowering operator*. These operators turn out to be powerful tools, which will help us simplify many calculations for harmonic oscillators. It will not matter too much if you forget the details of the derivations given in this section, but it is vitally important for you to understand how raising and lowering operators are used. The derivations are important in illustrating this.

5.3.1 Choosing appropriate variables

A first step towards solving the time-independent Schrödinger equation is to write it in a compact form by collecting together some of the constants that appear in it. Recalling that $a = \sqrt{\hbar/m\omega_0}$, we have

$$\frac{\hbar^2}{2m} = \frac{\hbar\omega_0}{2}\left(\frac{\hbar}{m\omega_0}\right) = \frac{\hbar\omega_0}{2}\,a^2$$

and

$$\tfrac{1}{2}m\omega_0^2 = \frac{\hbar\omega_0}{2}\left(\frac{m\omega_0}{\hbar}\right) = \frac{\hbar\omega_0}{2}\,\frac{1}{a^2}.$$

These two relationships allow us to write the time-independent Schrödinger equation for a harmonic oscillator (Equation 5.15) in the form

$$\widehat{\mathrm{H}}\,\psi_n(x) = E_n\,\psi_n(x),$$

where the Hamiltonian operator can be written as

$$\widehat{\mathrm{H}} = \frac{1}{2}\left(\frac{x^2}{a^2} - a^2\,\frac{\partial^2}{\partial x^2}\right)\hbar\omega_0. \qquad (5.24)$$

This consists of two terms: the first (proportional to x^2) is the potential energy term and the second (proportional to $\partial^2/\partial x^2$) is the kinetic energy term. For generality, we have used partial derivatives, but these can always be replaced by ordinary derivatives when $\widehat{\mathrm{H}}$ acts on a function that depends only on x, such as $\psi_n(x)$.

This way of writing the Hamiltonian operator is convenient because both x and a have the dimensions of length, so the term in round brackets is dimensionless, while the factor $\hbar\omega_0$ has the dimensions of energy. This shows that a quantum-mechanical harmonic oscillator is characterized by a length, a, and an energy, $\hbar\omega_0$. It is hardly surprising that the energy eigenfunctions are all functions of x/a, and the energy eigenvalues are all multiples of $\hbar\omega_0$.

5.3.2 Factorizing the Hamiltonian operator

The next step towards solving the time-independent Schrödinger equation is based on the observation that the term in round brackets in Equation 5.24 looks like a difference between two squares. In ordinary algebra, we know that any difference between two squares can be factorized: if m and n are two numbers, we have

$$m^2 - n^2 = (m - n)(m + n).$$

Equation 5.24 appears to involve a similar expression; instead of numbers we now have operators that act on functions, but we can ask whether a similar factorization can be achieved. To investigate this, consider

$$\left(\frac{x}{a} - a\frac{\partial}{\partial x}\right)\left(\frac{x}{a} + a\frac{\partial}{\partial x}\right)f(x), \tag{5.25}$$

where $f(x)$ is any reasonable (i.e. smooth) function of x. Multiplying out the brackets gives

$$\frac{x^2}{a^2}f(x) - a^2\frac{\partial^2 f}{\partial x^2} - \frac{\partial}{\partial x}(xf(x)) + x\frac{\partial f}{\partial x}. \tag{5.26}$$

Notice that the third term requires us to differentiate the product $xf(x)$. Carrying out this differentiation and cancelling terms, we obtain

$$\left(\frac{x}{a} - a\frac{\partial}{\partial x}\right)\left(\frac{x}{a} + a\frac{\partial}{\partial x}\right)f(x) = \frac{x^2}{a^2}f(x) - a^2\frac{\partial^2 f}{\partial x^2} - f(x).$$

Hence

$$\left(\frac{x^2}{a^2} - a^2\frac{\partial^2}{\partial x^2}\right)f(x) = \left[\left(\frac{x}{a} - a\frac{\partial}{\partial x}\right)\left(\frac{x}{a} + a\frac{\partial}{\partial x}\right) + 1\right]f(x).$$

This is true for all smooth functions $f(x)$, so we can express the result as an identity between *operators*. Including the factor of $1/2$ that appears in the Hamiltonian, we have

$$\frac{1}{2}\left(\frac{x^2}{a^2} - a^2\frac{\partial^2}{\partial x^2}\right) = \frac{1}{2}\left(\frac{x}{a} - a\frac{\partial}{\partial x}\right)\left(\frac{x}{a} + a\frac{\partial}{\partial x}\right) + \frac{1}{2}. \tag{5.27}$$

You can see that our attempt at factorization almost works, but there is a remainder of $1/2$. We can also take the two factors in Equation 5.25 in the opposite order. An almost identical calculation leads to

$$\frac{1}{2}\left(\frac{x^2}{a^2} - a^2\frac{\partial^2}{\partial x^2}\right) = \frac{1}{2}\left(\frac{x}{a} + a\frac{\partial}{\partial x}\right)\left(\frac{x}{a} - a\frac{\partial}{\partial x}\right) - \frac{1}{2}, \tag{5.28}$$

so in this case the remainder is $-1/2$.

These 'factorizations' turn out to be very useful, so we will introduce a notation that represents them more compactly. We define two new operators:

$$\widehat{\mathrm{A}}^\dagger = \frac{1}{\sqrt{2}}\left(\frac{x}{a} - a\frac{\partial}{\partial x}\right), \tag{5.29}$$

$$\widehat{\mathrm{A}} = \frac{1}{\sqrt{2}}\left(\frac{x}{a} + a\frac{\partial}{\partial x}\right). \tag{5.30}$$

For reasons that will emerge shortly, $\widehat{\mathrm{A}}^\dagger$ is called a **raising operator** and $\widehat{\mathrm{A}}$ is called a **lowering operator**. It is conventional to place a dagger on the raising operator and to leave the lowering operator unadorned. Notice that the dagger on the raising operator (which looks a bit like a plus sign) is accompanied by a *minus* sign in the definition of the operator.

Using these definitions, together with Equation 5.27, the Hamiltonian operator can be expressed in the 'factorized' form

$$\widehat{\mathrm{H}} = \left(\widehat{\mathrm{A}}^\dagger \widehat{\mathrm{A}} + \tfrac{1}{2}\right) \hbar\omega_0. \tag{5.31}$$

Remember that $\widehat{\mathrm{A}}^\dagger$ and $\widehat{\mathrm{A}}$ are first-order differential operators (Equations 5.29 and 5.30), so we have expressed a second-order differential operator, $\widehat{\mathrm{H}}$, in terms of two first-order differential operators. You will soon see why this is useful.

First, we must make a very important point about the ordering of the operators $\widehat{\mathrm{A}}^\dagger$ and $\widehat{\mathrm{A}}$. Subtracting Equation 5.27 from Equation 5.28, we obtain

$$\widehat{\mathrm{A}}\,\widehat{\mathrm{A}}^\dagger - \widehat{\mathrm{A}}^\dagger \widehat{\mathrm{A}} = 1, \tag{5.32}$$

which, of course, means that

$$\widehat{\mathrm{A}}\,\widehat{\mathrm{A}}^\dagger f(x) \neq \widehat{\mathrm{A}}^\dagger \widehat{\mathrm{A}} f(x),$$

so the ordering of the two operators obviously matters.

In mathematical terms, the expression on the left-hand side of Equation 5.32 is called the **commutator** of $\widehat{\mathrm{A}}$ and $\widehat{\mathrm{A}}^\dagger$, and the whole equation is called a **commutation relation**. The fact that the commutator does not vanish is also expressed by saying that the two operators *do not commute* with one another. The concept of non-commuting operators is familiar in everyday life, where the order of activities is often vital. Superman puts his underpants on after the rest of his clothes, while I aim to do the reverse; the effects are not similar! In the present context, it means that you must take care when writing down Equation 5.31. The order $\widehat{\mathrm{A}}^\dagger \widehat{\mathrm{A}}$ is called the **normal ordering** of $\widehat{\mathrm{A}}^\dagger$ and $\widehat{\mathrm{A}}$. We shall always use normal ordering when expressing the Hamiltonian operator of a harmonic oscillator in terms of raising and lowering operators.

Figure 5.11 Paul Dirac (1902–1984) was one of the founders of quantum mechanics. Dirac won the Nobel prize for physics in 1933.

5.3.3 Obtaining the eigenvalues and eigenfunctions

We shall now present an argument that leads to the energy eigenvalues and eigenfunctions of a harmonic oscillator. The argument takes a surprising route, originally discovered by Dirac (Figure 5.11). Discovering the route took genius, but the individual steps are not difficult to follow.

The eigenvalues and eigenfunctions are found by solving the time-independent Schrödinger equation, $\widehat{\mathrm{H}}\psi = E\psi$. We use Equation 5.31 to express the Hamiltonian operator $\widehat{\mathrm{H}}$ in terms of raising and lowering operators. The time-independent Schrödinger equation then takes the form:

$$\left(\widehat{\mathrm{A}}^\dagger \widehat{\mathrm{A}} + \tfrac{1}{2}\right) \hbar\omega_0\, \psi(x) = E\, \psi(x). \tag{5.33}$$

The first rung on the energy ladder

Rather than trying to find all of the solutions of this equation at once, we begin by noting that Equation 5.33 would be satisfied if

$$\widehat{\mathrm{A}}\psi(x) = 0 \quad \text{and} \quad E = \tfrac{1}{2}\hbar\omega_0. \tag{5.34}$$

Writing out the equation $\widehat{\mathrm{A}}\psi(x) = 0$ in full, we see that

$$\frac{1}{\sqrt{2}}\left(\frac{x}{a} + a\frac{\partial}{\partial x}\right)\psi(x) = 0.$$

This gives the first-order differential equation

$$\frac{\mathrm{d}\psi}{\mathrm{d}x} = -\frac{x}{a^2}\,\psi(x), \tag{5.35}$$

which has the solution

$$\psi(x) = C_0\,\mathrm{e}^{-x^2/2a^2}, \tag{5.36}$$

where C_0 is a constant.

- Verify that $\psi(x)$ in Equation 5.36 does satisfy Equation 5.35.

○ Differentiating $\psi(x) = C_0\,\mathrm{e}^{-x^2/2a^2}$ with respect to x gives

$$\frac{\mathrm{d}\psi}{\mathrm{d}x} = \frac{-2x}{2a^2} \times C_0\,\mathrm{e}^{-x^2/2a^2} = -\frac{x}{a^2}\,\psi(x),$$

as required.

Our analysis has shown that $\psi(x) = C_0\,\mathrm{e}^{-x^2/2a^2}$ is an energy eigenfunction — a solution of the time-independent Schrödinger equation — with eigenvalue $\hbar\omega_0/2$. You may recognize it as the eigenfunction discussed earlier in Worked Example 5.1. We can be sure that this is the *ground-state* eigenfunction because it has no nodes; this is a characteristic property of ground-state energy eigenfunctions, not shared by those of other states. We therefore denote it by $\psi_0(x)$. Exercise 5.5 showed that the appropriate normalization constant is $C_0 = 1/(\sqrt{\pi}a)^{1/2}$, so

$$\psi_0(x) = \left(\frac{1}{\sqrt{\pi}a}\right)^{1/2}\mathrm{e}^{-x^2/2a^2} \tag{5.37}$$

is the normalized ground-state eigenfunction of a harmonic oscillator, and the corresponding energy eigenvalue is $E_0 = \hbar\omega_0/2$. This is the lowest rung on the ladder of energy levels.

Climbing the energy level ladder

We have found the ground-state eigenfunction and eigenvalue, but what about all the excited states? Fortunately, the raising operator can be used to find many more eigenfunctions. To see how this works, consider the function

$$\widehat{\mathrm{A}}^\dagger\psi_0(x) = \frac{1}{\sqrt{2}}\left(\frac{x}{a} - a\frac{\partial}{\partial x}\right)\psi_0(x).$$

We will show that this is also an eigenfunction of the Hamiltonian, with a higher eigenvalue than ψ_0. To accomplish this, we apply the Hamiltonian operator to $\widehat{\mathrm{A}}^\dagger\psi_0$, giving

$$\widehat{\mathrm{H}}\big[\widehat{\mathrm{A}}^\dagger\psi_0(x)\big] = \big(\widehat{\mathrm{A}}^\dagger\,\widehat{\mathrm{A}} + \tfrac{1}{2}\big)\hbar\omega_0\big[\widehat{\mathrm{A}}^\dagger\psi_0(x)\big].$$

Leaving the order of the operators $\widehat{\mathrm{A}}^\dagger$ and $\widehat{\mathrm{A}}$ undisturbed, and remembering that $\frac{1}{2}$ and $\hbar\omega_0$ are just numbers, we can carefully pull out a factor $\widehat{\mathrm{A}}^\dagger$ from terms on the right-hand side to obtain

$$\widehat{\mathrm{H}}\big[\widehat{\mathrm{A}}^\dagger\psi_0(x)\big] = \widehat{\mathrm{A}}^\dagger\big(\widehat{\mathrm{A}}\,\widehat{\mathrm{A}}^\dagger + \tfrac{1}{2}\big)\hbar\omega_0\,\psi_0(x).$$

Now, the operators $\widehat{A}$ and $\widehat{A}^\dagger$ in the term contained in round brackets are not in their normal order. This can be rectified by using the commutation relation $\widehat{A}\,\widehat{A}^\dagger - \widehat{A}^\dagger\,\widehat{A} = 1$ (Equation 5.32) to write

$$\widehat{A}\,\widehat{A}^\dagger = \widehat{A}^\dagger\,\widehat{A} + 1,$$

giving

$$\widehat{H}\big[\widehat{A}^\dagger\psi_0(x)\big] = \widehat{A}^\dagger\big(\widehat{A}^\dagger\,\widehat{A} + 1 + \tfrac{1}{2}\big)\hbar\omega_0\,\psi_0(x) = \widehat{A}^\dagger\big(\widehat{H} + \hbar\omega_0\big)\psi_0(x).$$

We already know that $\psi_0(x)$ is an eigenfunction of $\widehat{H}$ with eigenvalue $\hbar\omega_0/2$, so we conclude that

$$\widehat{H}\big[\widehat{A}^\dagger\psi_0(x)\big] = \widehat{A}^\dagger\big(\tfrac{1}{2}\hbar\omega_0 + \hbar\omega_0\big)\psi_0(x) = \tfrac{3}{2}\hbar\omega_0\big[\widehat{A}^\dagger\psi_0(x)\big],$$

which shows that $\widehat{A}^\dagger\psi_0(x)$ is an eigenfunction of the Hamiltonian, with eigenvalue $3\hbar\omega_0/2$.

The nice thing about this argument is that it can be generalized. Given *any* energy eigenfunction $\psi_n(x)$, with eigenvalue E_n, we can form a new function $\widehat{A}^\dagger\psi_n(x)$. Then, repeating the above argument, we find that

$$\widehat{H}\big[\widehat{A}^\dagger\psi_n(x)\big] = \widehat{A}^\dagger\big(\widehat{H} + \hbar\omega_0\big)\psi_n(x) = \big(E_n + \hbar\omega_0\big)\big[\widehat{A}^\dagger\psi_n(x)\big], \qquad (5.38)$$

which shows that

$\widehat{A}^\dagger\psi_n(x)$ is an eigenfunction of $\widehat{H}$ with eigenvalue $E_n + \hbar\omega_0$.

Similar arguments apply to the operator $\widehat{A}$, but in this case we have

$$\widehat{H}\big[\widehat{A}\,\psi_n(x)\big] = \big(E_n - \hbar\omega_0\big)\big[\widehat{A}\,\psi_n(x)\big], \qquad (5.39)$$

and so

$\widehat{A}\,\psi_n(x)$ is an eigenfunction of $\widehat{H}$ with eigenvalue $E_n - \hbar\omega_0$.

Exercise 5.8 Prove Equation 5.39, using a method similar to that used for $\widehat{A}^\dagger$, but making earlier use of the commutation relation between $\widehat{A}$ and $\widehat{A}^\dagger$. ■

Raising and lowering operators are also called **ladder operators**.

It should now be clear why $\widehat{A}^\dagger$ and $\widehat{A}$ are called *raising* and *lowering* operators. Figure 5.12 summarizes their effects: $\widehat{A}^\dagger$ climbs a rung on the ladder of energy levels, while $\widehat{A}$ steps down a rung. The descent stops at $\psi_0(x)$ because we know that $\widehat{A}\,\psi_0(x) = 0$, so the lowering operator generates no new eigenvalues below $E_0 = \hbar\omega_0/2$.

These discoveries are remarkable because they reveal a never-ending ladder of energy eigenvalues. Starting from the ground state, with energy $\hbar\omega_0/2$, and repeatedly applying the raising operator, we can produce an infinite set of energy levels, $\frac{1}{2}\hbar\omega_0$, $\frac{3}{2}\hbar\omega_0$, $\frac{5}{2}\hbar\omega_0$, ...

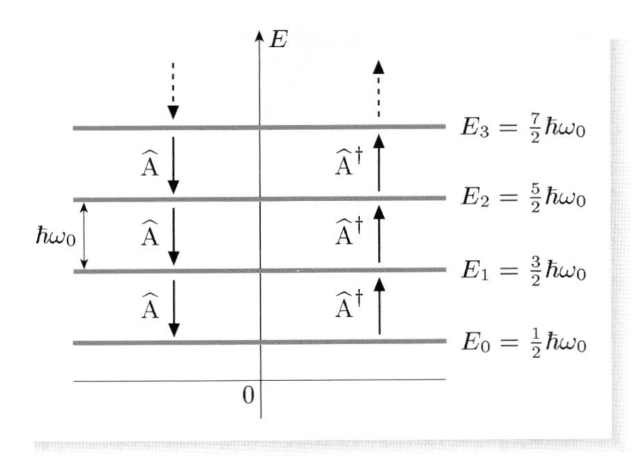

Figure 5.12 Climbing up and down the ladder of energy levels.

Can we be sure that this process generates *all* the energy levels? In fact, we can. To see why, let us start from an *arbitrary* energy level E, and repeatedly apply the *lowering* operator, $\widehat{\mathrm{A}}$. This generates energy levels $E - \hbar\omega_0$, $E - 2\hbar\omega_0, \ldots$. If this process continued forever, we would eventually reach energy levels below the ground state, which is absurd. The only way to avoid this infinite descent is for the lowering process to produce a function $\psi(x)$ for which $\widehat{\mathrm{A}}\,\psi(x) = 0$. Then, further applications of the lowering operator produce no new functions and the descent ceases. But we have already investigated the equation $\widehat{\mathrm{A}}\,\psi(x) = 0$ and seen that its only solution is the ground-state eigenfunction, with eigenvalue $\frac{1}{2}\hbar\omega_0$. Since the process of descending from E in steps of $\hbar\omega_0$ must eventually reach $\frac{1}{2}\hbar\omega_0$, it follows that all the eigenvalues are a whole number of $\hbar\omega_0$ steps above $\frac{1}{2}\hbar\omega_0$. In other words, all the energy levels are given by

$$E_n = (n + \tfrac{1}{2})\hbar\omega_0, \quad \text{where } n = 0, 1, 2, \ldots,$$

which confirms Equation 5.16.

Finally, let us briefly consider the energy eigenfunctions. We have shown that the raising and lowering operators allow us to move between neighbouring eigenfunctions, but nothing in our analysis has shown that these operators take a *normalized* eigenfunction and produce another *normalized* eigenfunction. In fact, it turns out that these operators do not preserve normalization. Detailed calculations show that if $\psi_{n-1}(x)$ and $\psi_n(x)$ are *normalized* energy eigenfunctions, then

$$\widehat{\mathrm{A}}^\dagger \psi_{n-1}(x) = \sqrt{n}\,\psi_n(x), \tag{5.40}$$

$$\widehat{\mathrm{A}}\,\psi_n(x) = \sqrt{n}\,\psi_{n-1}(x). \tag{5.41}$$

Note that, in both cases, the coefficient of the final eigenfunction involves the *higher* of the two quantum numbers being linked.

These equations make sense because they imply that

$$\widehat{\mathrm{A}}^\dagger \widehat{\mathrm{A}}\,\psi_n(x) = \widehat{\mathrm{A}}^\dagger \sqrt{n}\,\psi_{n-1}(x) = \sqrt{n}\sqrt{n}\,\psi_n(x) = n\,\psi_n(x),$$

which is necessary to obtain the correct energy eigenvalues:

$$\widehat{\mathrm{H}}\,\psi_n(x) = \left(\widehat{\mathrm{A}}^\dagger \widehat{\mathrm{A}} + \tfrac{1}{2}\right)\hbar\omega_0\,\psi_n(x) = \left(n + \tfrac{1}{2}\right)\hbar\omega_0\,\psi_n(x).$$

Because the operator $\widehat{\mathrm{A}}^\dagger \widehat{\mathrm{A}}$ has the quantum numbers $n = 0, 1, 2, \ldots$ as its eigenvalues, it is often called the **number operator**.

To find the *normalized* eigenfunction $\psi_n(x)$, we rewrite Equation 5.40 in the form

$$\psi_n(x) = \frac{1}{\sqrt{n}}\,\widehat{\mathrm{A}}^\dagger\psi_{n-1}(x), \tag{5.42}$$

and repeatedly apply this result to give

$$\psi_n(x) = \left(\frac{1}{2^n n!}\right)^{1/2}\left(\frac{x}{a} - a\frac{\partial}{\partial x}\right)^n \psi_0(x).$$

Inserting the ground-state eigenfunction (Equation 5.37), we finally obtain

$$\psi_n(x) = \left(\frac{1}{2^n n!\sqrt{\pi}a}\right)^{1/2}\left(\frac{x}{a} - a\frac{\partial}{\partial x}\right)^n \mathrm{e}^{-x^2/2a^2}. \tag{5.43}$$

The effect of repeatedly applying the operator $(x/a - a\,\partial/\partial x)$ is to leave the Gaussian term $\mathrm{e}^{-x^2/2a^2}$ unchanged, and to bring down increasingly complicated polynomials in front of it; these are precisely the Hermite polynomials mentioned in Section 5.2.

You could use Equation 5.43 to find the first few energy eigenfunctions, and compare your answers with Table 5.1. However, this is not really necessary. One of the great advantages of raising and lowering operators is that they allow us to explore the quantum properties of harmonic oscillators without using any explicit formulae for the energy eigenfunctions; this is the subject of the next section.

5.4 Quantum properties of oscillators

Raising and lowering operators have an importance beyond the derivation of energy eigenvalues and eigenfunctions. In this section, we shall use them to help understand the quantum-mechanical behaviour of harmonic oscillators. We shall look at the expectation values and uncertainties of various observable quantities, including position, momentum, kinetic energy and potential energy, in different stationary states. You will also see why the selection rule mentioned in Section 5.2.2 applies to harmonic oscillators.

5.4.1 Expectation values and uncertainties

If a system is in a stationary state $\Psi_n(x,t) = \psi_n(x)\,\mathrm{e}^{-\mathrm{i}E_nt/\hbar}$, the expectation value of any observable O is

$$\begin{aligned}\langle O\rangle &= \int_{-\infty}^{\infty} \Psi_n^*(x,t)\,\widehat{\mathrm{O}}\,\Psi_n(x,t)\,\mathrm{d}x\\ &= \int_{-\infty}^{\infty}\left(\psi_n^*(x)\,\mathrm{e}^{\mathrm{i}E_nt/\hbar}\right)\widehat{\mathrm{O}}\left(\psi_n(x)\,\mathrm{e}^{-\mathrm{i}E_nt/\hbar}\right)\mathrm{d}x\\ &= \int_{-\infty}^{\infty}\psi_n^*(x)\,\widehat{\mathrm{O}}\,\psi_n(x)\,\mathrm{d}x,\end{aligned}$$

and so depends on the normalized energy eigenfunction $\psi_n(x)$. The eigenfunction is independent of time, and so is the expectation value.

For a harmonic oscillator, we could evaluate such integrals by inserting the appropriate eigenfunctions. However, you have seen that these eigenfunctions are messy combinations of normalization constants, polynomials and Gaussian functions, so evaluating the integrals would be an arduous task. Fortunately, raising and lowering operators can save us a great deal of trouble, as we shall now show.

The key point is that position and momentum operators can be expressed in terms of raising and lowering operators. Combining Equations 5.29 and 5.30 and recalling that $\widehat{\mathrm{x}} = x$ and $\widehat{\mathrm{p}}_x = -\mathrm{i}\hbar\,\partial/\partial x$, we obtain

$$\widehat{\mathrm{x}} = \frac{a}{\sqrt{2}}(\widehat{\mathrm{A}} + \widehat{\mathrm{A}}^\dagger), \tag{5.44}$$

$$\widehat{\mathrm{p}}_x = \frac{-\mathrm{i}\hbar}{\sqrt{2}a}(\widehat{\mathrm{A}} - \widehat{\mathrm{A}}^\dagger). \tag{5.45}$$

Note that the denominator of Equation 5.45 is $\sqrt{2}a$, not $\sqrt{2a}$.

Consequently, the expectation value of position can be rewritten as

$$\langle x \rangle = \frac{a}{\sqrt{2}} \int_{-\infty}^{\infty} \psi_n^*(x)\,(\widehat{\mathrm{A}} + \widehat{\mathrm{A}}^\dagger)\,\psi_n(x)\,\mathrm{d}x. \tag{5.46}$$

The advantage of doing this is that $\widehat{\mathrm{A}}$ and $\widehat{\mathrm{A}}^\dagger$ have known, simple effects. They are lowering and raising operators, so we must have

$$(\widehat{\mathrm{A}} + \widehat{\mathrm{A}}^\dagger)\,\psi_n(x) = \alpha\,\psi_{n-1}(x) + \beta\,\psi_{n+1}(x), \tag{5.47}$$

where α and β are constants. (In fact, $\alpha = \sqrt{n}$ and $\beta = \sqrt{n+1}$, from Equations 5.40 and 5.41, but this detail is not needed.) We only need to substitute Equation 5.47 into Equation 5.46 to obtain

$$\langle x \rangle = \frac{a}{\sqrt{2}} \int_{-\infty}^{\infty} \psi_n^*(x)\,\left[\alpha\,\psi_{n-1}(x) + \beta\,\psi_{n+1}(x)\right]\mathrm{d}x.$$

Then, using the fact that $\psi_n(x)$ is orthogonal to both $\psi_{n-1}(x)$ and $\psi_{n+1}(x)$, we immediately see that $\langle x \rangle = 0$. An almost identical argument applies to $\langle p_x \rangle$, which is also equal to zero.

Worked Example 5.2

Essential skill

Calculating expectation values using raising and lowering operators

Show that a harmonic oscillator in a stationary state with quantum number n has

$$\langle x^2 \rangle = \left(n + \tfrac{1}{2}\right) a^2. \tag{5.48}$$

Solution

We begin by using Equation 5.44 to write

$$\langle x^2 \rangle = \frac{a^2}{2} \int_{-\infty}^{\infty} \psi_n^*(x)\,(\widehat{\mathrm{A}} + \widehat{\mathrm{A}}^\dagger)^2\,\psi_n(x)\,\mathrm{d}x.$$

It pays to think carefully about the effect of the operators. Taking care to maintain their ordering, we have

$$\begin{aligned}(\widehat{\mathrm{A}} + \widehat{\mathrm{A}}^\dagger)^2 &= (\widehat{\mathrm{A}} + \widehat{\mathrm{A}}^\dagger)(\widehat{\mathrm{A}} + \widehat{\mathrm{A}}^\dagger)\\ &= \widehat{\mathrm{A}}\,\widehat{\mathrm{A}} + \widehat{\mathrm{A}}^\dagger\,\widehat{\mathrm{A}}^\dagger + \widehat{\mathrm{A}}^\dagger\,\widehat{\mathrm{A}} + \widehat{\mathrm{A}}\,\widehat{\mathrm{A}}^\dagger.\end{aligned}$$

Consider what each term in this sum does to $\psi_n(x)$. The first two terms convert $\psi_n(x)$ into *different* eigenfunctions ($\psi_{n-2}(x)$ and $\psi_{n+2}(x)$). Because $\psi_n(x)$ is orthogonal to these eigenfunctions, we can safely drop these terms from the integral, leaving only

$$\langle x^2 \rangle = \frac{a^2}{2} \int_{-\infty}^{\infty} \psi_n^*(x) \left(\widehat{A}^\dagger \widehat{A} + \widehat{A}\,\widehat{A}^\dagger\right) \psi_n(x)\, dx.$$

The surviving terms have *equal* numbers of raising and lowering operators. Our tactic now is to express everything in terms of the number operator, $\widehat{A}^\dagger \widehat{A}$, because we know that this has a simple effect on $\psi_n(x)$. Using the commutation relation $\widehat{A}\,\widehat{A}^\dagger - \widehat{A}^\dagger \widehat{A} = 1$, we obtain

$$\widehat{A}\,\widehat{A}^\dagger = \widehat{A}^\dagger \widehat{A} + 1,$$

so

$$\left(\widehat{A}^\dagger \widehat{A} + \widehat{A}\,\widehat{A}^\dagger\right) \psi_n(x) = \left(2\widehat{A}^\dagger \widehat{A} + 1\right) \psi_n(x) = (2n+1)\, \psi_n(x),$$

and

$$\langle x^2 \rangle = \frac{a^2}{2}(2n+1) \int_{-\infty}^{\infty} \psi_n^*(x)\, \psi_n(x)\, dx = \left(n + \tfrac{1}{2}\right) a^2,$$

as required.

Exercise 5.9 Show that a harmonic oscillator in a stationary state with quantum number n has an expectation value of p_x^2 given by

$$\langle p_x^2 \rangle = \left(n + \tfrac{1}{2}\right) \frac{\hbar^2}{a^2}. \tag{5.49}$$

■

The two results just derived for $\langle x^2 \rangle$ and $\langle p_x^2 \rangle$ allow us to find the expectation values of potential and kinetic energy in a stationary state with quantum number n. For the potential energy,

$$\langle E_{\text{pot}} \rangle = \langle \tfrac{1}{2} m\omega_0^2 x^2 \rangle = \tfrac{1}{2} m\omega_0^2 \langle x^2 \rangle = \tfrac{1}{2}\left(n + \tfrac{1}{2}\right) m\omega_0^2 a^2,$$

and for the kinetic energy,

$$\langle E_{\text{kin}} \rangle = \frac{\langle p_x^2 \rangle}{2m} = \tfrac{1}{2}\left(n + \tfrac{1}{2}\right) \frac{\hbar^2}{ma^2}.$$

The ratio of the kinetic and potential energy expectation values is

$$\frac{\langle E_{\text{kin}} \rangle}{\langle E_{\text{pot}} \rangle} = \frac{\hbar^2}{m^2\omega_0^2 a^4} = 1,$$

where we have used the fact that $a = \sqrt{\hbar/m\omega_0}$. Hence the expectation value of the kinetic energy is equal to the expectation value of the potential energy. This is the quantum-mechanical version of the result illustrated in Figure 5.3, where the classical kinetic and potential energies, averaged over one cycle of oscillation, were the same.

We can also use our expressions for $\langle x^2 \rangle$ and $\langle p_x^2 \rangle$ to find the uncertainties in position and momentum. Because $\langle x \rangle$ and $\langle p_x \rangle$ are both equal to zero, we have

$$\Delta x = \sqrt{\langle x^2 \rangle - \langle x \rangle^2} = \sqrt{\langle x^2 \rangle} = \sqrt{n + \tfrac{1}{2}}\, a, \tag{5.50}$$

$$\Delta p_x = \sqrt{\langle p_x^2 \rangle - \langle p_x \rangle^2} = \sqrt{\langle p_x^2 \rangle} = \sqrt{n + \tfrac{1}{2}}\, \frac{\hbar}{a}. \tag{5.51}$$

Multiplying these uncertainties together gives

$$\Delta x\, \Delta p_x = \left(n + \tfrac{1}{2}\right)\hbar. \tag{5.52}$$

Three points are worth noting:

1. Equation 5.52 is consistent with the Heisenberg uncertainty principle, which requires that $\Delta x\, \Delta p_x \geq \hbar/2$ in any state.
2. The ground state of a harmonic oscillator ($n = 0$) has the minimum possible uncertainty product, $\Delta x\, \Delta p_x = \hbar/2$.
3. Highly-excited states (with $n \gg 1$) have uncertainty products that are much larger than $\hbar/2$, but this is fully consistent with the uncertainty principle, which is only an *inequality*.

5.4.2 A selection rule

Finally, we can use raising and lowering operators to gain an insight into the selection rule mentioned in Section 5.2.2. To take a definite case, we consider a vibrating diatomic molecule, such as HCl. In this molecule, one of the atoms (Cl) carries a slightly negative charge, while the other (H) carries a slightly positive charge, so the molecule can be thought of as an oscillating electric dipole. In classical physics, we would explain the absorption or emission of light from the molecule by treating it like a tiny radio aerial.

A very different picture applies in quantum physics, where the molecule has a set of equally-spaced vibrational energy levels. Neighbouring levels are separated by $\hbar\omega_0$, where ω_0 is the classical angular frequency of vibration. If there were no restrictions on transitions between the energy levels, a highly-excited molecule would emit photons of energies $\hbar\omega_0$, $2\hbar\omega_0$, $3\hbar\omega_0$, and so on. But this is not what is observed.

If we ignore extremely faint spectral lines, we only get photons of energy $\hbar\omega_0$, corresponding to transitions between neighbouring levels. A transition that is accompanied by the absorption or emission of a photon is generally called a **radiative transition**, so we can also say that the radiative transitions in a harmonic oscillator link *neighbouring* levels. This is the selection rule we mentioned earlier. Where does it come from?

Radiative transitions between vibrational levels of CO_2 are a cause for serious concern; via the greenhouse effect, they are responsible for the largest man-made contribution to global warming.

At the end of the course, we will describe the quantum theory of light interacting with matter, and you will see that, under most conditions, the rate of transitions from one state to another obeys the rule

$$\text{rate of transitions} \propto \left| \int_{-\infty}^{\infty} \psi_n^*(x)\, x\, \psi_m(x)\, \mathrm{d}x \right|^2, \tag{5.53}$$

where $\psi_m(x)$ and $\psi_n(x)$ are the energy eigenfunctions of the states involved. We cannot justify this formula here but, taking it on trust, we can trace the origins of the selection rule.

Exercise 5.10 Show that the integral in Equation 5.53 is equal to zero unless $n = m \pm 1$. ■

This establishes the selection rule: in a harmonic oscillator, radiative transitions link *neighbouring* energy levels. It is possible to observe this selection rule in a very direct way. Most experiments in quantum mechanics involve data from a vast number of atoms or molecules, but it has recently become possible to observe the behaviour of single particles. Figure 5.13 shows the results of an experiment that continuously monitors the energy of an electron trapped in a harmonic potential energy well. It is observed that the electron always jumps from one energy level to the next, never changing its quantum number by more than one. If you have ever doubted the reality of quantum jumps between discrete states, here is firm evidence!

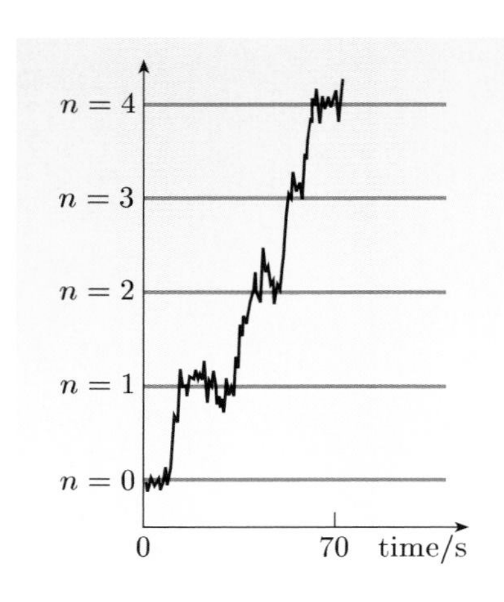

Figure 5.13 A single electron trapped in a potential energy well where its Hamiltonian is that of a harmonic oscillator. The electron performs quantum jumps between *neighbouring* energy levels.

Summary of Chapter 5

Section 5.1 Classical simple harmonic motion involves sinusoidal oscillations in response to a restoring force that is proportional to the displacement of a particle from an equilibrium position. The angular frequency of the oscillation is $\omega_0 = \sqrt{C/m}$, where C is the force constant and m is the mass of the particle.

The internal motion of a two-particle system, composed of particles of masses m_1 and m_2, can be analyzed by considering the motion of a single particle with the *reduced* mass $\mu = m_1 m_2/(m_1 + m_2)$.

Section 5.2 The time-independent Schrödinger equation for a harmonic oscillator takes the form

$$-\frac{\hbar^2}{2m}\frac{\mathrm{d}^2\psi_n}{\mathrm{d}x^2} + \tfrac{1}{2}m\omega_0^2 x^2\psi_n = E_n\psi_n.$$

The energy eigenvalues form an evenly-spaced ladder

$$E_n = (n + \tfrac{1}{2})\hbar\omega_0 \quad \text{for } n = 0, 1, 2, \ldots,$$

but transitions that involve the absorption or emission of photons are effectively restricted to those between neighbouring energy levels.

The corresponding energy eigenfunctions take the form

$$\psi_n(x) = C_n H_n(x/a)\,\mathrm{e}^{-x^2/2a^2},$$

where C_n is a normalization constant, $H_n(x/a)$ is an nth order Hermite polynomial in x/a, and $\mathrm{e}^{-x^2/2a^2}$ is a Gaussian function. The constant $a = \sqrt{\hbar/m\omega_0}$ is the length parameter that characterizes the oscillator.

The energy eigenfunction $\psi_n(x)$ has n nodes; it is even when n is even, and odd when n is odd. The energy eigenfunctions form an orthonormal set.

Section 5.3 The Hamiltonian operator for a harmonic oscillator can be expressed as

$$\widehat{\mathrm{H}} = \left(\widehat{\mathrm{A}}^\dagger\,\widehat{\mathrm{A}} + \tfrac{1}{2}\right)\hbar\omega_0,$$

where $\widehat{\mathrm{A}}^\dagger$ and $\widehat{\mathrm{A}}$ are raising and lowering operators. These operators convert any energy eigenfunction into a neighbouring eigenfunction. If the eigenfunctions are normalized, then

$$\widehat{\mathrm{A}}\,\psi_n(x) = \sqrt{n}\,\psi_{n-1}(x) \quad \text{and} \quad \widehat{\mathrm{A}}^\dagger\psi_{n-1}(x) = \sqrt{n}\,\psi_n(x).$$

By using this representation, and repeatedly applying $\widehat{\mathrm{A}}^\dagger$ to the ground state, it is possible to generate all the eigenfunctions and eigenvalues.

Section 5.4 Expectation values and uncertainties of x and p_x are conveniently calculated by expressing $\widehat{\mathrm{x}}$ and $\widehat{\mathrm{p}}_x$ in terms of the raising and lowering operators, and using the commutation relation $\widehat{\mathrm{A}}\,\widehat{\mathrm{A}}^\dagger - \widehat{\mathrm{A}}^\dagger\,\widehat{\mathrm{A}} = 1$. The selection rule for radiative transitions in vibrating molecules can also be derived using raising and lowering operators.

Achievements from Chapter 5

After studying this chapter you should be able to:

5.1 Explain the meanings of the newly defined (emboldened) terms and symbols, and use them appropriately.

5.2 Compare and contrast the quantum and classical behaviour of harmonic oscillators.

5.3 Write down the Schrödinger equation and the time-independent Schrödinger equation for a harmonic oscillator.

5.4 Recall the energy eigenvalues of a harmonic oscillator, and describe, in general terms, the properties of the energy eigenfunctions.

5.5 Verify that given functions are energy eigenfunctions of a harmonic oscillator, and determine their eigenvalues.

5.6 Show that if $\psi_n(x)$ is an energy eigenfunction with eigenvalue E_n, then $\widehat{\mathrm{A}}^\dagger\psi_n(x)$ is another energy eigenfunction, with eigenvalue $E_n + \hbar\omega_0$.

5.7 Calculate expectation values and uncertainties using raising and lowering operators.

5.8 Describe how the vibrations of a diatomic molecule can be modelled quantum-mechanically.

Chapter 6 Wave packets and motion

Introduction

In the real world, electrons, cannonballs and planets move, molecules and atomic nuclei vibrate and jellies wobble. We claim that quantum mechanics is a comprehensive theory, with Newtonian mechanics applying as a special case for relatively large and heavy objects, yet so far we have only studied the frozen world of stationary states.

Schrödinger's equation describes how quantum systems develop in time. However, soon after introducing this equation, we focused attention on a special type of solution known as a *stationary state*. Each stationary state corresponds to a definite energy, but the time-dependence of a stationary state is exceedingly dull. In a stationary state, the probability density for position is independent of time; whatever happened to the *motion* that Schrödinger's equation was supposed to describe?

To describe motion in quantum mechanics, we need to consider solutions of Schrödinger's equation known as *wave packets*, each wave packet being a *linear combination* of different stationary states. These solutions are not themselves stationary, and you will see that they can describe motion.

Wave packets have other uses as well. By comparing the motion of wave packets with the motion of classical particles, we can gain some insight into the relationship between quantum physics and classical physics. We seem to have two worlds: an everyday world and an utterly different quantum world. We feel that there must be some correspondence, some smooth transition, between these two realms, but what is its nature? Wave packets help to answer this question. Finally, wave packets provide the key to solving the problem of the continuum. You may recall that earlier chapters concentrated on *bound* stationary states because the unbound stationary states in the continuum (corresponding, for example, to the ionized states of atoms) cannot be normalized and so cannot represent any physically realizable state. You will see that the state of a particle in the continuum is best described by a wave packet.

The chapter is organized as follows. First we compare stationary states and wave packets, looking at examples of wave packets in a harmonic oscillator (Section 6.1). In such an oscillator, you will see that the expectation values of position and momentum depend sinusoidally on time, just like the position and momentum variables of a classical oscillator. This is part of a general result, known as Ehrenfest's theorem, which provides an important link between quantum mechanics and classical mechanics. Ehrenfest's theorem is discussed in Section 6.2. Finally, we consider the problem of predicting motion. Given an initial wave packet, you will see how the time-evolution of the wave packet can be predicted. Section 6.3 considers wave packets in a simple harmonic oscillator and Section 6.4 considers wave packets that describe free particles; you will see how these wave packets spread with time.

There are two multimedia sequences associated with this chapter: one supports your study of wave packets in wells, and the other explores free-particle wave packets. The first sequence can be studied at the end of Section 6.3, or they can both be studied at the end of the chapter.

6.1 Time-dependence in the quantum world

There is nothing in the everyday world corresponding to the stationary states studied in previous chapters. Electrons in an old-fashioned TV tube are emitted by a cathode and accelerated towards a screen, where they give up their energy with the emission of light: they move! So do the ions flowing through your brain, giving rise to thoughts and emotions. If quantum mechanics is a comprehensive theory, it must be capable of describing moving particles. This section will confirm that stationary states cannot describe motion, before introducing wave packets, which do provide the appropriate description of motion in quantum mechanics.

6.1.1 The frozen world of stationary states

A stationary state is a solution of Schrödinger's equation in which the wave function is a product of a function of position and a function of time: in one dimension, $\Psi(x,t) = \psi(x)T(t)$. In Chapter 2 we substituted this product function into Schrödinger's equation and used the method of separation of variables to show that $\psi(x)$ satisfies the time-independent Schrödinger equation at a fixed energy E, while $T(t)$ is a phase factor of the form $T(t) = \mathrm{e}^{-\mathrm{i}Et/\hbar}$. It follows that any stationary-state wave function can be written as

$$\Psi(x,t) = \psi(x)\mathrm{e}^{-\mathrm{i}Et/\hbar}. \tag{6.1}$$

The energy E is interpreted as the energy of the system in this state. This means that the state described by Equation 6.1 has the definite energy E; every measurement of energy in this stationary state yields the value E.

For present purposes, the most important point about stationary states is their simple time-dependence. The wave function of a stationary state *does* depend on time, but only via a phase factor $\mathrm{e}^{-\mathrm{i}Et/\hbar}$. Now, it is a general principle of quantum mechanics that an overall phase factor, multiplying the whole wave function, has no physical significance, so we should not expect the physical properties of a stationary state to depend on time.

To illustrate this point, consider the probability of finding the particle in a small interval δx, centred on the point x, in the stationary state described by Equation 6.1. As you have seen before, we can use Born's rule to obtain

$$\begin{aligned}\text{probability} &= \Psi^*(x,t)\Psi(x,t)\,\delta x \\ &= \psi^*(x)\mathrm{e}^{+\mathrm{i}Et/\hbar} \times \psi(x)\mathrm{e}^{-\mathrm{i}Et/\hbar}\,\delta x \\ &= \psi^*(x)\psi(x)\,\delta x,\end{aligned}$$

which is independent of time. In this sense, we can say that the stationary state does not describe motion at all.

Stationary states are static in other respects too. In a stationary state, the *expectation value* of an observable O is given by the sandwich integral

$$\begin{aligned}\langle O\rangle &= \int_{-\infty}^{\infty} \Psi^*(x,t)\widehat{\mathrm{O}}\Psi(x,t)\,\mathrm{d}x \\ &= \int_{-\infty}^{\infty} \left(\psi(x)\mathrm{e}^{-\mathrm{i}Et/\hbar}\right)^* \widehat{\mathrm{O}} \left(\psi(x)\mathrm{e}^{-\mathrm{i}Et/\hbar}\right)\,\mathrm{d}x,\end{aligned}$$

where $\widehat{\mathrm{O}}$ is the operator corresponding to O.

In general, the operator $\widehat{\mathrm{O}}$ does not depend on time, so it does not affect the time-dependent phase factors, which can be taken outside the integral. We therefore have

$$\begin{aligned}\langle O\rangle &= \mathrm{e}^{+\mathrm{i}Et/\hbar}\mathrm{e}^{-\mathrm{i}Et/\hbar}\int_{-\infty}^{\infty}\psi^*(x)\widehat{\mathrm{O}}\psi(x)\,\mathrm{d}x\\ &= \int_{-\infty}^{\infty}\psi^*(x)\widehat{\mathrm{O}}\psi(x)\,\mathrm{d}x,\end{aligned}$$

which is independent of time. So, the expectation values of any observable — position, momentum, kinetic energy, or whatever — remains constant in time in any stationary state. To describe motion, and the obvious fact that the expectation values of observable quantities can vary in time, we must look beyond stationary states.

6.1.2 The dynamic world of wave packets

Schrödinger's equation is linear, so if $\Psi_1(x,t)$, $\Psi_2(x,t)$, $\ldots\Psi_n(x,t)$ are all solutions of Schrödinger's equation for a given system, and a_1, a_2, $\ldots a_n$ are complex constants, then the linear combination

$$\Psi(x,t) = a_1\Psi_1(x,t) + a_2\Psi_2(x,t) + \ldots + a_n\Psi_n(x,t)$$

is also a solution of Schrödinger's equation for the system. This is the principle of superposition: a linear combination of stationary-state wave functions satisfies Schrödinger's equation. A normalized linear combination of two or more stationary-state wave functions is called a **wave packet**.

For simplicity, we will begin by considering a wave packet of just two stationary states:

$$\Psi(x,t) = a_1\psi_1(x)\mathrm{e}^{-\mathrm{i}E_1t/\hbar} + a_2\psi_2(x)\mathrm{e}^{-\mathrm{i}E_2t/\hbar}, \tag{6.2}$$

where $E_2 \neq E_1$. This wave function cannot be written as a product of a function of position and a function of time, so it is not a stationary state. If the energy of the system is measured in the state described by Equation 6.2, one or other of the values E_1 or E_2 will be obtained, and the energy of the system is indefinite to this extent. However, we can say that the probability of obtaining energy E_i is

This follows from the coefficient rule introduced in Section 4.2.

$$p_i = \left|a_i\mathrm{e}^{-\mathrm{i}E_it/\hbar}\right|^2 = \left|a_i\right|^2.$$

The coefficients a_i may therefore be called *probability amplitudes* for energy. The fact that an energy measurement must yield either E_1 or E_2 leads to the condition that

$$|a_1|^2 + |a_2|^2 = 1. \tag{6.3}$$

You will see shortly that Equation 6.3 ensures that $\Psi(x,t)$ is normalized.

We now come to a central idea in this chapter. For the wave packet in Equation 6.2, the probability density for position is

$$\begin{aligned}\Psi^*(x,t)\Psi(x,t) &= \left[a_1^*\psi_1^*(x)\mathrm{e}^{+\mathrm{i}E_1t/\hbar} + a_2^*\psi_2^*(x)\mathrm{e}^{+\mathrm{i}E_2t/\hbar}\right]\\ &\quad\times\left[a_1\psi_1(x)\mathrm{e}^{-\mathrm{i}E_1t/\hbar} + a_2\psi_2(x)\mathrm{e}^{-\mathrm{i}E_2t/\hbar}\right].\end{aligned}$$

Multiplying out the brackets and using the fact that $e^A e^B = e^{A+B}$, we obtain

$$\begin{aligned}\Psi^*(x,t)\Psi(x,t) &= |a_1|^2\psi_1^*(x)\psi_1(x) + |a_2|^2\psi_2^*(x)\psi_2(x)\\ &\quad + a_1^* a_2 \psi_1^*(x)\psi_2(x)e^{-i(E_2-E_1)t/\hbar}\\ &\quad + a_2^* a_1 \psi_2^*(x)\psi_1(x)e^{i(E_2-E_1)t/\hbar},\end{aligned} \tag{6.4}$$

where the first two terms come from multiplying corresponding terms in the two brackets, and the second two terms are cross-product terms, obtained by multiplying the first term in one bracket by the second term in the other bracket. These cross-product terms are important in our story because they are time-dependent; they ensure that the probability density depends on time, and so account for motion in wave mechanics. Motion is restored!

Exercise 6.1 Show that the linear combination of stationary-state wave functions in Equation 6.2 is normalized, with $\int_{-\infty}^{\infty} \Psi^*(x,t)\Psi(x,t)\,dx = 1$, provided that the coefficients a_1 and a_2 satisfy Equation 6.3. ■

6.1.3 A wave packet in a harmonic well

To take a definite example, we will now look more closely at a wave packet in a specific system, the harmonic oscillator. Our system consists of a particle of mass m that is subject to a potential energy function $V(x) = \frac{1}{2}Cx^2 = \frac{1}{2}m\omega_0^2 x^2$, where $\omega_0 = \sqrt{C/m}$ is the classical angular frequency corresponding to the classical period of oscillation, $T = 2\pi/\omega_0$.

In quantum mechanics, the energy levels of the harmonic oscillator are given by $E_n = (n + \frac{1}{2})\hbar\omega_0$, where $n = 0, 1, 2, \ldots$ and the first two energy eigenfunctions are

$$\psi_0(x) = \left(\frac{1}{\sqrt{\pi}\,a}\right)^{1/2} e^{-x^2/2a^2} \quad \text{and} \tag{6.5}$$

$$\psi_1(x) = \left(\frac{1}{2\sqrt{\pi}\,a}\right)^{1/2} \frac{2x}{a} e^{-x^2/2a^2}, \tag{6.6}$$

See Table 5.1 on page 134.

where $a = \sqrt{\hbar/m\omega_0}$, is the characteristic length parameter of the oscillator.

We shall look at a specific wave packet with equal coefficients for the first and second stationary-state wave functions:

$$\Psi_A(x,t) = \frac{1}{\sqrt{2}}\left[\psi_0(x)e^{-i\omega_0 t/2} + \psi_1(x)e^{-3i\omega_0 t/2}\right]. \tag{6.7}$$

Remember that $E_0 = \frac{1}{2}\hbar\omega_0$ and $E_1 = \frac{3}{2}\hbar\omega_0$.

Consider what happens at some particular point x. Since the two components have different time factors, there are times when they have the same sign at x, and other times when they have opposite signs. In effect, they sometimes add together and sometimes subtract, leading to a sum that seems to 'slosh' from side to side in the well, as shown in Figure 6.1. This figure represents $\Psi_A(x,t)$ at two convenient times, $t = 0$ in the top panel, and $t = T/2$ in the lower panel, where T is the classical period of the oscillation. The real and imaginary parts of the wave function are shown, together with the probability density $|\Psi_A(x,t)|^2$.

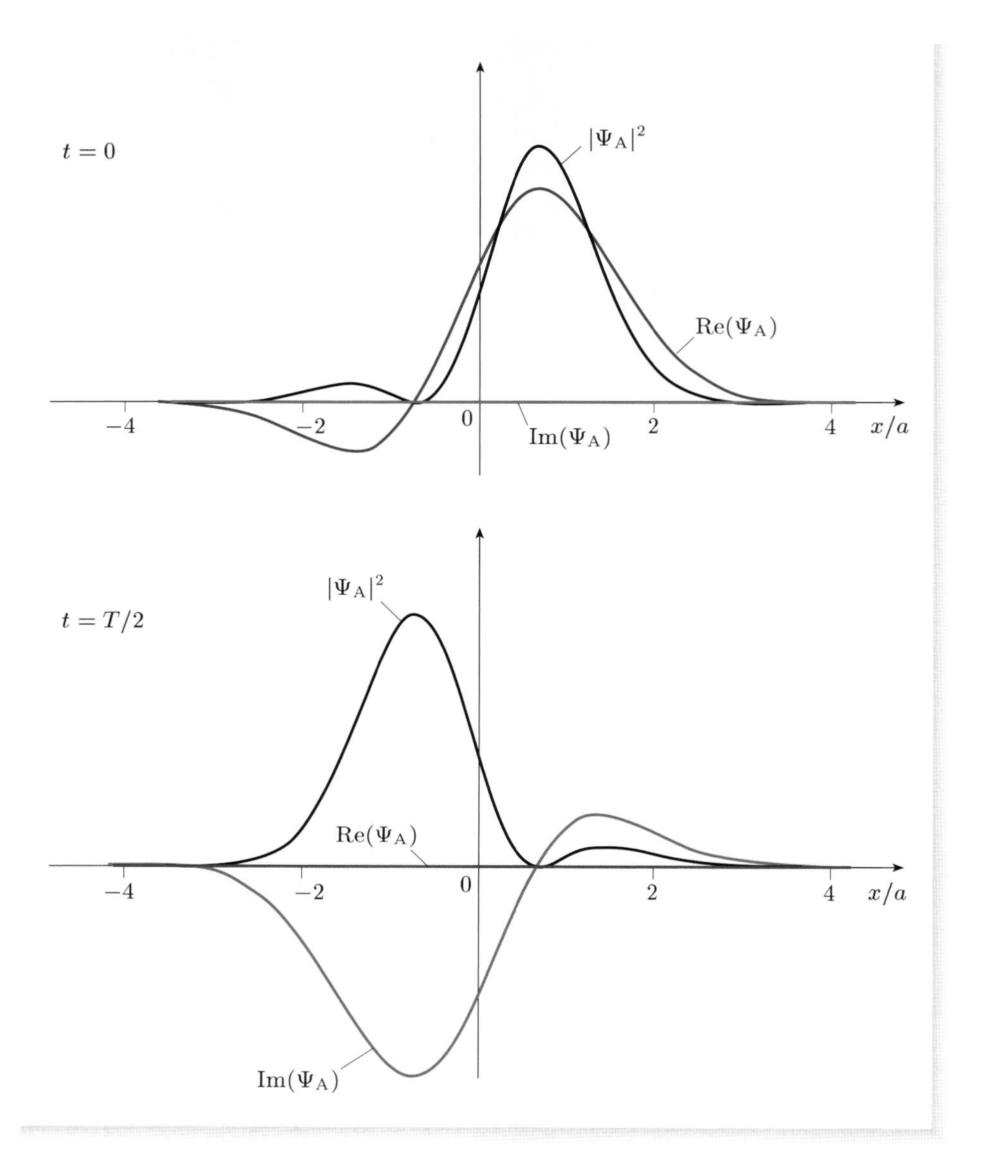

Figure 6.1 $|\Psi_A(x,t)|^2$ and the real and imaginary parts of $\Psi_A(x,t)$ for $t = 0$ and for $t = T/2$, where T is the classical period, $2\pi/\omega_0$. Note that the imaginary part is zero at $t = 0$ and the real part is zero at $t = T/2$. The graphs are plotted as a function of x/a where a is the characteristic length parameter of the harmonic oscillator.

At time $t = 0$, both the time-dependent phase factors in Equation 6.7 are equal to 1, so the wave function is

$$\Psi_A(x,0) = \frac{1}{\sqrt{2}}\left[\psi_0(x) + \psi_1(x)\right]. \tag{6.8}$$

At time $t = T/2 = \pi/\omega_0$, the phase factors are

$$e^{-i\omega_0 t/2} = e^{-i\pi/2} = -i \quad \text{and} \quad e^{-3i\omega_0 t/2} = e^{-3i\pi/2} = +i,$$

so the wave function is

$$\Psi_A(x,T/2) = \frac{i}{\sqrt{2}}\left[-\psi_0(x) + \psi_1(x)\right]. \tag{6.9}$$

Note the overall factor of i in the second equation; this makes no contribution to the probability density $|\Psi_A(x,T/2)|^2$; a more significant fact is that in Equation 6.8 the $\psi_0(x)$ and $\psi_1(x)$ terms are *added*, but in Equation 6.9 they are *subtracted*. The result is the 'sloshing' from side to side shown in Figure 6.1. You can see that the probability of finding the particle on the right-hand side of the well is large at $t = 0$ and the probability of finding the particle on the left-hand side is much smaller. Half a classical period later, exactly the reverse is true: the particle is much more likely to be found on the left-hand side of the well. The probability density $|\Psi|^2$ oscillates periodically, and is no longer independent of time as it is for stationary states. *Wave packets restore motion to quantum mechanics.*

Exercise 6.2 Evaluate $\Psi_A(x, T)$ in the same form as Equations 6.8 and 6.9. Verify that the sign of Ψ_A is reversed after one period. Also show that the sign is restored after two complete periods.

Exercise 6.3 Use Equations 6.5 and 6.6 to derive explicit expressions for $\Psi_A(x, 0)$ and $\Psi_A(x, T/2)$. Give a brief qualitative explanation as to why these expressions are consistent with what you see in Figure 6.1. ■

Expectation values

Figure 6.1 suggests that the time-dependence of the wave packet has at least some vestiges of the oscillatory behaviour of, say a mass on the end of a spring. Although the probability density $|\Psi_A(x,t)|^2$ contains complete information about this motion, it is more convenient to consider a single quantity — the expectation value of position — and see how this depends on time. For the given wave packet, the expectation value of position is given by

$$\langle x \rangle = \int_{-\infty}^{\infty} \Psi_A^*(x,t)\, \widehat{\mathrm{x}}\, \Psi_A(x,t)\, \mathrm{d}x. \tag{6.10}$$

Using

$$\Psi_A^*(x,t) = \frac{1}{\sqrt{2}} \left[\psi_0^*(x) \mathrm{e}^{+\mathrm{i}\omega_0 t/2} + \psi_1^*(x) \mathrm{e}^{+3\mathrm{i}\omega_0 t/2} \right]$$

$$\Psi_A(x,t) = \frac{1}{\sqrt{2}} \left[\psi_0(x) \mathrm{e}^{-\mathrm{i}\omega_0 t/2} + \psi_1(x) \mathrm{e}^{-3\mathrm{i}\omega_0 t/2} \right]$$

and combining phase factors, we see that

$$\begin{aligned}\langle x \rangle = \tfrac{1}{2} \int_{-\infty}^{\infty} \psi_0^*(x)\, \widehat{\mathrm{x}}\, \psi_0(x)\, \mathrm{d}x + \tfrac{1}{2} \int_{-\infty}^{\infty} \psi_1^*(x)\, \widehat{\mathrm{x}}\, \psi_1(x)\, \mathrm{d}x \\ + \tfrac{1}{2} \mathrm{e}^{-\mathrm{i}\omega_0 t} \int_{-\infty}^{\infty} \psi_0^*(x)\, \widehat{\mathrm{x}}\, \psi_1(x)\, \mathrm{d}x + \tfrac{1}{2} \mathrm{e}^{+\mathrm{i}\omega_0 t} \int_{-\infty}^{\infty} \psi_1^*(x)\, \widehat{\mathrm{x}}\, \psi_0(x)\, \mathrm{d}x.\end{aligned}$$

The functions $\psi_0(x)$ and $\psi_1(x)$ happen to be real in this case, but we retain the stars because complex conjugation is needed in general; this would matter if the eigenfunctions were complex.

Fortunately, the remaining integrals can be evaluated without too much effort. The trick is to remember, from Chapter 5, that the position operator in a simple harmonic oscillator can be expressed as

$$\widehat{\mathrm{x}} = \frac{a}{\sqrt{2}} (\widehat{\mathrm{A}} + \widehat{\mathrm{A}}^\dagger), \tag{6.11}$$

where $\widehat{\mathrm{A}}$ and $\widehat{\mathrm{A}}^\dagger$ are lowering and raising operators. The lowering operator converts ψ_n into a multiple of ψ_{n-1} (with $\widehat{\mathrm{A}}\psi_0 = 0$ as a special case) while the raising operator converts ψ_n into a multiple of ψ_{n+1}. The precise relationships for *normalized* energy eigenfunctions are:

$$\widehat{\mathrm{A}}\psi_n(x) = \sqrt{n}\,\psi_{n-1}(x) \tag{6.12}$$

$$\widehat{\mathrm{A}}^\dagger\psi_n(x) = \sqrt{n+1}\,\psi_{n+1}(x). \tag{6.13}$$

When we substitute Equation 6.11 into our expression for the expectation value of position, we get various integrals involving $\widehat{\mathrm{A}} + \widehat{\mathrm{A}}^\dagger$. Before carrying out the substitution, we shall evaluate the integrals that will be needed. Using Equations 6.12 and 6.13, together with the orthonormality of the energy eigenfunctions, we obtain

$$\int_{-\infty}^{\infty} \psi_0^*(x)\,(\widehat{\mathrm{A}} + \widehat{\mathrm{A}}^\dagger)\,\psi_0(x)\,\mathrm{d}x = \int_{-\infty}^{\infty} \psi_0^*(x)\,\psi_1(x)\,\mathrm{d}x = 0$$

$$\int_{-\infty}^{\infty} \psi_1^*(x)\,(\widehat{\mathrm{A}} + \widehat{\mathrm{A}}^\dagger)\,\psi_1(x)\,\mathrm{d}x = \int_{-\infty}^{\infty} \psi_1^*(x)\left[\psi_0(x) + \sqrt{2}\psi_2(x)\right]\,\mathrm{d}x = 0$$

$$\int_{-\infty}^{\infty} \psi_0^*(x)\,(\widehat{\mathrm{A}} + \widehat{\mathrm{A}}^\dagger)\,\psi_1(x)\,\mathrm{d}x = \int_{-\infty}^{\infty} \psi_0^*(x)\left[\psi_0(x) + \sqrt{2}\psi_2(x)\right]\,\mathrm{d}x = 1$$

$$\int_{-\infty}^{\infty} \psi_1^*(x)\,(\widehat{\mathrm{A}} + \widehat{\mathrm{A}}^\dagger)\,\psi_0(x)\,\mathrm{d}x = \int_{-\infty}^{\infty} \psi_1^*(x)\,\psi_1(x)\,\mathrm{d}x = 1.$$

Finally, substituting Equation 6.11 into our expression for $\langle x\rangle$ and using these results, we conclude that

$$\langle x\rangle = \frac{a}{\sqrt{2}}\left[\tfrac{1}{2}\mathrm{e}^{-\mathrm{i}\omega_0 t} + \tfrac{1}{2}\mathrm{e}^{\mathrm{i}\omega_0 t}\right] = \frac{a}{\sqrt{2}}\cos(\omega_0 t). \tag{6.14}$$

● Express this last equation in words.

❍ For the given wave packet, the expectation value of x undergoes sinusoidal motion with amplitude $a/\sqrt{2}$ and with the angular frequency ω_0 of the corresponding classical harmonic oscillator.

A similar calculation can be carried out for the expectation value of the momentum of a particle in a state described by the wave packet $\Psi_{\mathrm{A}}(x,t)$. As explained in Chapter 5, the momentum operator can also be expressed in terms of raising and lowering operators of the harmonic oscillator:

$$\widehat{\mathrm{p}}_x = \frac{-\mathrm{i}\hbar}{\sqrt{2}\,a}(\widehat{\mathrm{A}} - \widehat{\mathrm{A}}^\dagger). \tag{6.15}$$

We can therefore follow the same steps as before, noting that $\widehat{\mathrm{A}}^\dagger$ now appears with a minus sign. The only difference this produces is to give

$$\int_{-\infty}^{\infty} \psi_1^*(x)\,(\widehat{\mathrm{A}} - \widehat{\mathrm{A}}^\dagger)\,\psi_0(x)\,\mathrm{d}x = -1.$$

You may choose to follow through all the details for yourself or, if you are short of time, simply accept the final result which is

Remember

$$\sin x = \frac{\mathrm{e}^{\mathrm{i}x} - \mathrm{e}^{-\mathrm{i}x}}{2\mathrm{i}}.$$

$$\langle p_x\rangle = \frac{-\mathrm{i}\hbar}{\sqrt{2}\,a}\left[\tfrac{1}{2}\mathrm{e}^{-\mathrm{i}\omega_0 t} - \tfrac{1}{2}\mathrm{e}^{\mathrm{i}\omega_0 t}\right] = -\frac{\hbar}{\sqrt{2}a}\sin(\omega_0 t). \tag{6.16}$$

Here, again, we see something reminiscent of classical behaviour: the expectation value of momentum oscillates sinusoidally with the angular frequency ω_0 of the corresponding classical harmonic oscillator.

Exercise 6.4 Show that the expectation values x and p_x, given by Equations 6.14 and 6.16 satisfy

$$\langle p_x \rangle = m \frac{\mathrm{d}\langle x \rangle}{\mathrm{d}t}.$$

■

6.1.4 More general harmonic-oscillator wave packets

So far, we have considered a specific wave packet, with equal coefficients for the first two stationary states of a harmonic oscillator. You have seen that the probability density for position sloshes from one side of the well to the other with a period equal to T, the classical period of oscillation. Moreover, the expectation values of position and momentum vary sinusoidally, in close analogy with classical simple harmonic motion. It is natural to ask whether these features are true for all wave packets describing the state of a particle in a simple harmonic well.

A general wave packet describing a particle in a harmonic oscillator is of the form

$$\Psi(x,t) = \sum_{n=0}^{\infty} a_n \Psi_n(x,t) \quad \text{where} \quad \sum_{n=0}^{\infty} |a_n|^2 = 1. \tag{6.17}$$

In many cases, only a finite number of the a_n are non-zero, but Equation 6.17 covers all possibilities. Since $E_n = (n + \frac{1}{2})\hbar\omega_0$, the harmonic-oscillator stationary states are

$$\Psi_n(x,t) = \psi_n(x)\mathrm{e}^{-\mathrm{i}E_n t/\hbar} = \psi_n(x)\mathrm{e}^{-(2n+1)\mathrm{i}\omega_0 t/2}.$$

Now, it is helpful to remember that $\psi_0(x), \psi_2(x), \psi_4(x), \ldots$ are even functions of x, while $\psi_1(x), \psi_3(x), \psi_5(x), \ldots$ are odd functions of x. This allows us to write the wave function in Equation 6.17 as the sum of even and odd parts:

$$\Psi(x,t) = \Psi_{\text{even}}(x,t) + \Psi_{\text{odd}}(x,t),$$

where

$$\Psi_{\text{even}}(x,t) = a_0\psi_0(x)\mathrm{e}^{-\mathrm{i}\omega_0 t/2} + a_2\psi_2(x)\mathrm{e}^{-5\mathrm{i}\omega_0 t/2} + \ldots$$

$$\Psi_{\text{odd}}(x,t) = a_1\psi_1(x)\mathrm{e}^{-3\mathrm{i}\omega_0 t/2} + a_3\psi_3(x)\mathrm{e}^{-7\mathrm{i}\omega_0 t/2} + \ldots.$$

At time $t = T/2 = \pi/\omega_0$, the phase factors in the *even* part of the wave packet become

$$\mathrm{e}^{-\mathrm{i}\pi/2} = \mathrm{e}^{-5\mathrm{i}\pi/2} = \ldots = -\mathrm{i},$$

while the phase factors in the *odd* part of the wave packet become

$$\mathrm{e}^{-3\mathrm{i}\pi/2} = \mathrm{e}^{-7\mathrm{i}\pi/2} = \ldots = +\mathrm{i}.$$

We therefore have

$$\Psi_{\text{even}}(x,T/2) = -\mathrm{i}\Psi_{\text{even}}(x,0) \quad \text{and} \quad \Psi_{\text{odd}}(x,T/2) = +\mathrm{i}\Psi_{\text{odd}}(x,0),$$

and hence

$$\begin{aligned}\Psi(x,T/2) &= -\mathrm{i}\left[\Psi_{\text{even}}(x,0) - \Psi_{\text{odd}}(x,0)\right] \\ &= -\mathrm{i}\left[\Psi_{\text{even}}(-x,0) + \Psi_{\text{odd}}(-x,0)\right] \\ &= -\mathrm{i}\Psi(-x,0),\end{aligned}$$

giving

$$\left|\Psi(x, T/2)\right|^2 = \left|\Psi(-x, 0)\right|^2,$$

which is exactly what we discovered in Exercise 6.3 for the simple wave packet made up with just the first two energy eigenfunctions.

This shows that, no matter what the initial form of the wave packet, the probability density at time $t = T/2$ becomes a reflected version of the probability density at $t = 0$. If the wave packet was concentrated mainly on the right-hand side of the well at time $t = 0$, it becomes concentrated mainly on the left-hand side of the well at time $t = T/2$, where T is the classical period of oscillation.

A similar analysis shows that the wave function changes sign after a time T, so that $\Psi(x, T) = -\Psi(x, 0)$, but this change in sign has no physical consequences because $|\Psi(x, T)|^2 = |\Psi(x, 0)|^2$. After two classical periods, the wave function returns to its initial value: $\Psi(x, 2T) = \Psi(x, 0)$. We therefore conclude that the probability density describing the position of the particle generally sloshes from one side of the well to the other, endlessly repeating this motion at the classical period $T = 2\pi/\omega_0$. This typical motion is shown in Figure 6.2, which shows the real and imaginary parts of the wave function, together with the probability density $|\Psi|^2$, at a number of different times during the classical period.

Figure 6.3a shows the time-dependence of $\langle x \rangle$, and Figure 6.3b shows the time-dependence of $\langle p_x \rangle$, for the wave packet shown in Figure 6.2. Both these quantities vary sinusoidally with time, in agreement with our findings for the two-component wave packet discussed earlier. Remember from Equations 6.14 and 6.16 that $\langle x \rangle$ and $\langle p_x \rangle$ have the same period, T, as the classical oscillator. Comparing the two parts of this figure, you can see that $\langle p_x \rangle$ is zero when $\langle x \rangle$ is at one of its extreme values. This is analogous to the motion of a pendulum which comes instantaneously to rest at the extreme points of its swing.

The error bars in Figure 6.3a indicate the uncertainties in position at selected times, while those in Figure 6.3b show the uncertainties in momentum. You can see that Δx is largest when Δp_x is smallest. By comparing Figures 6.2 and 6.3, you can also see that Δx is largest when the wave packet is visibly broadest. It is as if the wave packet is breathing in and out as it moves to and fro.

It is interesting to note that wave packets can also breathe in and out *without* moving to and fro. This happens if the wave function is initially symmetric (or antisymmetric) about the centre of the well. Such wave functions always remain symmetric (or antisymmetric) and so do not oscillate from one side of the well to the other. For such a wave packet, the expectation values of position and momentum are permanently equal to zero.

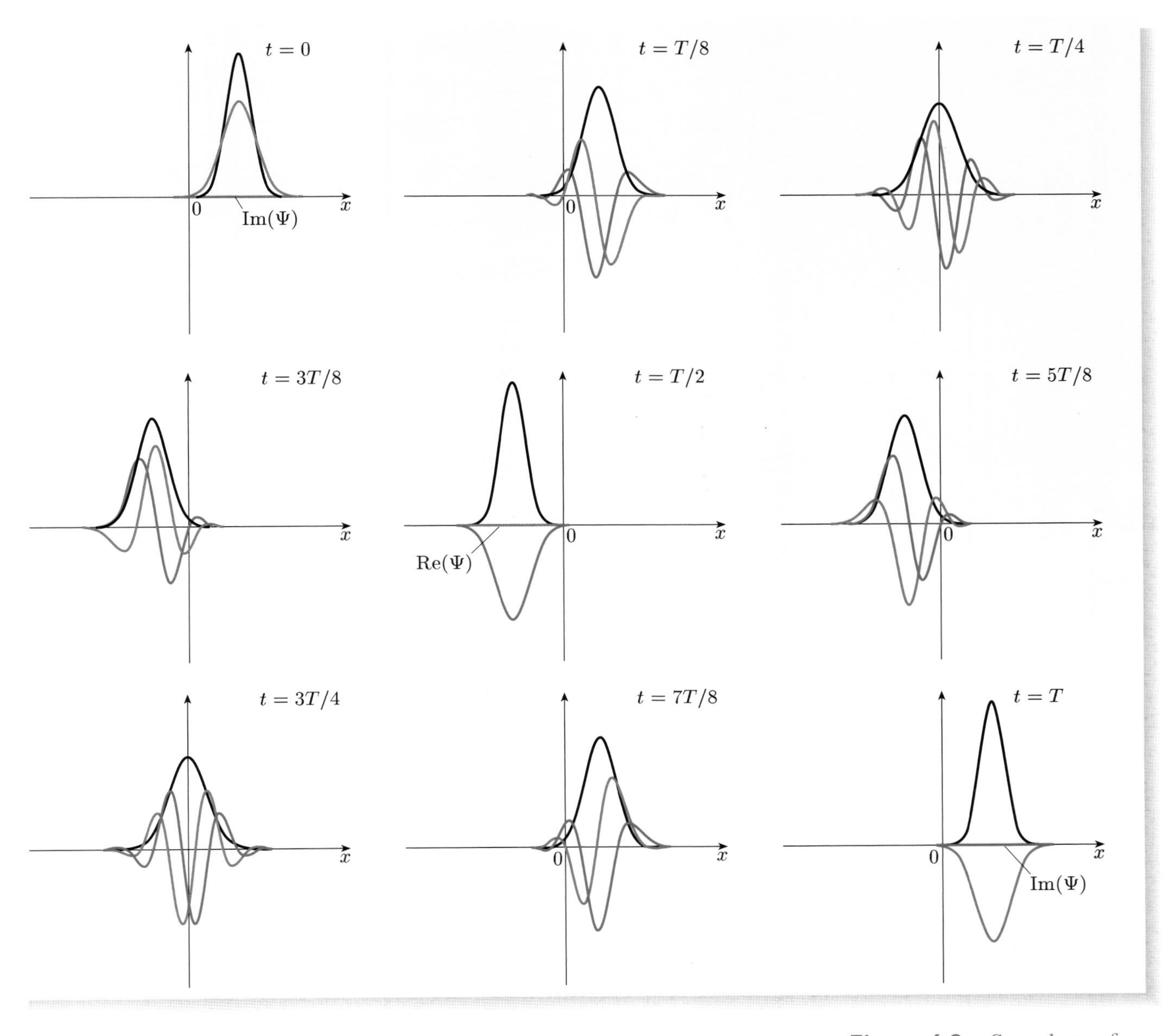

Figure 6.2 Snapshots of a wave packet in a harmonic well at times $t = 0$, $t = T/8$, $t = T/4$, $t = 3T/8$, $t = T/2$, $t = 5T/8$, $t = 3T/4$, $t = 7T/8$ and $t = T$, where $T = 2\pi/\omega_0$ is the classical period of oscillation. The real part of the wave function is shown in blue, the imaginary part in green and the probability density $|\Psi|^2$ in black. The wave function changes sign after one classical period, but this has no observable consequences.

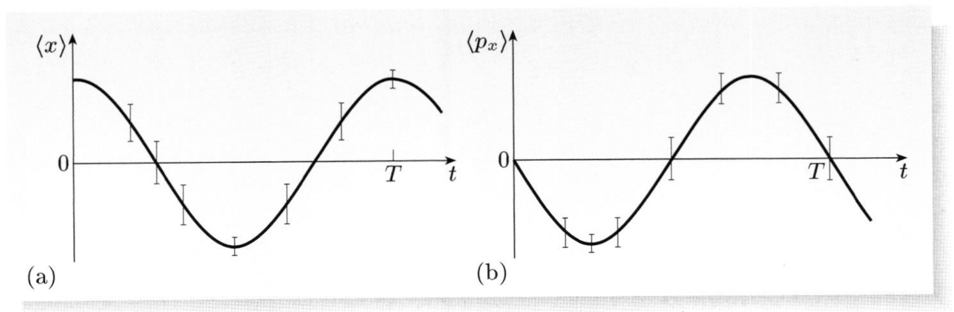

Figure 6.3 The time-dependence of (a) the expectation value $\langle x \rangle$ of position and (b) the expectation value $\langle p_x \rangle$ of momentum for the wave packet of Figure 6.2. At selected times the uncertainties Δx and Δp_x are indicated by error bars.

Figure 6.4 Paul Ehrenfest (1880–1933). Ehrenfest was a student of Boltzmann and a close friend of Einstein. He proved his famous theorem in 1927.

Exercise 6.5 A wave packet in a harmonic potential energy well takes the form

$$\Psi(x,t) = a_1\psi_1(x)\mathrm{e}^{-3\mathrm{i}\omega_0 t/2} + a_3\psi_3(x)\mathrm{e}^{-7\mathrm{i}\omega_0 t/2},$$

where a_1 and a_3 are constants. Show that, at all times, the probability density $|\Psi(x,t)|^2$ is symmetrical about the centre of the well and the expectation value of position is equal to zero. What type of motion does this describe? ■

6.2 Ehrenfest's theorem and the classical limit

You have seen several examples of wave packets that describe the motion of particles in a harmonic potential energy well. These examples suggest that the *expectation values* of position and momentum vary sinusoidally with time (as in Figure 6.3) or remain constant in time (as in Exercise 6.5). This behaviour is intriguing because we know that a classical harmonic oscillator has *values* of position and momentum that vary sinusoidally in time, or remain constant.

This section will introduce an important theorem due to Paul Ehrenfest (Figure 6.4) which sheds light on this apparent coincidence. Later in the course, we shall prove Ehrenfest's theorem from general quantum-mechanical principles. It would be too much of a diversion to include the proof in this chapter, so we shall simply state the result here, and concentrate on its interpretation. You will see that Ehrenfest's theorem provides a key link from the quantum world to the more familiar everyday world of classical physics.

6.2.1 Ehrenfest's theorem

Ehrenfest's theorem is a general statement about the rates of change of $\langle x\rangle$ and $\langle p_x\rangle$ in quantum-mechanical systems. It can be stated as follows:

Ehrenfest's theorem

If a particle of mass m is in a state described by a normalized wave function in a system with a potential energy function $V(x)$, the expectation values of position and momentum of the particle obey the equations

$$\textbf{E1:}\quad \frac{\mathrm{d}\langle x\rangle}{\mathrm{d}t} = \frac{\langle p_x\rangle}{m}$$

$$\textbf{E2:}\quad \frac{\mathrm{d}\langle p_x\rangle}{\mathrm{d}t} = -\left\langle \frac{\partial V}{\partial x}\right\rangle.$$

So far, in this chapter, we have concentrated on wave packets in harmonic oscillators, but it is important to note that Ehrenfest's theorem is a completely general result, valid throughout quantum mechanics. Moreover, it is an exact result — it does not rely on any approximations.

Perhaps the most obvious feature of Ehrenfest's theorem is that it looks very like Newtonian mechanics. The first Ehrenfest equation (which we label **E1**) can be rearranged to give

$$\langle p_x \rangle = m\frac{\mathrm{d}\langle x \rangle}{\mathrm{d}t}, \tag{6.18}$$

which is similar to the familiar Newtonian equation, $p_x = m\,\mathrm{d}x/\mathrm{d}t$. The second Ehrenfest equation (which we label **E2**) can be expressed as

$$\frac{\mathrm{d}\langle p_x \rangle}{\mathrm{d}t} = \langle F_x \rangle, \tag{6.19}$$

where we have defined $F_x = -\partial V/\partial x$, in line with classical physics. So **E2** is the quantum-mechanical analogue of Newton's second law equating the rate of change of momentum to the applied force. However, there is one important difference between **E1** and **E2** and Newtonian mechanics. Ehrenfest's theorem involves *expectation values*, while the corresponding Newtonian equations deal with the *actual values* of position, momentum and force. In quantum mechanics the description is necessarily less precise because the particle cannot have precise values of both position and momentum, according to the uncertainty principle.

We shall examine the precise relationship between Ehrenfest's theorem and Newtonian mechanics shortly. Before doing so, it is worth looking at some examples of Ehrenfest's theorem in action, beginning with the case that has been discussed at length in previous sections — a particle of mass m in a harmonic potential energy well $V(x) = \frac{1}{2}Cx^2 = \frac{1}{2}m\omega_0^2x^2$. In this case, we have

$$\frac{\partial V}{\partial x} = \frac{\partial}{\partial x}\left(\tfrac{1}{2}m\omega_0^2x^2\right) = m\omega_0^2x. \tag{6.20}$$

Differentiating **E1** with respect to time and then using **E2** and Equation 6.20 we obtain

$$\frac{\mathrm{d}^2\langle x \rangle}{\mathrm{d}t^2} = \frac{1}{m}\frac{\mathrm{d}\langle p_x \rangle}{\mathrm{d}t} = -\frac{1}{m}\left\langle \frac{\partial V}{\partial x} \right\rangle = -\frac{1}{m}\langle m\omega_0^2x \rangle = -\omega_0^2\langle x \rangle,$$

which gives

$$\frac{\mathrm{d}^2\langle x \rangle}{\mathrm{d}t^2} + \omega_0^2\langle x \rangle = 0. \tag{6.21}$$

This is the usual equation of motion for a classical harmonic oscillator, *but* applied to $\langle x \rangle$ rather than to x. The general solution of this differential equation can be written as

$$\langle x \rangle = A\sin(\omega_0 t + \phi), \tag{6.22}$$

where A and ϕ are arbitrary constants. The expectation value of the momentum then follows immediately from **E1**:

$$\langle p_x \rangle = m\frac{\mathrm{d}\langle x \rangle}{\mathrm{d}t} = m\omega_0 A\cos(\omega_0 t + \phi). \tag{6.23}$$

Equations 6.22 and 6.23 confirm the sinusoidal time-dependences of $\langle x \rangle$ and $\langle p_x \rangle$ found in the particular wave packet of Equations 6.14 and 6.16, and the more general case illustrated in Figure 6.3. You may recall that symmetric wave packets in a harmonic potential energy well do not move to and fro, but breathe in and out with $\langle x \rangle$ and $\langle p_x \rangle$ permanently equal to zero. Such behaviour is also consistent with Ehrenfest's theorem; Equations 6.22 and 6.23 continue to apply, but with the constant A equal to zero.

Exercise 6.6 A particle of mass m is subject only to the downward force of gravity. With the x-axis pointing vertically upwards, the potential energy function is $V(x) = mgx$, where g is the magnitude of the acceleration due to gravity. What can be said about the expectation value of x for this particle? ■

From the above examples, you might be tempted to suppose that the time-dependences of $\langle x \rangle$ and $\langle p_x \rangle$ in quantum physics are always identical to the time-dependences of $x(t)$ and $p_x(t)$ in classical physics. If so, you will need to think again! To show that this need not be so, we consider a particle of mass m subject to a potential energy function $V(x) = Ax^4$, where A is a constant. In this case,

$$\frac{\partial V}{\partial x} = \frac{\partial}{\partial x}(Ax^4) = 4Ax^3,$$

and Ehrenfest's theorem gives

$$\frac{\mathrm{d}^2\langle x \rangle}{\mathrm{d}t^2} = \frac{1}{m}\frac{\mathrm{d}\langle p_x \rangle}{\mathrm{d}t} = -\frac{1}{m}\left\langle \frac{\partial V}{\partial x} \right\rangle = -\frac{4A}{m}\langle x^3 \rangle.$$

Now, there is no reason to suppose that $\langle x^3 \rangle$ is the same as $\langle x \rangle^3$. In general, these quantities are different, though the precise details will depend on the nature of the wave packet. This means that $\langle x \rangle$ satisfies a different differential equation from $x(t)$ in classical physics. so there is no reason to think that the time-dependences of $\langle x \rangle$ and $x(t)$ will be exactly the same.

6.2.2 The classical limit of quantum mechanics

Quantum mechanics is believed to be a more accurate description of the world than classical mechanics. This is certainly true for systems on the scale of atoms and nuclei, for which classical physics runs into severe difficulties. However, we know that classical physics still provides an excellent description of macroscopic phenomena, on the scale of footballs, cars or planets. No one would think of solving Schrödinger's equation to predict the motion of a planet; it is much easier, and perfectly satisfactory, to use classical physics for this. Nevertheless, a point of principle remains: if quantum physics is a truly universal theory, it should be able to explain *all* phenomena — the motion of a planet as well as the motion of an electron. This suggests that the predictions of quantum mechanics should reduce to those of classical mechanics in a suitable 'classical limit'. In rough terms, this **classical limit** corresponds to bodies that are very much larger than atoms.

Ehrenfest's theorem helps explain how this classical limit is approached. Although **E1** and **E2** look very similar to the basic equations of Newtonian mechanics, you have seen that there is one important difference: Ehrenfest's equations involve expectation values while Newton's equations do not. In the classical limit, this difference must become unimportant. In taking the classical limit we need to make some approximations. The approximations are not part of Ehrenfest's theorem, but they are needed to bring the predictions of Ehrenfest's theorem into line with those of classical physics.

The first point to note is that classical physics assumes that a particle has a definite position and momentum at each instant in time. In the classical limit, the uncertainties of position and momentum associated with quantum-mechanical wave packets must therefore be negligible. This is not hard to achieve: a wave packet describing the centre of mass of a pendulum bob could have $\Delta x \simeq 10^{-14}$ m and $\Delta p_x \simeq 10^{-19}$ kg m s^{-1}. These uncertainties are consistent with the Heisenberg uncertainty principle, but we cannot measure the position or momentum of the bob to such precisions so, from a practical point of view, the uncertainties are insignificant.

We also need to consider the expectation value $\langle \partial V/\partial x \rangle$. Combining **E1** and **E2** we obtain

$$m\frac{\mathrm{d}^2\langle x \rangle}{\mathrm{d}t^2} = -\left\langle \frac{\partial V}{\partial x} \right\rangle. \tag{6.24}$$

For a very narrow wave packet, centred on the point $x = X$ at time t, the expectation value $\langle x \rangle$ is essentially equal to X, and this can be interpreted classically as 'the position of the particle'. Writing the partial derivative $\partial V/\partial x$ as $V'(x)$, the right-hand side of Equation 6.24 is given by the sandwich integral

$$-\left\langle \frac{\partial V}{\partial x} \right\rangle = -\int_{-\infty}^{\infty} \Psi^*(x,t)\, V'(x)\, \Psi(x,t)\, \mathrm{d}x.$$

Because the wave packet is very narrow, only the region around $x = X$ makes a significant contribution to this integral. Provided that $V'(x)$ varies slowly enough across this region, we can approximate $V'(x)$ by the *constant* value $V'(X)$, and then take this constant value outside the integral. Then, because the wave function is normalized we obtain

$$-\left\langle \frac{\partial V}{\partial x} \right\rangle = -V'(X) \int_{-\infty}^{\infty} \Psi^*(x,t)\Psi(x,t)\, \mathrm{d}x = -V'(X).$$

This can be thought of as the force on the particle at position X. Equation 6.24 is then interpreted as Newton's second law: mass $\times$ acceleration $=$ applied force. The important point to remember is that:

> The classical limit applies to wave packets with narrow spreads of position and momentum subject to potential energy functions for which $\partial V/\partial x$ varies slowly over the width of the wave packet.

These conditions are generally met for macroscopic objects (which have very short de Broglie wavelengths) but are seldom valid for microscopic particles such as atoms or electrons.

6.3 Predicting the motion of a wave packet

We now turn to consider the problem of predicting the time-development of a wave packet. This is analogous to the classical problem of releasing a particle with some initial position and velocity, and predicting its future motion.

To take a definite case, we shall continue to discuss a particle of mass m in a harmonic potential energy well. Suppose that, at time $t = 0$, the state of the particle is described by a linear combination of harmonic-oscillator energy eigenfunctions

$$\Psi(x,0) = a_0\psi_0(x) + a_1\psi_1(x) + \ldots \tag{6.25}$$

where $\psi_n(x)$ is an energy eigenfunction of the harmonic oscillator with eigenvalue $E_n = (n + \frac{1}{2})\hbar\omega_0$. Then, we can ask what the wave function will be at some later time t. We shall assume (correctly) that there is a unique answer to this question. In other words, if we know the wave function at time $t = 0$, we assume that the time-development of the wave function is completely determined by Schrödinger's equation.

The assumption of a unique answer is liberating. It means that, if we can write down a wave function $\Psi(x, t)$ that (i) satisfies Schrödinger's equation for the given system and (ii) is equal to the known form of the wave function at time $t = 0$, then this must be *the* wave function of the system — there can be no other. Now, it is easy to see that the wave function

$$\Psi(x, t) = a_0\psi_0(x)e^{-iE_0t/\hbar} + a_1\psi_1(x)e^{-iE_1t/\hbar} + \ldots \tag{6.26}$$

is a solution of Schrödinger's equation. This is because each term in the sum is a stationary-state wave function that satisfies Schrödinger's equation and, by the principle of superposition, any linear combination of solutions of Schrödinger's equation is itself a solution. What is more, if we substitute $t = 0$ in Equation 6.26, we recover Equation 6.25, so the particular linear combination of stationary-state wave functions in Equation 6.26 is equal to the known wave function at time $t = 0$. Since conditions (i) and (ii) are met, uniqueness allows us to conclude that the wave function in Equation 6.26 describes the state of the particle at all times.

Now let us consider a slightly different case. Suppose that, at time $t = 0$, the state of the particle is described by the wave function $\Psi(x, 0) = f(x)$, where $f(x)$ is a known function, such as that in Figure 6.5.

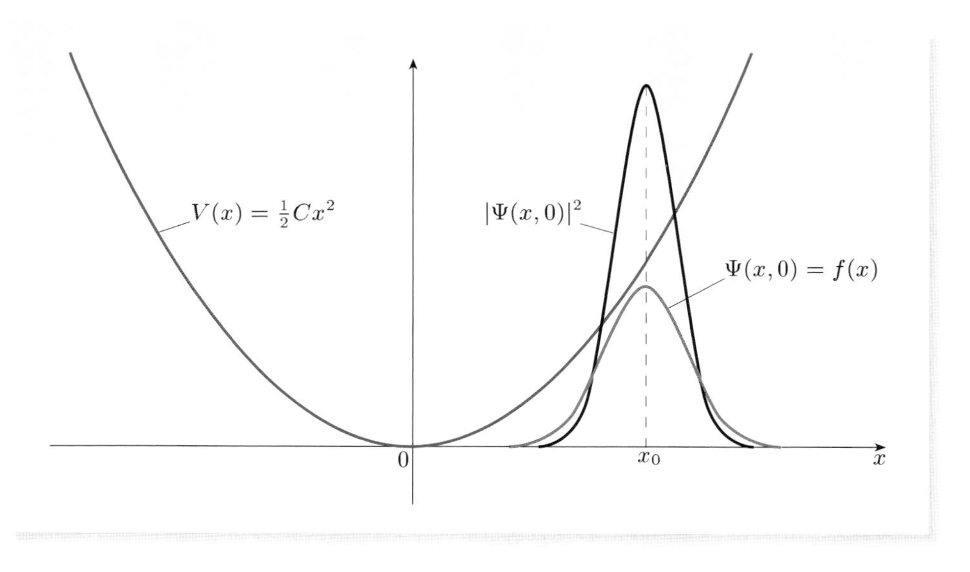

Figure 6.5 The wave function at $t = 0$ is $\Psi(x, 0) = f(x)$, centred at point x_0.

How can we predict the time-development of the wave function in this case? We do this by assuming that it is possible to write $f(x)$ as a linear combination of energy eigenfunctions. That is, we write

$$f(x) = a_0\psi_0(x) + a_1\psi_1(x) + \ldots = \sum_{i=0}^{\infty} a_i\psi_i(x). \tag{6.27}$$

Then, by the above argument, the wave function we are looking for is given by Equation 6.26. The only snag is, we do not know what the coefficients $a_0, a_1, \ldots$ are.

- What is the physical significance of the coefficient a_j?

- It is the probability amplitude for obtaining the energy eigenvalue E_j in the initial state $\Psi(x, 0) = f(x)$. The probability of obtaining the energy E_j is $|a_j|^2$.

To find the coefficients $a_0, a_1, \ldots$ we can use a method introduced in Chapter 4. We multiply both sides of Equation 6.27 by $\psi_j^*(x)$ and integrate over all x. This gives

$$\int_{-\infty}^{\infty} \psi_j^*(x)\, f(x)\, \mathrm{d}x = \sum_{i=0}^{\infty} a_i \int_{-\infty}^{\infty} \psi_j^*(x)\, \psi_i(x)\, \mathrm{d}x. \tag{6.28}$$

We then use the fact that the energy eigenfunctions are orthonormal. Expressed in terms of the Kronecker delta symbol, this means that

$$\int_{-\infty}^{\infty} \psi_j^*(x)\, \psi_i(x)\, \mathrm{d}x = \delta_{ji}. \tag{6.29}$$

Inserting this result into Equation 6.28 and noting that the Kronecker delta kills off all the terms on the right-hand side *except the term with* $i = j$, we obtain

$$\int_{-\infty}^{\infty} \psi_j^*(x)\, f(x)\, \mathrm{d}x = a_j.$$

Turning this equation around, replacing j by i, and recalling that $f(x) = \Psi(x, 0)$, we conclude that

$$a_i = \int_{-\infty}^{\infty} \psi_i^*(x)\, \Psi(x, 0)\, \mathrm{d}x. \tag{6.30}$$

You may recognize this equation. It tells us that the probability amplitude for measuring energy E_i in the initial state is given by the *overlap integral* of the ith energy eigenfunction with the initial wave function. In the context of our present discussion, however, the importance of this result is that it gives us a way of finding the coefficients needed in the expansion of the wave function.

We have now solved the following basic problem in quantum mechanics: *If a system starts in a given state, described by an initial wave function* $\Psi(x, 0) = f(x)$, *how does the wave function evolve in the future*? The answer is given by the following three-step procedure:

1. Expand the initial wave function as a sum of energy eigenfunctions (Equation 6.27).
2. Use the overlap integral to find the coefficients a_i in the sum (Equation 6.30).
3. Construct the wave function at an arbitrary time t by including the appropriate time-dependent phase factors, (Equation 6.26). The energy eigenfunctions in Step 1 are replaced by stationary-state wave functions to give

$$\Psi(x, t) = \sum_{i=0}^{\infty} a_i\, \psi_i(x)\, \mathrm{e}^{-\mathrm{i}E_i t/\hbar}. \tag{6.31}$$

The energy eigenfunctions used in Step 1 are, of course, those appropriate to the system under study — products of polynomials and Gaussian functions for a harmonic oscillator, sinusoidal functions for an infinite square well, and so on. In general, the task of evaluating the coefficients is lengthy, and is not something you will be asked to do. However, the coefficients can be evaluated quite painlessly using a computer, and this is how data for the graphs in Figure 6.2 were calculated.

Finally, one mathematical detail must be cleared up. We have *assumed* that it is possible to expand the initial wave function $f(x)$ in terms of the appropriate energy eigenfunctions. You might wonder whether this assumption is justified. It turns out that any reasonable function $f(x)$ can be expanded as a linear combination of energy eigenfunctions, although infinitely many terms may be needed in general. Mathematicians have proved this fact in various cases, and it also makes good physical sense because the coefficients in the sum have the physical meaning of being probability amplitudes for the various possible energy eigenvalues. In mathematical language, the set of energy eigenfunctions is said to form a **complete set** — which is just another way of saying that any reasonable function can be expanded in terms of them.

Exercise 6.7 We stated earlier that if a wave packet in a harmonic potential energy well is initially symmetric about the centre of the well, it will remain symmetric at all times. Use Equations 6.26 and 6.30 to confirm this fact. ■

Comparison with classical physics

In classical physics, the initial state of a particle is specified by its initial position and momentum. Newton's laws then allow us to deduce the position and momentum of the particle at all future times. For example, we can move a pendulum bob to one side and release it from rest; Newton's laws will predict the subsequent to-and-fro motion.

In quantum mechanics, the initial state of a particle is specified by giving its initial wave function. Schrödinger's equation then allows us to deduce the wave function of the particle at all future times. The methods used above may have disguised the fact that Schrödinger's equation is involved, but this is only because we have found a crafty way of solving Schrödinger's equation, using the principle of superposition to represent the solution as a linear combination of stationary-state wave functions.

It is interesting to compare the ways in which the initial conditions are specified. For a particle in one dimension, Newtonian mechanics requires two real numbers, $x(0)$ and $p_x(0)$, but quantum mechanics requires much more information — a function $\Psi(x, 0)$, which may be complex. It is worth reflecting on why this is so. All that can happen to a classical particle is that it can move from one point to another, but a quantum-mechanical wave packet undergoes a more complicated evolution in which the whole probability distribution shifts with time, moving, spreading and possibly developing peaks and troughs; this more complicated time-development is related to more complicated initial conditions.

In classical physics, we can release a particle at rest from an initial position $x = x_0$, and then calculate its subsequent motion in a harmonic well. What is the quantum version of this calculation? The closest we can get is to consider a narrow wave packet that is initially centred on the point $x = x_0$ (as in Figure 6.5). To correspond to starting from rest, we assume that the expectation value of momentum is equal to zero at the instant of release. We can then use the methods outlined above to find the coefficients of different harmonic-oscillator energy eigenfunctions in the initial wave packet, and construct a linear combination of stationary-state wave functions that tells us how the wave packet develops in time. In the classical limit, when the wave packet has extremely narrow spreads of position and momentum, the detailed shape of the wave packet becomes unimportant and all that really matters is $\langle x \rangle$ and $\langle p_x \rangle$. In this limit, the predictions of quantum physics become indistinguishable from those of classical physics, as you saw earlier in the discussion of Ehrenfest's theorem.

One detail is worth further thought. How can we ensure that an initial wave function has $\langle p_x \rangle = 0$, corresponding to the classical notion of releasing a particle from rest? One way of achieving this is to take the initial wave function to be real, as we shall now show. Let the initial wave function be $\Psi(x, 0) = f(x)$, where $f(x)$ is a normalized *real* function. Then the initial expectation value of momentum is

$$\begin{aligned}\langle p_x \rangle &= \int_{-\infty}^{\infty} \Psi^*(x,0)\Big(-\mathrm{i}\hbar\frac{\partial}{\partial x}\Big)\Psi(x,0)\,\mathrm{d}x \\ &= -\mathrm{i}\hbar \int_{-\infty}^{\infty} f(x)\frac{\partial f(x)}{\partial x}\,\mathrm{d}x,\end{aligned}$$

where we have used the fact that $f(x)$ is real to take $f^*(x) = f(x)$. Since $f(x)$ is a function of x only, the partial derivative can be replaced by $\mathrm{d}f/\mathrm{d}x$. Now we can rewrite the integral as

$$\langle p_x \rangle = -\mathrm{i}\hbar \int_{-\infty}^{\infty} \tfrac{1}{2}\frac{\mathrm{d}}{\mathrm{d}x}\left[f^2(x)\right]\mathrm{d}x$$

which, on integration, gives

$$\langle p_x \rangle = -\frac{\mathrm{i}\hbar}{2}\left[f^2(x)\right]_{-\infty}^{\infty} = 0,$$

because $f^2(x)$ must go to zero at both $+\infty$ and $-\infty$ in order for $f(x)$ be normalized. It therefore follows that $\langle p_x \rangle = 0$ in any state described by a real-valued wave function.

Choosing the initial wave function to be real is only one way of ensuring that the wave packet starts from rest. In the case of a harmonic oscillator, *any* wave packet that does not contain contributions from any *neighbouring* energy eigenfunctions (such as $\psi_i(x)$ and $\psi_{i+1}(x)$) has zero expectation value of momentum, as the following exercise illustrates.

Exercise 6.8 Show that the expectation value of momentum is equal to zero in a state described by the wave function

$$\Psi(x, 0) = a_2\psi_2(x) + a_5\psi_5(x),$$

where $\psi_2(x)$ and $\psi_5(x)$ are energy eigenfunctions in a harmonic oscillator, and a_2 and a_5 are complex constants. ■

Computer simulations: wave packets in wells

Now is the ideal time to study the software package *Wave packets in wells*. This is designed to help you visualize the wave packets we have been describing. Part of this package reviews the topics covered so far, but you will also be able to explore some additional topics in a fairly open-ended way. Here is a brief summary of some things you can do with this package:

- Set up your own combinations of harmonic oscillator eigenfunctions to explore how wave packets, probability densities and expectation values of position and momentum vary with time in a harmonic oscillator.
- Study a special type of harmonic-oscillator wave packet (known as a *coherent-state wave packet*) which behaves very much like a classical particle.
- Compare the behaviour of wave packets in a harmonic oscillator potential well with those in an infinite square well. There are many similarities, but the differences include the fact that $\langle x \rangle$ and $\langle p_x \rangle$ in an infinite square well do not always vary sinusoidally in time.

6.4 Free-particle wave packets

Finally, we consider a wave packet that describes a **free particle**, i.e. one that is not subject to any force. At first sight, you might think that it would be trivial to describe the motion of a free particle, but this is not always so in quantum mechanics. We face two special difficulties: (i) the allowed energies of a free particle form a continuum and (ii) the energy eigenfunctions of a free particle cannot be normalized. Nevertheless, you will see that the approach used earlier to describe wave packets in a harmonic oscillator can be adapted to free particles, provided we judiciously replace sums by integrals. One of the things we will be able to do is explain how electrons passing through a narrow slit give rise to a characteristic single-slit diffraction pattern.

6.4.1 Free-particle stationary states

Before constructing wave packets for a free particle, we will first look at the building blocks from which these wave packets are made — the energy eigenfunctions and de Broglie wave functions of free particles. The Hamiltonian operator for a free particle in one dimension is

$$\widehat{\mathrm{H}} = -\frac{\hbar^2}{2m}\frac{\partial^2}{\partial x^2},$$

and the corresponding time-independent Schrödinger equation is

$$-\frac{\hbar^2}{2m}\frac{\mathrm{d}^2\psi}{\mathrm{d}x^2} = E\psi(x). \tag{6.32}$$

The energy eigenfunctions are functions that satisfy this equation together with the required boundary conditions (i.e. not diverging as $x \to \pm\infty$). There are no satisfactory solutions for $E < 0$, but solutions exist for all $E \geq 0$.

The solutions to Equation 6.32 can be expressed in various alternative forms. From your experience with an infinite square well, you might expect these solutions to be expressed as sines or cosines. However, for the purpose of describing a free particle, it is much more convenient to write the solutions as complex exponentials. That is, we take

$$\psi(x) = D\mathrm{e}^{\mathrm{i}kx},$$

where D is a constant and k can have any real value from $-\infty$ to $+\infty$. The advantage of writing the eigenfunctions in this form is that they are eigenfunctions of momentum as well as energy.

Worked Example 6.1

Essential skill

Verifying properties of momentum eigenfunctions

(a) Verify that $\psi(x) = D\mathrm{e}^{\mathrm{i}kx}$ is an eigenfunction of momentum and determine its eigenvalue.

(b) Show that $\psi(x)$ is also an eigenfunction of the free-particle Hamiltonian operator and determine the corresponding eigenvalue, E.

(c) What is the degree of degeneracy of each energy level?

(d) Show explicitly that $\psi(x)$ cannot be normalized.

Solution

(a) Applying the momentum operator to the given function,

$$-\mathrm{i}\hbar\frac{\partial}{\partial x}D\mathrm{e}^{\mathrm{i}kx} = -\mathrm{i}\hbar(\mathrm{i}k)D\mathrm{e}^{\mathrm{i}kx} = \hbar kD\mathrm{e}^{\mathrm{i}kx},$$

so $D\mathrm{e}^{\mathrm{i}kx}$ is an eigenfunction of momentum with eigenvalue $\hbar k$.

(b) Applying the free-particle Hamiltonian operator to the given function,

$$-\frac{\hbar^2}{2m}\frac{\partial^2}{\partial x^2}D\mathrm{e}^{\mathrm{i}kx} = -\frac{\hbar^2}{2m}(\mathrm{i}k)^2D\mathrm{e}^{\mathrm{i}kx} = \frac{\hbar^2k^2}{2m}D\mathrm{e}^{\mathrm{i}kx},$$

so $D\mathrm{e}^{\mathrm{i}kx}$ is an eigenfunction of the free-particle Hamiltonian operator with eigenvalue $E = \hbar^2k^2/2m$; this is just the kinetic energy associated with momentum $\hbar k$.

(c) There are two values of k for each energy eigenvalue $E = \hbar^2k^2/2m$. These values are $k_1 = \sqrt{2mE}/\hbar$ and $k_2 = -\sqrt{2mE}/\hbar$. They correspond to two *different* energy eigenfunctions, $D\mathrm{e}^{\mathrm{i}k_1x}$ and $D\mathrm{e}^{\mathrm{i}k_2x}$, characterized by momenta of equal magnitudes but opposite directions along the x-axis. So each energy level is doubly-degenerate.

(d) We evaluate

$$\int_{-\infty}^{\infty} |D\mathrm{e}^{\mathrm{i}kx}|^2\,\mathrm{d}x = \int_{-\infty}^{\infty} |D|^2\,\mathrm{d}x$$

which is infinite. There is no choice of D that will yield a finite (and unity-valued) integral, so the function cannot be normalized.

Although the function De^{ikx} cannot be normalized, it is helpful to choose a particular value for the constant D. While this cannot help with the normalization, you will see that it simplifies some equations that will appear later. We choose

$$\psi_k(x) = \frac{1}{\sqrt{2\pi}}\, e^{ikx}. \tag{6.33}$$

The index k runs over a continuous range of real values, from $-\infty$ to $+\infty$. The function $\psi_k(x)$ is an eigenfunction of momentum with eigenvalue $\hbar k$, and it is also an eigenfunction of the free-particle Hamiltonian operator (an energy eigenfunction) with eigenvalue $\hbar^2 k^2/2m$. For brevity, we will generally refer to $\psi_k(x)$ as a **momentum eigenfunction**.

The stationary-state wave function corresponding to the eigenfunction $\psi_k(x)$ is

$$\Psi_k(x,t) = \psi_k(x) e^{-iE_k t/\hbar} \quad \text{where} \quad E_k = \frac{\hbar^2 k^2}{2m}.$$

Using Equation 6.33, and combining exponentials, this can also be written as

$$\Psi_k(x,t) = \frac{1}{\sqrt{2\pi}}\, e^{i(kx-\omega_k t)} \quad \text{where} \quad \omega_k = \frac{E_k}{\hbar} = \frac{\hbar k^2}{2m}.$$

Apart from the factor $1/\sqrt{2\pi}$, you should recognize this as a de Broglie wave function with wave number k and angular frequency ω_k, corresponding to momentum $\hbar k$ and energy $\hbar\omega_k$. The wavelength of the de Broglie wave is given by $\lambda = 2\pi/k$, so high wave numbers correspond to short wavelengths.

The trouble with de Broglie wave functions is that they cannot be normalized, so we cannot define a sensible probability density for position, let alone describe how this probability density changes when a particle moves. Put bluntly, this means that de Broglie waves are not acceptable wave functions for a single particle. However, you will soon see that we can construct normalized wave packets, using de Broglie waves as building blocks.

6.4.2 Constructing a free-particle wave packet

For a harmonic oscillator, we constructed a wave packet by taking a linear combination of harmonic-oscillator stationary states:

$$\Psi(x,t) = \sum_{i=0}^{\infty} a_i \psi_i(x) e^{-iE_i t/\hbar},$$

where $\psi_i(x)$ is the ith harmonic-oscillator energy eigenfunction, with eigenvalue E_i. By the principle of superposition, this wave packet satisfies Schrödinger's equation for the harmonic oscillator.

Now we shall do something very similar for a free particle, by taking a linear combination of free-particle stationary states (de Broglie waves). In fact, it does not help to take a discrete sum of de Broglie wave functions because such a sum would not produce a normalizable wave function. Instead, we integrate over a *continuous* set of de Broglie wave functions, each labelled by the continuous index k which represents the wave number. In other words, we consider the wave packet

$$\Psi(x,t) = \int_{-\infty}^{\infty} A(k)\, \psi_k(x)\, e^{-iE_k t/\hbar}\, dk,$$

where $A(k)$ is a complex function and $\psi_k(x)$ is a momentum eigenfunction, which is also an energy eigenfunction with energy eigenvalue E_k. Using Equation 6.33, and combining exponentials, the wave packet can be expressed as

$$\Psi(x,t) = \frac{1}{\sqrt{2\pi}} \int_{-\infty}^{\infty} A(k)\, \mathrm{e}^{\mathrm{i}(kx - E_k t/\hbar)}\, \mathrm{d}k, \tag{6.34}$$

which is a linear superposition of de Broglie wave functions. Because each de Broglie wave function is a stationary-state solution of the free-particle Schrödinger equation, this linear combination must also satisfy the free-particle Schrödinger equation, by the principle of superposition.

Exercise 6.9 Given that the wave function $\Psi(x,t)$ in Equation 6.34 satisfies

$$-\frac{\hbar^2}{2m}\frac{\partial^2\Psi}{\partial x^2} = \frac{1}{\sqrt{2\pi}} \int_{-\infty}^{\infty} A(k) \left[\frac{\hbar^2 k^2}{2m}\right] \mathrm{e}^{\mathrm{i}(kx - E_k t/\hbar)}\, \mathrm{d}x$$

$$\mathrm{i}\hbar\frac{\partial\Psi}{\partial t} = \frac{1}{\sqrt{2\pi}} \int_{-\infty}^{\infty} A(k) \left[E_k\right] \mathrm{e}^{\mathrm{i}(kx - E_k t/\hbar)}\, \mathrm{d}x,$$

show that $\Psi(x,t)$ satisfies the free-particle Schrödinger equation. ■

Now something rather wonderful happens. There is a famous mathematical result known as *Plancherel's theorem* that we can exploit. The proof of this theorem is difficult, and we shall not attempt any justification here. Suffice it to say that the proof is a piece of pure mathematics that was given in 1910, before quantum mechanics had been established. In the notation of Equation 6.34, **Plancherel's theorem** tells us that

Some authors call this *Parseval's theorem*.

$$\int_{-\infty}^{\infty} |\Psi(x,0)|^2\, \mathrm{d}x = \int_{-\infty}^{\infty} |A(k)|^2\, \mathrm{d}k. \tag{6.35}$$

Consequently, if we choose the function $A(k)$ in such a way that

$$\int_{-\infty}^{\infty} |A(k)|^2\, \mathrm{d}k = 1, \tag{6.36}$$

then our wave packet will be normalized at $t = 0$; as always, this normalization is preserved at later times. Such a choice of $A(k)$ is always possible and it completely solves the problem of non-normalizable de Broglie wave functions. To describe a free particle in quantum mechanics we must use a wave packet of the form given in Equation 6.34 with a function $A(k)$ that satisfies Equation 6.36. This normalized wave packet does not have a definite momentum or a definite energy, but has a spread in both of these quantities.

6.4.3 Interpreting $A(k)$ as a momentum amplitude

What is the physical significance of the function $A(k)$ introduced above? To explore this question, it is helpful to recall the interpretation of a harmonic-oscillator wave packet:

$$\Psi(x,t) = \sum_{i=0}^{\infty} a_i\, \psi_i(x)\, \mathrm{e}^{-\mathrm{i}E_i t/\hbar}.$$

In this case, the coefficient a_i is interpreted as the *probability amplitude* for obtaining the value E_i in the given wave packet, which means that the *probability* of obtaining the energy E_i is $|a_i|^2$.

Now, a very similar interpretation can be given to $A(k)$, but with a slight difference because we are dealing with a continuous variable k, rather than a discrete index i. In fact we can say that $|A(k)|^2\,\delta k$ is the probability of obtaining a wave number in a narrow range of width δk, centred on k. We are usually more interested in momentum than wave number, but these two quantities are related by the simple formula $p_x = \hbar k$. Hence, the probability of finding the wave number to be in a narrow range δk, centred on k is the same as the probability of finding the momentum p_x to be in a narrow range $\hbar\,\delta k$, centred on $\hbar k$. We can therefore make the following statement:

Born's rule for momentum

For a free-particle wave packet, the probability of finding the momentum of the particle to lie in a small range $\hbar\,\delta k$, centred on $\hbar k$, is $|A(k)|^2\,\delta k$.

There is a strong similarity between this rule and Born's rule for the wave function, which interprets $|\Psi(x,t)|^2\,\delta x$ as the probability of finding the particle in a small range δx, centred on x. For this reason, $A(k)$ is called the **momentum amplitude**. This term emphasizes the central role of $|A(k)|^2$ in determining the probability distribution of momentum. Note, however, that $|A(k)|^2$ is really a probability density *per unit wave number*. The probability density *per unit momentum* is $|A(k)|^2/\hbar$, which when multiplied by the small momentum range $\hbar\,\delta k$, gives the probability $|A(k)|^2\,\delta k$.

$A(k)$ is also called the **momentum wave function**.

While the expectation value $\langle p_x \rangle$ can always be found by evaluating the usual sandwich integral

$$\langle p_x \rangle = \int_{-\infty}^{+\infty} \Psi^*(x,t)\,\widehat{\mathrm{p}}_x\,\Psi(x,t)\,\mathrm{d}x,$$

it is also possible to calculate $\langle p_x \rangle$ directly from $A(k)$, based on Born's rule for momentum given above. The following exercise shows how this is done.

Exercise 6.10 How would you express the expectation values of the momentum and kinetic energy of a free-particle wave packet in terms of the momentum amplitude $A(k)$? Do these expectation values depend on time? ■

6.4.4 Predicting free-particle motion

It is easy to predict the motion of a free particle in classical physics, but the corresponding task is less straightforward in quantum mechanics. Given a wave function $\Psi(x,0) = f(x)$ that describes the initial state of a free particle at time $t = 0$, we need to predict how this wave function evolves in time. The method we shall use is similar to that outlined earlier for harmonic-oscillator wave packets.

1. For a harmonic oscillator, our first step was to express the initial wave function as a linear combination of harmonic-oscillator energy

eigenfunctions (Equation 6.25). For a free particle, we expand the initial wave function $\Psi(x,0) = f(x)$ as a linear combination of momentum eigenfunctions (which are also energy eigenfunctions):

$$\begin{aligned}\Psi(x,0) &= \int_{-\infty}^{\infty} A(k)\,\psi_k(x)\,\mathrm{d}k \\ &= \frac{1}{\sqrt{2\pi}}\int_{-\infty}^{\infty} A(k)\,\mathrm{e}^{\mathrm{i}kx}\,\mathrm{d}k. \qquad (6.37)\end{aligned}$$

An integral, rather than a sum, is needed here because k can have any real value between $-\infty$ and $+\infty$.

2. For a harmonic oscillator, the coefficients a_i were determined by using an overlap integral (Equation 6.30). The coefficients $A(k)$ in a free-particle wave packet can be determined in a very similar way. It turns out that

$$\begin{aligned}A(k) &= \int_{-\infty}^{\infty} \psi_k^*(x)\,\Psi(x,0)\,\mathrm{d}x \\ &= \frac{1}{\sqrt{2\pi}}\int_{-\infty}^{\infty} \Psi(x,0)\,\mathrm{e}^{-\mathrm{i}kx}\,\mathrm{d}x. \qquad (6.38)\end{aligned}$$

3. Finally, as for a harmonic oscillator, we construct the wave function $\Psi(x,t)$ by inserting the appropriate time-dependent phase factors. This has the effect of replacing the momentum eigenstates in Equation 6.37 by de Broglie wave functions:

$$\begin{aligned}\Psi(x,t) &= \int_{-\infty}^{\infty} A(k)\,\psi_k(x)\,\mathrm{e}^{-\mathrm{i}E_kt/\hbar}\,\mathrm{d}k \\ &= \frac{1}{\sqrt{2\pi}}\int_{-\infty}^{\infty} A(k)\,\mathrm{e}^{\mathrm{i}kx}\,\mathrm{e}^{-\mathrm{i}E_kt/\hbar}\,\mathrm{d}k \\ &= \frac{1}{\sqrt{2\pi}}\int_{-\infty}^{\infty} A(k)\,\mathrm{e}^{\mathrm{i}(kx-E_kt/\hbar)}\,\mathrm{d}k, \qquad (6.39)\end{aligned}$$

where $E_k = \hbar^2k^2/2m$.

The steps in this procedure can all be rigourously justified using a branch of mathematics called *Fourier analysis*, created at the beginning of the nineteenth century by the French mathematician Jean Baptiste Fourier (Figure 6.6). Mathematicians have shown that the set of functions $\mathrm{e}^{\mathrm{i}kx}$ is *complete*, so any reasonable function $f(x)$ can be expanded in terms of them, as in Equation 6.37. They have also shown that Equation 6.38 is a *mathematical* consequence of Equation 6.37, independent of any physical interpretation. We follow the standard mathematical convention of calling the right-hand side of Equation 6.38 the **Fourier transform** of $\Psi(x,0)$ and calling the right hand-side of Equation 6.37 the **inverse Fourier transform** of $A(k)$. Using this terminology we can state that:

Figure 6.6 Jean Baptiste Fourier (1768–1830) invented Fourier analysis to predict the flow of heat.

- The momentum amplitude $A(k)$ is the Fourier transform of the initial wave function $\Psi(x,0)$.
- The initial wave function $\Psi(x,0)$ is the inverse Fourier transform of the momentum amplitude $A(k)$.

Exercise 6.11 Use Equations 6.37 and 6.38 to suggest why it was sensible to include a factor $1/\sqrt{2\pi}$ in our definition of the momentum eigenfunction (Equation 6.33). ■

6.4.5 Wave-packet spreading

As an example of the method outlined above, let's consider a free particle whose initial state is described by a normalized *Gaussian function*

$$\Psi(x,0) = f(x) = \left(\frac{1}{\sqrt{\pi}a}\right)^{1/2} e^{-x^2/2a^2}, \tag{6.40}$$

sketched in Figure 6.7a. The significance of the constant a is that it gives a typical width for the initial wave packet (in fact, $\Delta x = a/\sqrt{2}$).

The initial wave packet is real-valued, so the initial value of $\langle p_x \rangle$ is equal to zero. Because a free particle experiences no forces, Ehrenfest's theorem tells us that

$$\frac{d\langle p_x \rangle}{dt} = -\left\langle \frac{\partial V}{\partial x} \right\rangle = 0,$$

so the expectation value of momentum remains permanently equal to zero for this wave packet. The wave packet cannot move bodily through space, but it can expand. This important phenomenon is called **wave-packet spreading**.

To predict how the wave packet spreads, we first calculate $A(k)$, using Equation 6.38 to take the Fourier transform of $f(x)$. The integration is quite tricky, so we shall just quote the answer:

$$A(k) = \left(\frac{a}{\sqrt{\pi}}\right)^{1/2} e^{-a^2k^2/2}.$$

This function is sketched in Figure 6.7b. It is interesting to note that $A(k)$ is also a Gaussian function, but of typical width $1/a$. So there is a reciprocal relation between the widths of $f(x)$ and $A(k)$ as can be seen in Figures 6.7c and d, where we show an alternative $\Psi(x,0)$ with a smaller Δx and with a correspondingly more spread-out $A(k)$. The narrower the spread of the initial wave packet $f(x)$, the broader the spread of the momentum amplitude function $A(k)$, and vice versa. This is nothing other than the Heisenberg uncertainty principle in action.

Our initial wave function (Equation 6.40) has a typical width of order a. To see how the width of the wave packet grows with time, we put the function $A(k)$ into Equation 6.39 to find $\Psi(x,t)$, the wave function of the free particle at any given time t. The integrals are again tricky, so we shall just illustrate the final result. Figure 6.8 shows the wave packet at three different times after $t = 0$; the steady spreading of the wave packet is evident.

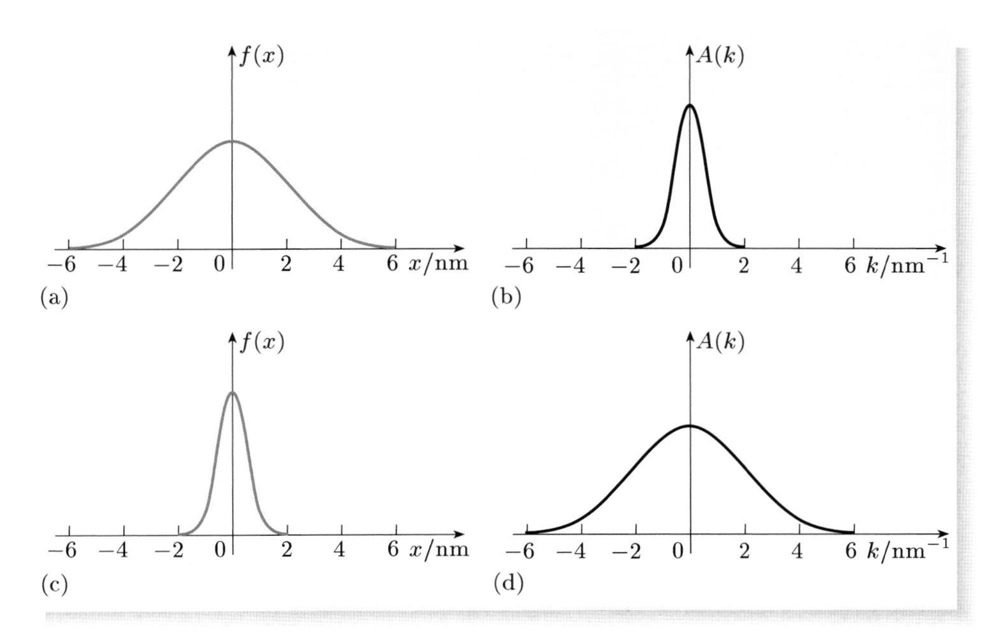

Figure 6.7 (a) A normalized Gaussian function $f(x)$, representing the initial state of a free particle. (b) The corresponding momentum amplitude function $A(k)$, found by taking the Fourier transform of $f(x)$. (c) Another Gaussian with a smaller Δx. (d) The momentum amplitude function $A(k)$ corresponding to (c).

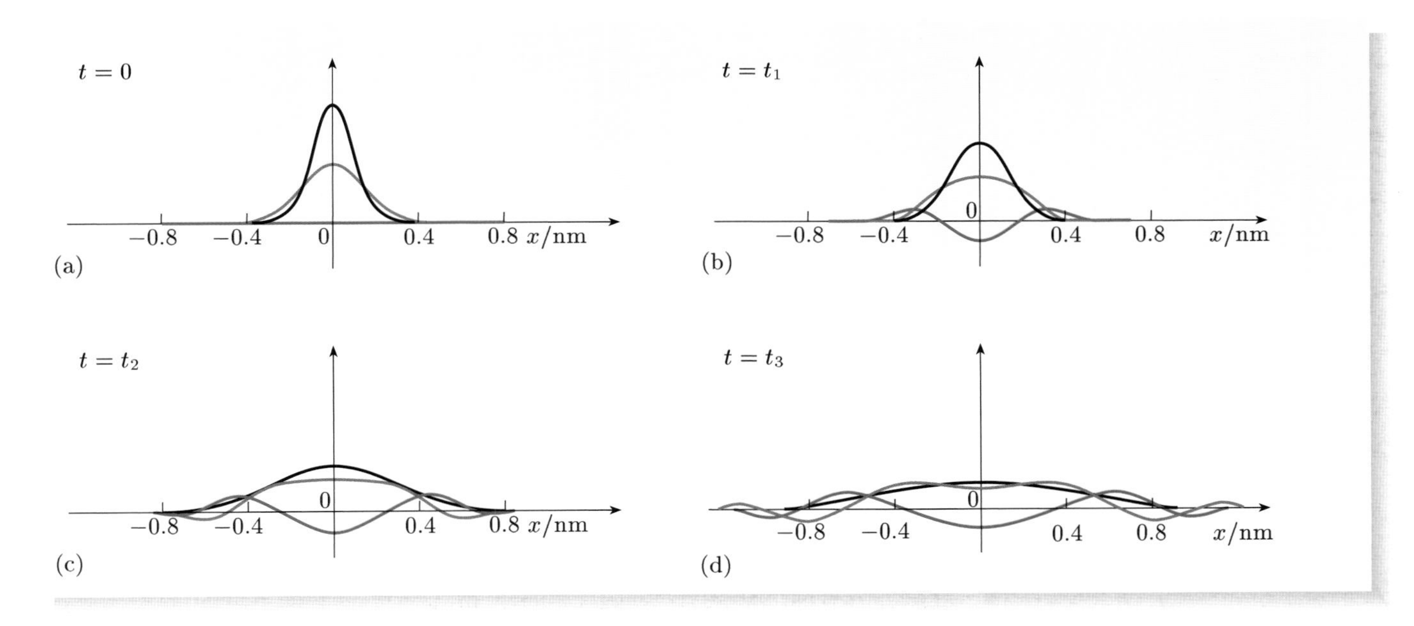

Figure 6.8 A free-particle wave packet shown at $t = 0$ and at three later times, with $0 < t_1 < t_2 < t_3$. The initial spread of the wave packet has been chosen to be comparable to the size of an atom. The real part of the wave function is shown in blue, the imaginary part in green and the probability density $|\Psi|^2$ in black.

Although the wave function develops real and imaginary parts, both of which have lots of wiggles, the modulus of the wave function turns out to be yet another Gaussian function

$$|\Psi(x,t)| = \left(\frac{1}{\sqrt{\pi}b}\right)^{1/2} \mathrm{e}^{-x^2/2b^2},$$

with a width

$$b = \left(a^2 + \frac{\hbar^2 t^2}{a^2 m^2}\right)^{1/2} \tag{6.41}$$

that *increases with time*. The corresponding uncertainty in position is $\Delta x = b/\sqrt{2}$. In the limit of large t, when the wave packet has spread far beyond its initial width a, we can make the approximation

$$\Delta x = \frac{b}{\sqrt{2}} \simeq \frac{\hbar}{\sqrt{2}am}\,t, \tag{6.42}$$

so the width of the wave packet eventually becomes proportional to time.

There is nothing mysterious about wave-packet spreading; it merely reflects the initial uncertainty in the velocity of the particle. The initial wave function corresponds to a spread of various possible momenta, with a momentum distribution determined by $|A(k)|^2$, and this momentum distribution remains constant for a free-particle wave packet because there are no forces to change it. Corresponding to the spread in possible momenta, there is a spread in possible velocities, $\Delta v_x = \Delta p_x/m$. This spread in velocities gives rise to an uncertainty in the position of the particle that increases with time.

Ignoring the initial width of the wave packet, the uncertainty in position can be estimated to be

$$\Delta x \simeq \Delta v_x\, t = \frac{\Delta p_x}{m}\,t. \tag{6.43}$$

The wave packet in Equation 6.40 is Gaussian in shape, just like the ground state of a harmonic oscillator. Although we are not dealing with a harmonic oscillator here, the fact that our wave function has the *same shape* as the harmonic-oscillator ground state allows us to use the result

$$\Delta p_x = \frac{\hbar}{\sqrt{2}a} \tag*{(Eqn 5.51)}$$

obtained near the end of the Chapter 5. Combining Equations 6.43 and 5.51, we then recover Equation 6.42.

It is worth noting that *decreasing* the spatial extent of the initial wave packet *increases* the spread of momenta and therefore increases the rate at which the wave packet spreads. If you want to construct a wave packet that remains compact for a long time, you should not start out with one that is too narrow.

Exercise 6.12 A free electron is described at $t = 0$ by a Gaussian wave packet with $\Delta x = 1\,\mathrm{mm}$. Show that 5 hours later $\Delta x \approx 1\,\mathrm{km}$. ■

6.5 Beyond free particles

You have seen that the function $|A(k)|^2$ tells us about the distribution of momentum in a free-particle wave packet. But free particles are a very special case. It also makes sense to talk about the distribution of momentum in other types of wave packet — a harmonic-oscillator wave packet, for example. So far, we have not said how such a momentum distribution can be calculated. For completeness, we now tie up this loose end.

A free particle has an unchanging distribution of momentum, but this is not true for a particle subject to external forces. You saw earlier that the expectation value of momentum varies sinusoidally in a harmonic-oscillator wave packet. Nevertheless, we can extend Born's rule for momentum to cover all particles, provided we use a time-dependent momentum amplitude $A(k,t)$. In general, the probability of finding the momentum at time t to lie in a small range $\hbar\,\delta k$, centred on $\hbar k$ is $|A(k,t)|^2\,\delta k$.

The only remaining issue is specifying an appropriate function $|A(k,t)|^2$. For free particles, we know that this is done by taking the Fourier transform of the initial wave function. Equation 6.38 shows that

$$|A(k)|^2 = \left|\frac{1}{\sqrt{2\pi}}\int_{-\infty}^{\infty}\Psi(x,0)\mathrm{e}^{-ikx}\,\mathrm{d}x\right|^2 \quad \text{for a free particle.}$$

This equation generalizes to non-free particles, simply by using the wave function at the appropriate time:

$$|A(k,t)|^2 = \left|\frac{1}{\sqrt{2\pi}}\int_{-\infty}^{\infty}\Psi(x,t)\mathrm{e}^{-ikx}\,\mathrm{d}x\right|^2. \tag{6.44}$$

In general, this leads to a time-dependent distribution of momentum, reflecting the fact that external forces cause the momentum distribution to change.

Figure 6.9 shows an example of a momentum distribution calculated by this method. It shows $|A(k,t)|^2$ for the harmonic-oscillator wave packet of Figure 6.2 at $t = T/8$. At the instant shown, a measurement of momentum would give a negative value with overwhelming probability, corresponding to the classical picture of a particle moving back towards the origin. Nevertheless, there is a small but non-vanishing probability of finding a positive momentum, corresponding to a particle that is moving *away* from the origin. Not for the first (or last) time, quantum mechanics confounds our classical intuition.

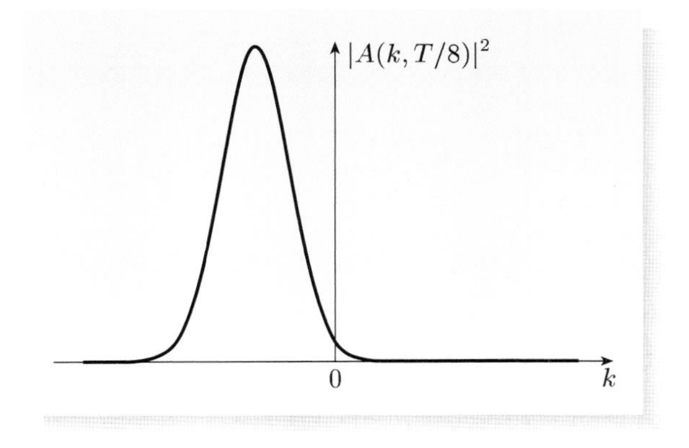

Figure 6.9 A graph of $|A(k,t)|^2$ against k for the harmonic-oscillator wave packet of Figure 6.2 at time $t = T/8$.

Computer simulations: free-particle wave packets

Now is a good time to study the software package *Free-particle wave packets*. Here is a brief summary of some things you can learn from this package:

- Free-particle wave packets with $\langle p_x \rangle = 0$ spread at a rate inversely related to their initial width.
- Free-particle wave packets with $\langle p_x \rangle \neq 0$ spread in a similar way, but with a steadily moving 'centre of mass'.
- The process of single-slit diffraction of an electron wave can be analyzed, starting with a 'top-hat' initial wave function representing an electron that has just passed through the slit. The subsequent spreading of this wave packet produces a probability distribution for finding the electron at any given point on the screen.
- Moving beyond free-particle wave packets we can consider a wave-packet that encounters a potential energy ramp. The momentum distribution then changes with time in response to the external force.

Summary of Chapter 6

Section 6.1 A wave packet is a normalized linear combination of two or more stationary-state wave functions. Because of the superposition principle, wave packets satisfy Schrödinger's equation. Unlike stationary states, wave packets have probability densities that change in time, and so can describe motion.

In a harmonic oscillator, the probability density associated with a wave packet can oscillate to-and-fro with the classical frequency. Other types of motion, such as symmetrical breathing in and out are also possible. The expectation values of position and momentum are either permanently equal to zero, or oscillate sinusoidally with time.

Section 6.2 Ehrenfest's theorem states that

$$\frac{\mathrm{d}\langle x \rangle}{\mathrm{d}t} = \frac{\langle p_x \rangle}{m} \quad \text{and} \quad \frac{\mathrm{d}\langle p_x \rangle}{\mathrm{d}t} = -\left\langle \frac{\partial V}{\partial x} \right\rangle .$$

This theorem is always true. The classical limit corresponds to a wave packet with narrow spreads of position and momentum, subject to a force that varies slowly over the width of the wave packet. In this limit, the predictions of quantum mechanics approach those of classical mechanics.

Section 6.3 Given an initial wave function $\Psi(x, 0)$, it is possible to predict the time-development of this wave function using a three-step procedure. We illustrated this for a harmonic oscillator. First, we expanded the initial state as a sum of energy eigenfunctions $\psi_i(x)$ of the oscillator. This is always possible because these eigenfunctions form a complete set. Then we determined the coefficients in the expansion, using the overlap integral

$$a_i = \int_{-\infty}^{\infty} \psi_i^*(x)\Psi(x, 0)\,\mathrm{d}x.$$

Finally, we introduced time-dependent phase factors into the sum, replacing each energy eigenstate $\psi_i(x)$ by the corresponding stationary-state wave function $\psi_i(x)\,\mathrm{e}^{-\mathrm{i}E_it/\hbar}$. This provides a solution of Schrödinger's equation that obeys the initial conditions, and so is the wave function we were looking for.

Section 6.4 Free-particle wave packets can be constructed by taking a continuous linear combination of de Broglie wave functions:

$$\Psi(x,t) = \frac{1}{\sqrt{2\pi}} \int_{-\infty}^{\infty} A(k)\mathrm{e}^{\mathrm{i}(kx - E_kt/\hbar)}\,\mathrm{d}k.$$

This wave packet is normalized provided that the function $A(k)$ obeys the condition $\int_{-\infty}^{\infty} |A(k)|^2\,\mathrm{d}k = 1$. The coefficients $A(k)$ are called momentum amplitudes. They are interpreted with the aid of Born's rule for momentum, which states that the probability of obtaining a momentum in a small interval of momentum $\hbar\,\delta k$, centred on $\hbar k$, is $|A(k)|^2\,\delta k$.

The time development of a free-particle wave packet can be predicted using an extension of the three-step procedure used for a harmonic oscillator. For a free particle, the overlap integral becomes a Fourier transform:

$$A(k) = \frac{1}{\sqrt{2\pi}} \int_{-\infty}^{\infty} \Psi(x,0)\mathrm{e}^{-\mathrm{i}kx}\,\mathrm{d}x.$$

By analyzing the motion of a free-particle wave packet, we predicted how a Gaussian packet spreads with time, and (in a computer simulation) how a single-slit diffraction pattern is formed.

Achievements from Chapter 6

After studying this chapter, you should be able to:

6.1 Explain the meanings of the newly-defined (emboldened) terms and symbols, and use them appropriately.

6.2 Give a general account of harmonic-oscillator wave packets and their role in accounting for motion in a harmonic oscillator.

6.3 Calculate the time-dependences of $\langle x\rangle$ and $\langle p_x\rangle$ for given harmonic-oscillator wave packets.

6.4 State Ehrenfest's theorem and give an account of its significance in relating quantum mechanics to classical mechanics.

6.5 Describe the procedure used to predict the time-development of a harmonic-oscillator wave function, given its value at time $t = 0$.

6.6 Describe how a free-particle wave packet is constructed, including the condition needed to ensure normalization.

6.7 Interpret momentum amplitudes and use them in Born's rule for momentum.

6.8 Give an account of the spreading of free-particle wave packets.

After studying the computer simulation packages, you should also be able to:

6.9 Compare and contrast the behaviour of wave packets in a harmonic potential energy well and an infinite square potential energy well.

6.10 Recognize that momentum amplitudes become time-dependent in wave packets that describe particles subject to forces.

Chapter 7 Scattering and tunnelling

Introduction

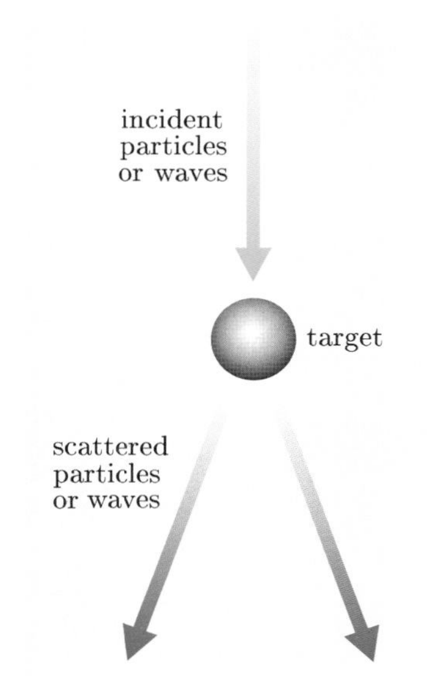

Figure 7.1 The phenomenon of scattering.

You have now met two approaches to describing the state of a system in wave mechanics. In cases where the probability distributions are independent of time a *stationary-state* approach can be used. In other cases, where probabilities are time-dependent and motion is really taking place, a *wave-packet* approach can be used. The two approaches are related but different. In many situations the choice of approach is obvious and straightforward, but that is not always the case, as you will soon see.

In this chapter we shall consider two physical phenomena of fundamental importance: *scattering* and *tunnelling*. Each will be treated using both a stationary-state approach *and* a wave-packet approach.

The phenomenon of **scattering** was an important topic in physics long before the development of wave mechanics. In its most general sense, scattering is a process in which incident particles (or waves) are affected by interaction with some kind of target, quite possibly another particle (Figure 7.1). The interaction can affect an incident particle in a number of ways; it may change its speed, direction of motion or state of internal excitation. Particles can even be created, destroyed or absorbed.

Figure 7.2 Red sunsets are a direct consequence of the scattering of sunlight.

It can be argued that almost everything we know about the world is learnt as a result of scattering. When we look at a non-luminous object such as a book or a building we see it because of the light that is scattered from its surface. The sky is blue because the particles in the Earth's atmosphere are more effective at scattering blue light (relatively short wavelengths) than yellow or red light (longer wavelengths). This is also the reason why sunsets are red (Figure 7.2). As the Sun approaches the horizon, its light has to traverse a lengthening path through the Earth's atmosphere; as a consequence shorter wavelengths are increasingly scattered out of the beam until all that remains is red.

Much of what we know about the structure of matter has been derived from scattering experiments. For example, the scattering of alpha particles from a gold foil, observed by Geiger and Marsden in 1909, led Rutherford to propose the first nuclear model of an atom. More recent scattering experiments, involving giant particle accelerators, have provided insight into the fundamental constituents of matter such as the quarks and gluons found inside protons and neutrons. Even our knowledge of cosmology, the study of the Universe as a whole, is deeply dependent on scattering. One of the main sources of precise cosmological information is the study of the *surface of last scattering* observed all around us at microwave wavelengths (Figure 7.3).

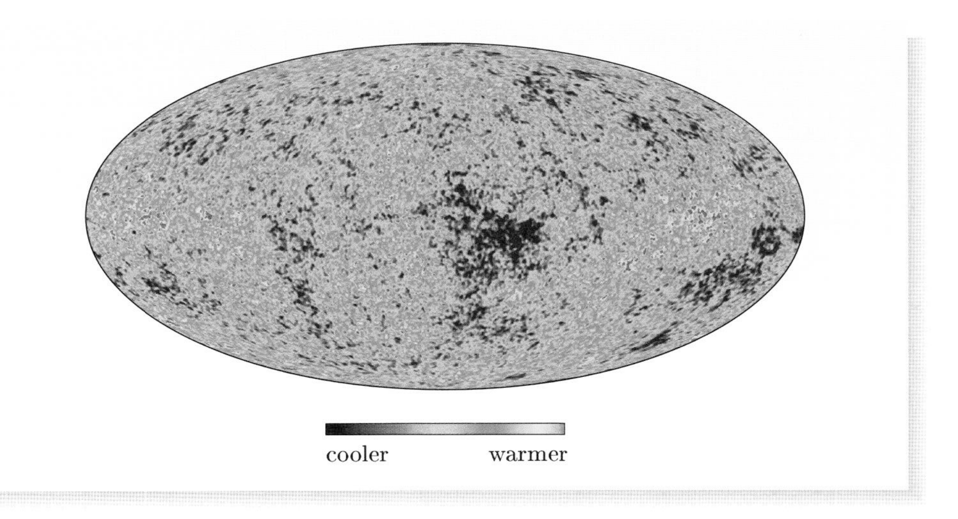

Figure 7.3 Microwave image of the surface of last scattering.

In our detailed discussions of scattering we shall not consider cases where the scattering changes the number or nature of the scattered particles, since that requires the use of *quantum field theory*, a part of quantum physics beyond the scope of this book. Rather, we shall mainly restrict ourselves to one-dimensional problems in which an incident beam or particle is either *transmitted* (allowed to pass) or *reflected* (sent back the way it came) as a result of scattering from a target. Moreover, that target will generally be represented by a fixed potential energy function, typically a finite well or a finite barrier of the kind indicated in Figure 7.4. Despite these restrictions, our discussion of quantum-mechanical scattering will contain many surprises. For instance, you will see that a finite potential energy barrier of height V_0 can reflect a particle of energy E_0, even when $E_0 > V_0$. Perhaps even more amazingly you will see that unbound particles can be reflected when they encounter a finite well.

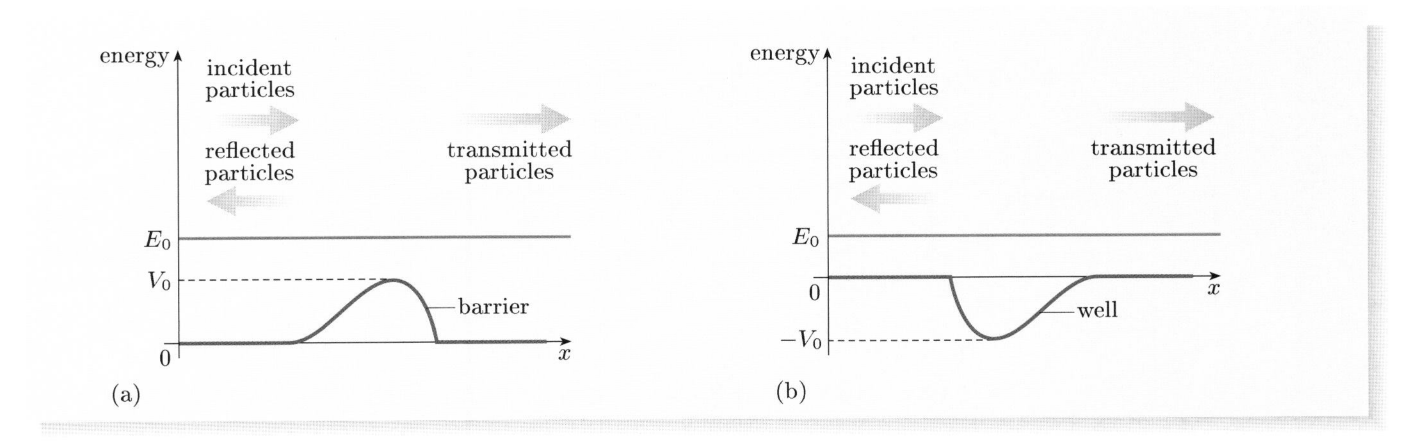

Figure 7.4 (a) Particles with energy $E_0 > V_0$, encountering a finite barrier of height V_0, have some probability of being reflected. (b) Similarly, unbound particles with energy $E_0 > 0$ can be reflected by a finite well.

The phenomenon of **tunnelling** is entirely quantum-mechanical with no analogue in classical physics. It is an extension of the phenomenon of *barrier penetration* that you met in earlier chapters. Barrier penetration involves the appearance of particles in *classically forbidden regions*. In cases of tunnelling, such as that shown in Figure 7.5, a particle with energy $E_0 < V_0$ can penetrate a potential energy barrier of height V_0, pass through the classically forbidden region within the barrier and have some finite probability of emerging into the classically allowed region on the far side.

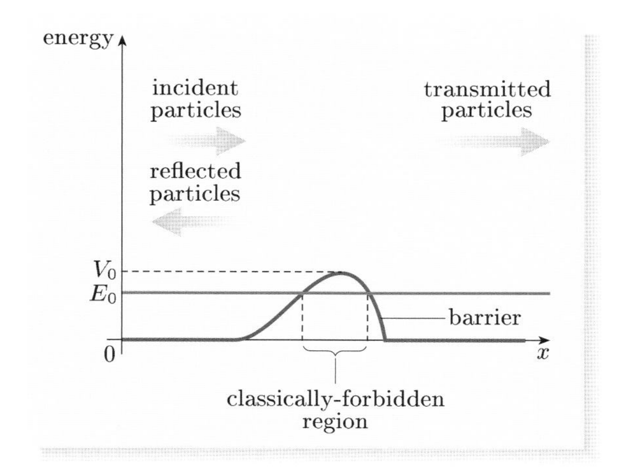

Figure 7.5 Particles with energy $E_0 < V_0$, encountering a finite barrier of height V_0, have some probability of being transmitted by tunnelling through the barrier. Such a process is forbidden in classical physics.

Tunnelling phenomena are common in many areas of physics. In this chapter you will see how tunnelling provides an explanation of the *alpha decay* of radioactive nuclei, and is also an essential part of the *nuclear fusion* processes by which stars produce light. Finally you will see how quantum tunnelling has allowed the development of instruments called *scanning tunnelling microscopes* (STMs) that permit the positions of individual atoms on a surface to be mapped in stunning detail.

7.1 Scattering: a wave-packet approach

This section discusses the scattering of a particle using wave packets of the type discussed in Chapter 6. We shall restrict attention to one dimension and suppose that the incident particle is initially free, described by a wave packet of the form

$$\Psi(x,t) = \frac{1}{\sqrt{2\pi}} \int_{-\infty}^{\infty} A(k) e^{i(kx - E_k t/\hbar)} \, dk. \tag{7.1}$$

This is a superposition of de Broglie waves, with the function $e^{i(kx-E_kt/\hbar)}$ corresponding to momentum $p_x = \hbar k$ and energy $E_k = \hbar^2k^2/2m$. The momentum amplitude function $A(k)$ determines the blend of de Broglie waves in the initial free wave packet, but when the wave packet encounters a change in the potential energy function, the blend of de Broglie waves changes, and scattering takes place.

7.1.1 Wave packets and scattering in one dimension

Figure 7.6 shows the scattering of a wave packet, incident from the left, on a target represented by a potential energy function of the form

$$V(x) = V_0 \quad \text{for } 0 \leq x \leq L \tag{7.2}$$

$$V(x) = 0 \quad \text{for } x < 0 \text{ and } x > L. \tag{7.3}$$

Potential energy functions of this type are called **finite square barriers**. They are simple idealizations of the more general kind of finite barrier shown in Figure 7.4a. The de Broglie waves that make up the wave packet extend over a range of energy and momentum values. In the case illustrated in Figure 7.6, the expectation value of the energy, $\langle E \rangle$, has a value E_0 that is greater than the height of the potential energy barrier V_0. The classical analogue of the process illustrated in Figure 7.6 would be a particle of energy E_0 scattering from a repulsive target. The unrealistically steep sides of the potential energy function imply that the encounter is sudden and impulsive — not like the encounter between two negatively charged particles for instance; there is no gradual slope for the incident particle to 'climb', nor for it to descend after the interaction. Still, in the classical case, the fact that E_0 is greater than V_0, implies that the incident particle has enough energy to overcome the resistance offered by the target, and it would be certain to emerge on the far side of it.

The quantum analysis tells a different story. Based on numerical solutions of Schrödinger's equation, the computer-generated results in Figure 7.6 show successive snapshots of $|\Psi|^2$, each of which represents the probability density of the particle at a particular instant. Examining the sequence of pictures it is easy to visualize the probability as a sort of fluid that flows from one part of the x-axis to another. Initially, as the wave packet approaches the barrier, the probability is concentrated in a single 'blob', flowing from left to right. Then, as the wave packet encounters the barrier, something odd starts to happen; the probability distribution develops closely-spaced peaks. These peaks are a consequence of reflection — part of the quantum wave packet passes through the barrier, but another part is reflected; the reflected part interferes with the part still advancing from the left and results in the spiky graph of $|\Psi|^2$. Eventually, however, the interference subsides and what remains are two distinct 'blobs' of probability density: one returning to the left, the other progressing to the right.

It's important to realize that the splitting of the wave packet illustrated in Figure 7.6 does *not* represent a splitting of the particle described by the wave packet. The normalization of the wave packet is preserved throughout the scattering process (the area under each of the graphs in Figure 7.6 is equal to 1); there is only ever one particle being scattered. The splitting of the wave packet simply indicates that, following the scattering, there are two distinct regions in which the particle might be found. In contrast to the certainty of transmission in the classical case, the quantum calculation predicts some probability of transmission, but also some probability of reflection. Indeterminacy is, of course, a characteristic feature of quantum mechanics.

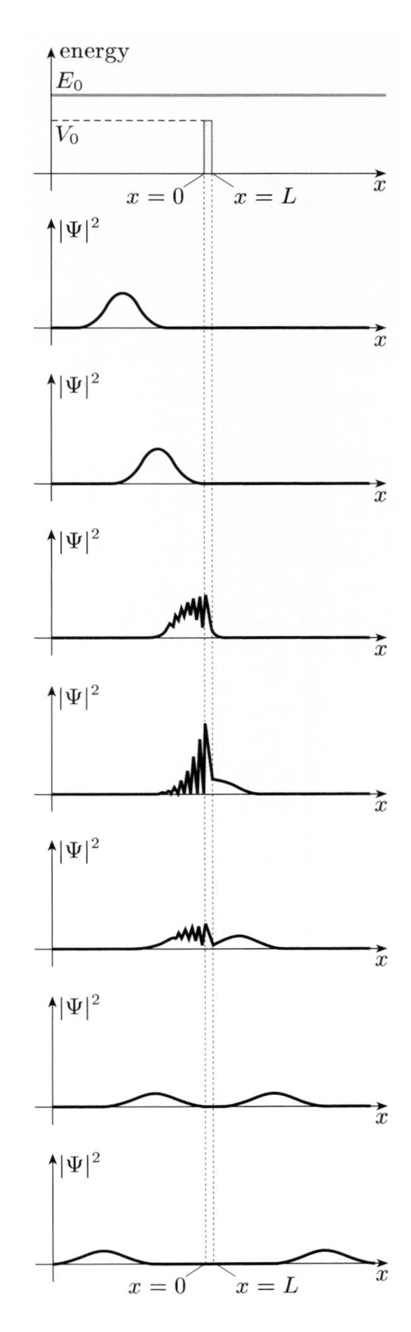

Figure 7.6 The scattering of a wave packet with $\langle E \rangle = E_0$ by a finite square barrier of height V_0 when $E_0 > V_0$. The probability density $|\Psi|^2$ is shown in a sequence of snapshots with time increasing from top to bottom. The barrier has been made narrow in this example but a greater width could have been chosen.

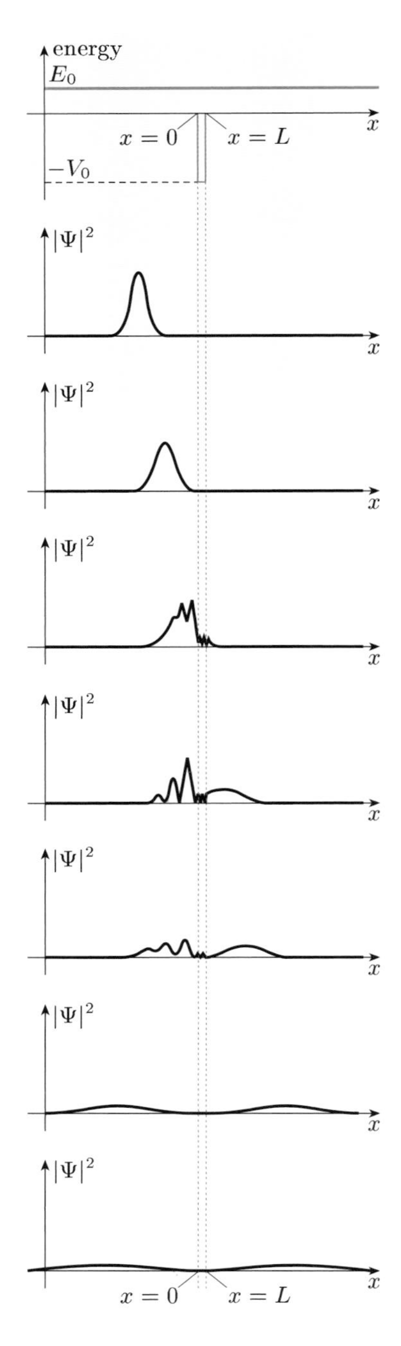

Figure 7.7 The scattering of a wave packet with $\langle E \rangle = E_0$ by a finite square well of depth V_0 when $E_0 > 0$. The probability density $|\Psi|^2$ is shown in a sequence of snapshots with time increasing from top to bottom. The well has been made narrow in this example but a greater width could have been chosen.

Exercise 7.1 Simply judging by eye, what are the respective probabilities of reflection and transmission as the final outcome of the scattering process shown in Figure 7.6? ■

The probability that a given incident particle is reflected is called the **reflection coefficient**, R, while the probability that it is transmitted is called the **transmission coefficient**, T. Since one or other of these outcomes must occur, we always have

$$R + T = 1. \tag{7.4}$$

The values of R and T that apply in any particular case of one-dimensional scattering can be worked out by examining the solution of the relevant Schrödinger equation that applies at a time long after the incident particle has encountered the target. For properly normalized wave packets, the values of R and T are found by measuring the areas under the graph of $|\Psi|^2$ that are located to the left and to the right of the target at that time.

Reflection and transmission coefficients can also describe scattering from an attractive target. The idealized case of a wave packet with $\langle E \rangle = E_0 > 0$ encountering a finite square well of depth V_0 and width L is shown in Figure 7.7. Again, the wave packet is incident from the left, and the wave packet is partly reflected, giving rise to interference effects. Eventually, however, the interference abates, leaving two distinct parts of the wave packet. The areas of these two parts determine the reflection and transmission coefficients.

Of course, wave mechanics is probabilistic, so although reflection and transmission coefficients can be calculated in any given situation, it is never possible to predict what will happen to any individual particle (unless R or T happen to be equal to zero in the situation considered). Indeed, the detection of the scattered particle, after the scattering has taken place, is exactly the kind of measurement that brings about the *collapse of the wave function* — a sudden, abrupt and unpredictable change that is not described by Schrödinger's equation. After such a collapse has occurred, the particle is localized on one side of the target or the other.

7.1.2 Simulations using wave packets

Now would be a good time to study the computer package *The scattering of wave packets in one dimension* that was used to generate the probability density plots shown in Figures 7.6 and 7.7. This package will allow you to specify a range of barriers and wells (not necessarily square), and to observe the way in which they scatter **Gaussian wave packets** (wave packets for which $|\Psi|^2$ is shaped like a Gaussian function).

The simulation is powerful and will reward detailed investigation. However, you should not spend too much time on it at this stage in your study of the chapter; thirty minutes should be sufficient for a first look. As you work through the package you should use its features, such as the ability to freeze the motion, or to advance the simulation step-by-step, to observe the following points:

- The spreading of the wave packet before, and after, it encounters the barrier or well (a consequence of the range of momentum values that contribute to the wave packet).

- The reflection of the wave packet by different parts of the barrier or well, and the way this leads to interference effects within the wave packet.
- The differences in behaviour between the real and imaginary parts of the wave packet and the way they combine to produce the probability density $|\Psi|^2$.
- The way in which the transmission and reflection coefficients depend on the energy of the incident wave packet and the parameters that define the barrier or well.

7.2 Scattering: a stationary-state approach

Scattering calculations using wave packets are so laborious that they are generally done numerically, using a computer. However, in many cases, scattering phenomena can be adequately treated using a procedure based on stationary states. This approach can give valuable insight into the scattering process without the need for computer simulations.

This section introduces the stationary-state approach to scattering. The discussion is mainly confined to one dimension, so a stationary-state solution to Schrödinger's equation can be written in the form $\Psi(x,t) = \psi(x)e^{-iEt/\hbar}$, where $\psi(x)$ satisfies the appropriate time-independent Schrödinger equation at energy E. The first challenge is to find a way of interpreting stationary-state solutions that makes them relevant to an inherently time-dependent phenomenon like scattering.

7.2.1 Stationary states and scattering in one dimension

The key idea of the stationary-state approach is to avoid treating individual particles, and to consider instead the scattering of a steady intense beam of particles, each particle having the same energy E_0. It is not possible to predict the exact behaviour of any individual particle but, if the incident beam is sufficiently intense, the result of the scattering will be reflected and transmitted beams with steady intensities that are determined by the reflection and transmission coefficients we are aiming to evaluate. Provided we consider the beams as a whole, nothing in this arrangement depends on time. A snapshot of the set-up taken at one time would be identical to a similar snapshot taken at another time. In contrast to the wave-packet approach, there are no moving 'blobs of probability density', so the whole process can be described in terms of stationary states.

For a one-dimensional beam, we define the **beam intensity** j to be the number of beam particles that pass a given point per unit time. We also define the **linear number density** n of the beam to be the number of beam particles per unit length. Then, thinking in classical terms for a moment, if all the particles in a beam have the same speed v, the beam intensity is given by $j = vn$. Specializing this relationship to the incident, reflected and transmitted beams, we have

$$j_{\text{inc}} = v_{\text{inc}}\, n_{\text{inc}}, \qquad j_{\text{ref}} = v_{\text{ref}}\, n_{\text{ref}}, \qquad j_{\text{trans}} = v_{\text{trans}}\, n_{\text{trans}}.$$

In the stationary-state approach, the reflection and transmission coefficients can be expressed in terms of beam intensity ratios, as follows:

$$R = \frac{j_{\text{ref}}}{j_{\text{inc}}} \quad \text{and} \quad T = \frac{j_{\text{trans}}}{j_{\text{inc}}}. \tag{7.5}$$

If all the incident particles are scattered, and no particles are created or destroyed, it must be the case that $j_{\text{inc}} = j_{\text{ref}} + j_{\text{trans}}$. Dividing both sides by j_{inc} and rearranging, gives $R + T = 1$, as expected from our earlier discussions.

We now need to relate these steady beam intensities to stationary-state solutions of the relevant Schrödinger equation. This requires some care, since we are familiar with the use of Schrödinger's equation to describe individual particles but not beams of particles. To make the steps in the analysis as clear as possible, we shall begin by considering a particularly simple kind of one-dimensional scattering target.

7.2.2 Scattering from a finite square step

The kind of one-dimensional scattering target we shall be concerned with in this subsection is called a **finite square step**. It can be represented by the potential energy function

$$V(x) = 0 \quad \text{for } x \leq 0, \tag{7.6}$$

$$V(x) = V_0 \quad \text{for } x > 0. \tag{7.7}$$

The finite square step (Figure 7.8) provides a simplified model of the potential energy function that confronts an electron as it crosses the interface between two homogeneous media. The discontinuous change in the potential energy at $x = 0$ is, of course, unrealistic, but this is the feature that makes the finite square step simple to treat mathematically. The fact that we are dealing with a square step means that we shall only have to consider *two* regions of the x-axis: Region 1 where $x \leq 0$, and Region 2 where $x > 0$.

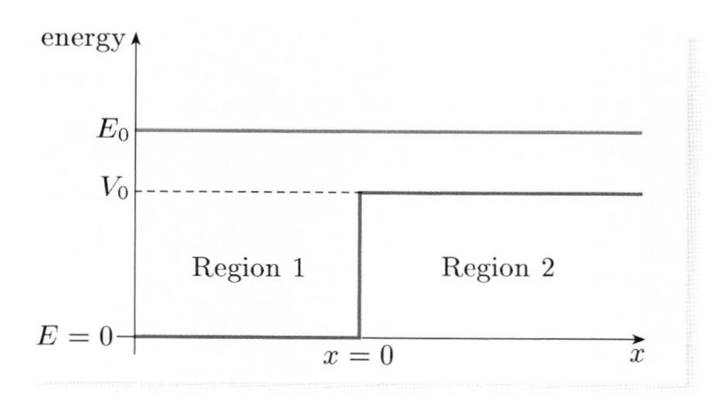

Figure 7.8 A finite square step of height $V_0 < E_0$.

Classically, when a finite square step of height V_0 scatters a rightward moving beam in which each particle has energy $E_0 > V_0$, each of the particles will continue moving to the right but will be suddenly slowed as it passes the point $x = 0$. The transmitted particles are slowed because, in the region $x > 0$, each particle has an increased potential energy, and hence a reduced kinetic energy. The intensity of each beam is the product of the linear number density and the speed of the particles in that beam. To avoid any accumulation of particles at the step, the incident and transmitted beams must have equal intensities; the slowing of the transmitted beam therefore implies that it has a greater linear number density than the incident beam.

Exercise 7.2 In general terms, how would you expect the outcome of the quantum scattering process to differ from the classical outcome? ■

As usual, we start our analysis by writing down the relevant Schrödinger equation:

$$i\hbar\frac{\partial\Psi(x,t)}{\partial t} = -\frac{\hbar^2}{2m}\frac{\partial^2\Psi(x,t)}{\partial x^2} + V(x)\Psi(x,t), \tag{7.8}$$

where $V(x)$ is the finite square step potential energy function given in Equations 7.6 and 7.7. We seek stationary-state solutions of the form $\Psi(x,t) = \psi(x)e^{-iE_0t/\hbar}$, where E_0 is the fixed energy of each beam particle. The task of solving Equation 7.8 then reduces to that of solving the time-independent Schrödinger equations

$$-\frac{\hbar^2}{2m}\frac{d^2\psi}{dx^2} = E_0\psi(x) \qquad \text{for } x \leq 0, \tag{7.9}$$

$$-\frac{\hbar^2}{2m}\frac{d^2\psi}{dx^2} + V_0\psi(x) = E_0\psi(x) \qquad \text{for } x > 0. \tag{7.10}$$

A simple rearrangement gives

$$\frac{d^2\psi}{dx^2} + \frac{2mE_0}{\hbar^2}\psi(x) = 0 \qquad \text{for } x \leq 0,$$

$$\frac{d^2\psi}{dx^2} + \frac{2m(E_0 - V_0)}{\hbar^2}\psi(x) = 0 \qquad \text{for } x > 0,$$

and it is easy to see that these equations have the general solutions

$$\psi(x) = Ae^{ik_1x} + Be^{-ik_1x} \qquad \text{for } x \leq 0 \text{ (Region 1)}, \tag{7.11}$$

$$\psi(x) = Ce^{ik_2x} + De^{-ik_2x} \qquad \text{for } x > 0 \text{ (Region 2)}, \tag{7.12}$$

where A, B, C and D are arbitrary complex constants, and the wave numbers in Region 1 and Region 2 are respectively

$$k_1 = \frac{\sqrt{2mE_0}}{\hbar} \qquad \text{and} \qquad k_2 = \frac{\sqrt{2m(E_0 - V_0)}}{\hbar}. \tag{7.13}$$

You may wonder why we have expressed these solutions in terms of complex exponentials rather than sines and cosines. The reason is that the individual terms in Equations 7.11 and 7.12 have simple interpretations in terms of the incident, reflected and transmitted beams. To see how this works, it is helpful to note that

$$\widehat{p}_x e^{\pm ikx} = -i\hbar\frac{\partial}{\partial x}e^{\pm ikx} = \pm\hbar k e^{\pm ikx}.$$

It therefore follows that terms proportional to e^{ikx} are associated with particles moving rightward at speed $\hbar k/m$, while terms proportional to e^{-ikx} are associated with particles moving leftward at speed $\hbar k/m$.

These directions of motion can be confirmed by writing down the corresponding stationary-state solutions, which take the form

$$\Psi(x,t) = Ae^{i(k_1x-\omega t)} + Be^{-i(k_1x+\omega t)} \qquad \text{for } x \leq 0, \tag{7.14}$$

$$\Psi(x,t) = Ce^{i(k_2x-\omega t)} + De^{-i(k_2x+\omega t)} \qquad \text{for } x > 0, \tag{7.15}$$

where $\omega = E_0/\hbar$. We can then identify terms of the form $e^{i(kx-\omega t)}$ as plane waves travelling in the positive x-direction, while terms of the form $e^{-i(kx+\omega t)}$ are plane waves travelling in the negative x-direction. None of these waves can be normalized, so they cannot describe individual particles, but you will see that they can describe steady beams of particles.

In most applications of wave mechanics, the wave function $\Psi(x,t)$ describes the state of a single particle, and $|\Psi(x,t)|^2$ represents the probability density for that particle. In the steady-state approach to scattering, however, it is assumed that the wave function $\Psi(x,t)$ describes steady beams of particles, with $|\Psi(x,t)|^2$ interpreted as the number of particles per unit length — that is, the linear number density of particles. We know that the wave function is not normalizable, and this corresponds to the fact that the steady beams extend indefinitely to the left and right of the step and therefore contain an infinite number of particles. This will not concern us, however, because we only need to know the linear number density of particles, and this is given by the square of the modulus of the wave function.

Looking at Equation 7.14, and recalling that the first term $Ae^{i(k_1x-\omega t)}$ represents a wave travelling in the positive x-direction for $x \leq 0$, we identify this term as representing the incident wave in Region 1 ($x \leq 0$). We can say that each particle in the beam travels to the right with speed $v_{\text{inc}} = \hbar k_1/m$, and that the linear number density of particles in the beam is

You will find further justification of this interpretation in the next subsection.

$$n_{\text{inc}} = \left|Ae^{i(k_1x-\omega t)}\right|^2 = \left|A\right|^2\left|e^{i(k_1x-\omega t)}\right|^2 = \left|A\right|^2.$$

Similarly, the second term on the right of Equation 7.14 can be interpreted as representing the reflected beam in Region 1 ($x \leq 0$). This beam travels to the left with speed $v_{\text{ref}} = \hbar k_1/m$ and has linear number density $n_{\text{ref}} = |B|^2$.

The first term on the right of Equation 7.15 represents the transmitted beam in Region 2 ($x > 0$). This beam travels to the right with speed $v_{\text{trans}} = \hbar k_2/m$ and has linear number density $n_{\text{trans}} = |C|^2$. The second term on the right of Equation 7.15 would represent a leftward moving beam in the region $x > 0$. On physical grounds, we do not expect there to be any such beam, so we ensure its absence by setting $D = 0$ in our equations.

Using these interpretations, we see that the beam intensities are:

$$j_{\text{inc}} = \frac{\hbar k_1}{m}|A|^2, \quad j_{\text{ref}} = \frac{\hbar k_1}{m}|B|^2 \quad \text{and} \quad j_{\text{trans}} = \frac{\hbar k_2}{m}|C|^2. \tag{7.16}$$

Expressions for the reflection and transmission coefficients then follow from Equation 7.5:

$$R = \frac{j_{\text{ref}}}{j_{\text{inc}}} = \frac{|B|^2}{|A|^2} = \left|\frac{B}{A}\right|^2, \tag{7.17}$$

$$T = \frac{j_{\text{trans}}}{j_{\text{inc}}} = \frac{k_2|C|^2}{k_1|A|^2} = \frac{k_2}{k_1}\left|\frac{C}{A}\right|^2. \tag{7.18}$$

It is worth noting that the expression for the transmission coefficient includes the wave numbers k_1 and k_2, which are proportional to the speeds of the beams in Regions 1 and 2. The wave numbers cancel in the expression for the reflection coefficient because the incident and reflected beams both travel in the same region.

To calculate R and T, we need to find the ratios B/A and C/A. To achieve this, we must eliminate unwanted arbitrary constants from our solutions to the time-independent Schrödinger equation. This can be done by requiring that the solutions satisfy the continuity boundary conditions introduced in Chapter 3:

- $\psi(x)$ is continuous everywhere.
- $\mathrm{d}\psi(x)/\mathrm{d}x$ is continuous where the potential energy function is finite.

The first of these conditions tells us that our two expressions for $\psi(x)$ must match at their common boundary $x = 0$. From Equations 7.11 and 7.12, we therefore obtain

$$A + B = C. \tag{7.19}$$

Taking the derivatives of Equations 7.11 and 7.12,

$$\frac{\mathrm{d}\psi}{\mathrm{d}x} = \mathrm{i}k_1 A\mathrm{e}^{\mathrm{i}k_1 x} - \mathrm{i}k_1 B\mathrm{e}^{-\mathrm{i}k_1 x} \quad \text{for } x \leq 0$$

$$\frac{\mathrm{d}\psi}{\mathrm{d}x} = \mathrm{i}k_2 C\mathrm{e}^{\mathrm{i}k_2 x} \quad \text{for } x > 0,$$

so requiring the continuity of $\mathrm{d}\psi/\mathrm{d}x$ at $x = 0$ implies that

$$\mathrm{i}k_1 A - \mathrm{i}k_1 B = \mathrm{i}k_2 C. \tag{7.20}$$

After some manipulation, Equations 7.19 and 7.20 allow us to express B and C in terms of A

$$B = \frac{k_1 - k_2}{k_1 + k_2} A \quad \text{and} \quad C = \frac{2k_1}{k_1 + k_2} A.$$

Combining these expressions with Equations 7.17 and 7.18, we finally obtain

$$R = \frac{(k_1 - k_2)^2}{(k_1 + k_2)^2}, \tag{7.21}$$

$$T = \frac{4k_1 k_2}{(k_1 + k_2)^2}. \tag{7.22}$$

Since $k_1 = \sqrt{2mE_0}/\hbar$ and $k_2 = \sqrt{2m(E_0 - V_0)}/\hbar$, where E_0 is the incident particle energy and V_0 is the height of the step, we have now managed to express R and T entirely in terms of given quantities. The transmission coefficient, T, is plotted against E_0/V_0 in Figure 7.9.

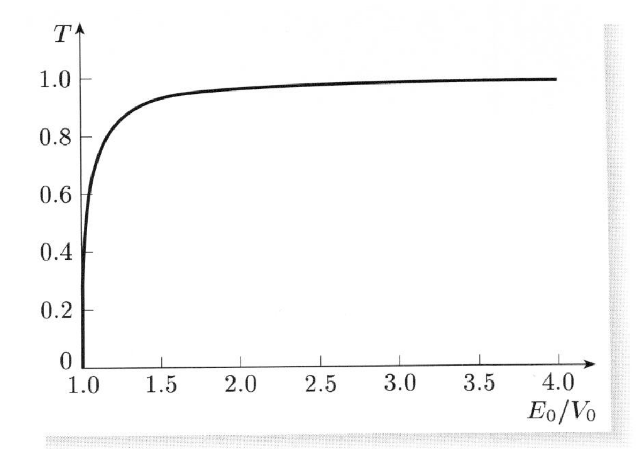

Figure 7.9 A graph of the transmission coefficient T against E_0/V_0 for a finite square step of height V_0.

The above results have been derived by considering a rightward moving beam incident on an upward step of the kind shown in Figure 7.8. However, almost identical calculations can be carried out for leftward moving beams or downward steps. Equations 7.21 and 7.22 continue to apply in all these cases, provided we take k_1 to be the wave number of the incident beam and k_2 to be the wave number of the transmitted beam.

The formulae for R and T are symmetrical with respect to an interchange of k_1 and k_2, so a beam of given kinetic energy, incident on a step of given magnitude, is reflected to the same extent *no matter whether the step is upwards or downwards*. This may seem strange, but you should note that the reflection is a purely quantum effect, and has nothing to do with any classical forces provided by the step.

Another surprising feature of Equation 7.21 is that R is independent of m and so does not vanish as the particle mass m becomes very large. However, we know from experience that macroscopic objects are *not* reflected by small changes in their potential energy function — you can climb upstairs without serious risk of being reflected! How can such everyday experiences be reconciled with wave mechanics?

This puzzle can be resolved by noting that our calculation assumes an *abrupt* step. Detailed quantum-mechanical calculations show that Equation 7.21 provides a good approximation to reflections from a diffuse step *provided that* the wavelength of the incident particles is much longer than the distance over which the potential energy function varies. For example, Equation 7.21 accurately describes the reflection of an electron with a wavelength of 1 nm from a finite step that varies over a distance of order 0.1 nm. However, macroscopic particles have wavelengths that are much shorter than the width of any realistic step, so the above calculation does not apply to them. Instead, these particles obey the classical limit (as outlined in Chapter 6) so they are not reflected to any noticeable extent.

Although we have been discussing the behaviour of beams of particles in this section, it is important to realize that these beams are really no more than a convenient fiction. The beams were simply used to provide a physical interpretation of de Broglie waves that could not be normalized. The crucial point is that we have arrived at explicit expressions for R and T, and we have done so using relatively simple stationary state-methods based on the time-independent Schrödinger equation rather than computationally complicated wave packets. Moreover, as you will see, the method we have used in this section can be generalized to other one-dimensional scattering problems.

Exercise 7.3 (a) Use Equations 7.21 and 7.22 to show that $R + T = 1$.
(b) Evaluate R and T in the case that $E_0 = 2V_0$, and confirm that their sum is equal to 1 in this case.

Exercise 7.4 Consider the case where $k_2 = k_1/2$.

(a) Express B and C in terms of A.

(b) Show that in the region $x > 0$, we have $|\Psi|^2 = 16|A|^2/9 =$ constant, while in the region $x \leq 0$, we have $|\Psi|^2 = |A|^2(10/9 + (2/3)\cos(2k_1x))$.

(c) What general physical phenomenon is responsible for the spatial variation of $|\Psi|^2$ to the left of the step?

(d) If the linear number density in the incident beam is $1.00 \times 10^{24}\,\mathrm{m}^{-1}$, what are the linear number densities in the reflected and transmitted beams?

Exercise 7.5 Based on the solution to Exercise 7.4, sketch a graph of $|\Psi|^2$ that indicates its behaviour both to the left and to the right of a finite square step. ■

7.2.3 Probability currents

The expressions we have derived for reflection and transmission coefficients were based on the assumption that the intensity of a beam is the product of the speed of its particles and their linear number density. This assumption seems very natural from the viewpoint of classical physics, but we should always be wary about carrying over classical ideas into quantum physics. In this section we shall establish a general quantum-mechanical formula for the beam intensity. The formula will be consistent with the assumptions made so far, but is also more general, applying in regions where a classical beam would not exist and for localized wave packets as well as steady beams.

At the heart of our analysis lies the idea that matter is conserved. Neglecting relativistic processes in which particles can be created or destroyed, the total number of particles remains fixed. This is built deep into the formalism of quantum mechanics: if the wave function describing a particle is normalized now, it will remain normalized forever because particles do not simply disappear. The conservation of particles applies locally as well as globally, so if the number of particles in a small region changes, this must be due to particles entering or leaving the region by crossing its boundaries. We shall now express this idea in mathematical terms.

Let us first consider the one-dimensional flow of a fluid along the x-axis. At each point, we define a fluid current $j_x(x,t)$ that represents the rate of flow of fluid particles along the x-axis. If the fluid is compressible, like air, this fluid current may vary in space and time.

Figure 7.10 shows a small one-dimensional region between x and $x+\delta x$. The number of particles in this region can be written as $n(x,t)\,\delta x$, where $n(x,t)$ is the linear number density of particles. The *change* in the number of particles in the region during a small time interval δt is then

$$\delta n(x,t)\,\delta x = j_x(x,t)\,\delta t - j_x(x+\delta x)\,\delta t, \tag{7.23}$$

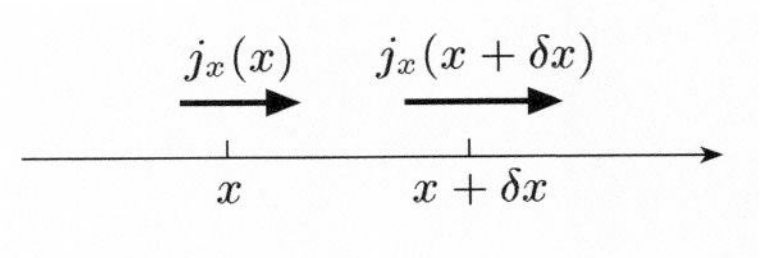

where, for flow in the positive x-direction, the first term on the right-hand side represents the number of particles *entering* the region from the left and the second term represents the number of particles *leaving* the region to the right. Rearranging Equation 7.23 gives

$$\frac{\delta n(x,t)}{\delta t} = -\frac{j_x(x+\delta x) - j_x(x,t)}{\delta x},$$

and, on taking the limit as δx and δt tend to zero we see that

$$\frac{\partial n(x,t)}{\partial t} = -\frac{\partial j_x(x,t)}{\partial x}. \tag{7.24}$$

This result is called the **equation of continuity** in one dimension. With a little care, it can be extended to quantum mechanics.

Figure 7.10 Any change in the number of particles in a small region is due to fluid currents that carry particles into or out of the region.

For a single-particle wave packet in quantum mechanics, the flowing quantity is *probability density*. This is evident from images of wave packets as 'blobs' of moving probability density (e.g. Figure 7.6). Now we know that probability density is represented in quantum mechanics by $\Psi^*\Psi$, so we should be able to construct the appropriate equation of continuity by examining the time derivative of this quantity.

Obviously, we have

$$\frac{\partial(\Psi^*\Psi)}{\partial t} = \Psi^*\frac{\partial\Psi}{\partial t} + \frac{\partial\Psi^*}{\partial t}\Psi, \tag{7.25}$$

where Schrödinger's equation dictates that

$$\mathrm{i}\hbar\frac{\partial\Psi}{\partial t} = -\frac{\hbar^2}{2m}\frac{\partial^2\Psi}{\partial x^2} + V(x)\Psi(x,t).$$

Dividing through by $\mathrm{i}\hbar$, the rate of change of the wave function is

$$\frac{\partial\Psi}{\partial t} = \frac{\mathrm{i}\hbar}{2m}\frac{\partial^2\Psi}{\partial x^2} - \frac{\mathrm{i}}{\hbar}V(x)\Psi(x,t). \tag{7.26}$$

Substituting this equation, and its complex conjugate, into Equation 7.25, we then obtain

The potential energy function $V(x)$ is real and cancels out.

$$\frac{\partial(\Psi^*\Psi)}{\partial t} = \frac{\mathrm{i}\hbar}{2m}\left(\Psi^*\frac{\partial^2\Psi}{\partial x^2} - \Psi\frac{\partial^2\Psi^*}{\partial x^2}\right),$$

and a further manipulation gives

$$\frac{\partial(\Psi^*\Psi)}{\partial t} = \frac{\partial}{\partial x}\left[\frac{\mathrm{i}\hbar}{2m}\left(\Psi^*\frac{\partial\Psi}{\partial x} - \Psi\frac{\partial\Psi^*}{\partial x}\right)\right].$$

This equation can be written in the form of an equation of continuity:

$$\frac{\partial n(x,t)}{\partial t} = -\frac{\partial j_x(x,t)}{\partial x}. \tag{7.27}$$

provided that we interpret $n(x,t)$ as the probability density $\Psi^*\Psi$, with a corresponding current

$$j_x(x,t) = -\frac{\mathrm{i}\hbar}{2m}\left(\Psi^*\frac{\partial\Psi}{\partial x} - \Psi\frac{\partial\Psi^*}{\partial x}\right). \tag{7.28}$$

In the one-dimensional situations we are considering, $j_x(x,t)$ is called the **probability current**. In one dimension, the probability density $n = \Psi^*\Psi$ is a probability per unit length, and therefore has the dimensions of $[\mathrm{L}]^{-1}$. It follows from Equation 7.27 that the probability current has dimensions of

$$[j_x] = [\mathrm{L}]^{-1} \times [\mathrm{T}]^{-1} \times [\mathrm{L}] = [\mathrm{T}]^{-1},$$

and therefore has the SI unit 'per second', as expected for a current of particles.

These ideas can be readily extended to steady beams of particles. Snapshots of a steady beam would not reveal any changes from one moment to the next but the beam nevertheless carries a steady flow of particles, just as a steadily flowing river carries a current of water. For a particle beam, $\Psi^*\Psi$ represents the linear number density of particles and the probability current is the rate of flow of particles in the positive x-direction. For a steady beam, described by a stationary-state wave function, $\Psi(x,t) = \psi(x)e^{-iEt/\hbar}$, the time-dependent phase factors cancel out in Equation 7.28, and the probability current can be written more simply as

$$j_x(x) = -\frac{i\hbar}{2m}\left(\psi^*\frac{d\psi}{dx} - \psi\frac{d\psi^*}{dx}\right), \tag{7.29}$$

which is independent of time. It is important to realize that j_x is a *signed* quantity; it is positive for a beam travelling in the positive x-direction, and negative for a beam travelling in the negative x-direction. This is unlike the beam intensity j introduced earlier, which is always positive. It is natural to define the beam intensity of a steady beam to be the *magnitude* of the probability current: $j = |j_x|$. It is this definition that gives us a way of calculating beam intensities without making unwarranted classical assumptions.

In fact, each beam intensity calculated using Equation 7.29 turns out to be precisely what we have always assumed — the product of a particle speed and a linear number density — as you can check by tackling Exercise 7.7 below. So our analysis adds rigour, but contains no surprises. However, the really significant feature of Equation 7.28 is its generality; it applies to single-particle wave packets as well as to steady beams, and (as you will see later) it will also apply in cases of tunnelling, where a classical beam does not exist.

Exercise 7.6 Show that $j_x(x,t)$ as defined in Equation 7.28 is a real quantity.

Exercise 7.7 Using the solutions to the Schrödinger equation that were obtained in the stationary-state approach to scattering from a finite square step, evaluate the probability current in the regions $x > 0$ and $x \leq 0$. Interpret your results in terms of the beam intensities in these two regions. ■

7.2.4 Scattering from finite square wells and barriers

The procedure used to analyze scattering from a finite square step can also be applied to scattering from finite square wells or barriers, or indeed to any combination of finite square steps, wells and barriers. The general procedure is as follows:

- Divide the x-axis into the minimum possible number of regions of constant potential energy.
- Write down the general solution of the relevant time-independent Schrödinger equation in each of these regions, remembering to use the appropriate value of the wave number k in each region and introducing arbitrary constants as necessary.
- Use continuity boundary conditions to determine all but one of the arbitrary constants. The one remaining constant is associated with the incident beam, which may enter from the left or the right.
- Obtain expressions for all the beam intensities relative to the intensity of the incident beam.
- Determine the reflection and transmission coefficients from ratios of beam intensities.

The best way to become familiar with this procedure is by means of examples and exercises. Here is a worked example involving a finite square well of the kind shown in Figure 7.11.

Essential skill

Solving scattering problems in one dimension

Worked Example 7.1

A particle of mass m with positive energy E_0 is scattered by a one-dimensional finite square well of depth V_0 and width L (shown in Figure 7.11). Derive an expression for the probability that the particle will be transmitted across the well.

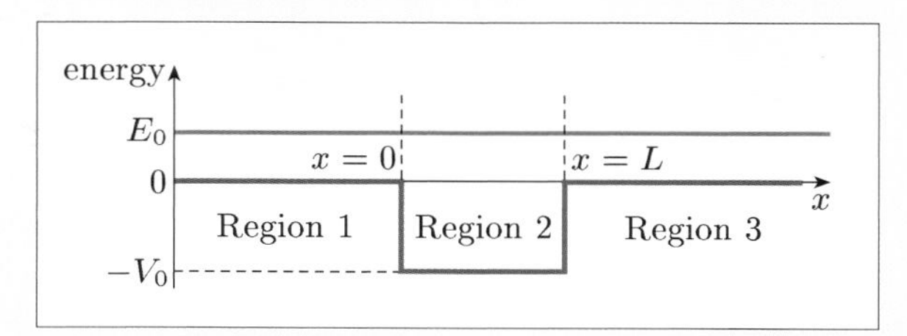

Figure 7.11 A finite square well of depth V_0 and width L. A particle of energy E_0 is scattered by the well.

Solution

Suppose the particle concerned to be part of an intense steady beam of identical particles, each having energy E_0 and each incident from the left on a well located between $x = 0$ and $x = L$.

Divide the x-axis into three regions: Region 1, $(x < 0)$, where $V(x) = 0$; Region 2, $(0 \le x \le L)$, where $V(x) = -V_0$; and Region 3, $(x > L)$, where $V(x) = 0$. In each region the time-independent Schrödinger equation takes the general form

$$-\frac{\hbar^2}{2m}\frac{d^2\psi}{dx^2} + V(x)\psi(x) = E_0\psi(x), \tag{7.30}$$

so the solution in each region is:

$$\psi(x) = Ae^{ik_1x} + Be^{-ik_1x} \quad \text{for } x < 0, \tag{7.31}$$

$$\psi(x) = Ce^{ik_2x} + De^{-ik_2x} \quad \text{for } 0 \le x \le L, \tag{7.32}$$

$$\psi(x) = Fe^{ik_1x} + Ge^{-ik_1x} \quad \text{for } x > L, \tag{7.33}$$

where A, B, C, D, F and G are arbitrary constants, and the wave numbers in Region 1 and Region 2 are

To avoid confusion with energy, we don't use the symbol E for an arbitrary constant in Equation 7.33.

$$k_1 = \frac{\sqrt{2mE_0}}{\hbar} \quad \text{and} \quad k_2 = \frac{\sqrt{2m(E_0 + V_0)}}{\hbar}. \tag{7.34}$$

Note that the wave number in Region 3 is the same as that in Region 1. This is because the potential energy function is the same in these two regions.

There is no leftward moving beam in Region 3, so we set $G = 0$. Since the potential energy function is finite everywhere, $\psi(x)$ and $d\psi/dx$ must be continuous everywhere. Continuity of $\psi(x)$ and $d\psi/dx$ at $x = 0$ implies that

$$A + B = C + D \tag{7.35}$$

$$ik_1A - ik_1B = ik_2C - ik_2D, \tag{7.36}$$

while continuity of $\psi(x)$ and $\mathrm{d}\psi/\mathrm{d}x$ at $x = L$ gives

$$Ce^{ik_2L} + De^{-ik_2L} = Fe^{ik_1L} \tag{7.37}$$

$$ik_2Ce^{ik_2L} - ik_2De^{-ik_2L} = ik_1Fe^{ik_1L}. \tag{7.38}$$

Since the wave numbers in Regions 1 and 3 are both equal to k_1, the intensities of the incident and transmitted beams are

$$j_{\text{inc}} = \frac{\hbar k_1}{2m}|A|^2 \quad \text{and} \quad j_{\text{trans}} = \frac{\hbar k_1}{2m}|F|^2, \tag{7.39}$$

and the required transmission coefficient is given by $T = |F|^2/|A|^2$.

The mathematical task is now to eliminate B, C and D from Equations 7.35 to 7.38 in order to find the ratio F/A. To achieve this, we note that the constant B only appears in the first two equations, so we take the opportunity of eliminating it immediately. Multiplying Equation 7.35 by ik_1 and adding the result to Equation 7.36 we obtain

$$2ik_1A = i(k_2 + k_1)C - i(k_2 - k_1)D. \tag{7.40}$$

Now we must eliminate C and D from the remaining equations. To eliminate D, we multiply Equation 7.37 by ik_2 and add the result to Equation 7.38. This gives

$$2ik_2e^{ik_2L}C = i(k_2 + k_1)Fe^{ik_1L},$$

$$\text{so} \quad C = \frac{k_2 + k_1}{2k_2}Fe^{ik_1L}e^{-ik_2L}. \tag{7.41}$$

Similarly, multiplying Equation 7.37 by ik_2 and subtracting the result from Equation 7.38 we see that

$$2ik_2e^{-ik_2L}D = i(k_2 - k_1)Fe^{ik_1L}$$

$$\text{so} \quad D = \frac{k_2 - k_1}{2k_2}Fe^{ik_1L}e^{ik_2L}. \tag{7.42}$$

Finally, substituting Equations 7.41 and 7.42 into Equation 7.40 and rearranging slightly we obtain

$$\frac{F}{A} = \frac{4k_1k_2e^{-ik_1L}}{(k_2 + k_1)^2e^{-ik_2L} - (k_2 - k_1)^2e^{ik_2L}},$$

so the transmission coefficient is given by

$$T = \left|\frac{F}{A}\right|^2 = \frac{16k_1^2k_2^2}{|(k_2 + k_1)^2e^{-ik_2L} - (k_2 - k_1)^2e^{ik_2L}|^2}. \tag{7.43}$$

This is the expression we have been seeking; it only involves k_1, k_2 and L, and both k_1 and k_2 can be written in terms of m, E_0 and V_0.

Although Equation 7.43 provides a complete answer to the worked example, it is expressed in a rather opaque form. After several pages of substitutions and manipulations (which are not a good investment of your time) it is possible to recast and simplify this formula. We just quote the final result:

$$T = \frac{4E_0(E_0 + V_0)}{4E_0(E_0 + V_0) + V_0^2 \sin^2(k_2 L)}. \tag{7.44}$$

Treating E_0 as an independent variable and V_0 as a given constant, this function can be displayed as a graph of T against E_0/V_0, as shown in Figure 7.12.

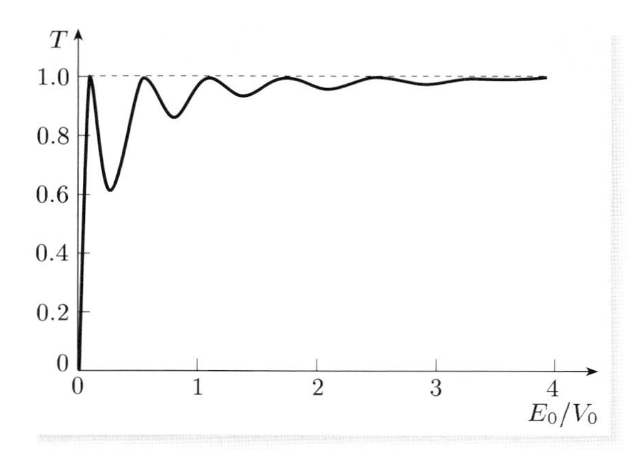

Figure 7.12 The transmission coefficient T for a particle of energy E_0 scattering from a finite square well of depth V_0, plotted against E_0/V_0. To plot this graph, we have taken the particle to have the mass of an electron and taken the well to have a depth of 8.6 eV and a width of 1.0 nm.

As the incident energy is increased significantly above 0 (the top of the well), transmission becomes relatively more likely, but there is generally some chance of reflection ($R = 1 - T \neq 0$), and for some energies the reflection probability may be quite high. However, Figure 7.12 also shows that there are some special incident particle energies at which $T = 1$, so that transmission is certain to occur. Although it is a rather poor use of terminology, these peaks of high transmission are usually called **transmission resonances**. Equation 7.44 shows that transmission resonances occur when $\sin(k_2 L) = 0$, that is when

$$k_2 L = N\pi \quad \text{for} \quad N = 1, 2, 3, \ldots. \tag{7.45}$$

Recalling the relationship $k = 2\pi/\lambda$ between wave number and wavelength, we can also express this condition as $N\lambda_2 = 2L$; in other words:

A transmission resonance occurs when a whole number of wavelengths occupies the full path length $2L$ of a wave that crosses the width of the well and is reflected back again.

This condition can be interpreted in terms of interference between waves reflected at $x = 0$ and $x = L$. The interference turns out to be destructive because reflection at the $x = 0$ interface is accompanied by a phase change of π. Of course, suppression of the reflected beam is accompanied by an enhancement of the transmitted beam. The effect is similar to one found in optics, where destructive interference between waves reflected from the front and back surfaces of a thin transparent film accounts for the success of the anti-reflective coatings on lenses and mirrors.

Before leaving the subject of scattering from a finite well there is one other point that deserves attention. This concerns the precise form of the functions $\psi(x)$ in the three regions we identified earlier. The value of $|A|^2$ is equal to the linear number density of particles in the incident beam. If we regard A as being known, the values of the constants B, C, D and F can be evaluated using the continuity boundary conditions and graphs of $\psi(x)$ can be drawn.

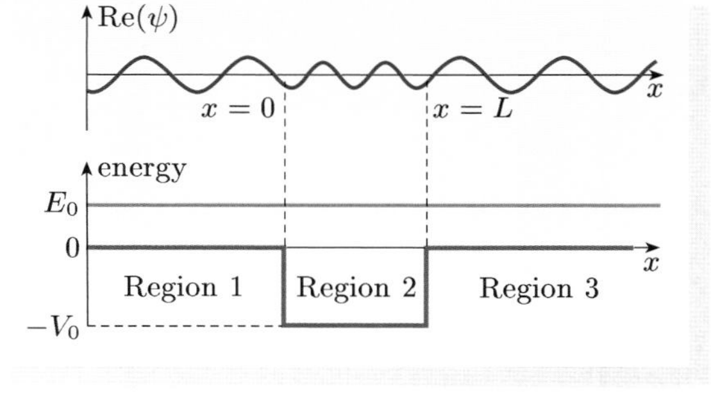

Figure 7.13 A typical example of the real part $\mathrm{Re}(\psi)$ of the function $\psi(x)$ for the kind of square well shown in Figure 7.11.

Since $\psi(x)$ is generally complex there are several possible graphs that might be of interest, including the real part of ψ, the imaginary part of ψ, and $|\psi|^2$. Figure 7.13 shows a typical plot of the real part of ψ for chosen values of m, E_0, V_0 and L; the following points should be noted:

1. In each region, $\psi(x)$ is a periodic function of x.
2. The wavelength is smaller inside the well than outside. The shorter wavelength corresponds to a higher wave number, and higher momentum, and echoes the classical increase in speed that a particle experiences as it enters the well.
3. The amplitude of the wave is smaller inside the well than outside. This is because the beam moves more rapidly inside the well, and has a lower linear number density there.

Exercise 7.8 With one modification, the stationary-state method that was applied to scattering from a finite square well can also be applied to scattering from a finite square barrier of height V_0 and width L, when $E_0 > V_0$. Describe the modification required, draw a figure analogous to Figure 7.13, but for the case of a square barrier, and comment on any differences between the two graphs.

Exercise 7.9 Draw two sketches to illustrate the form of the function $|\Psi(x,t)|^2$ in the case of the stationary-state approach to scattering from (a) a finite square well and (b) a finite square barrier. Comment on the time-dependence of $|\Psi(x,t)|^2$ in each case. ■

7.2.5 Scattering in three dimensions

Sophisticated methods have been developed to analyze scattering in three-dimensions. The complexity of these methods makes them unsuitable for inclusion in this book but it is appropriate to say something about the basic concepts involved.

In three dimensions, we are obliged to think in terms of scattering at a given angle, rather than in terms of one-dimensional reflection or transmission. We distinguish between the incident particles (some of which may be unaffected by the target) and the scattered particles which are affected by the target in some way (changing their direction of motion, energy or state of internal excitation). The detectors for the scattered particles are placed far from the target, well outside the range of interaction of the incident beam and the target, so the scattering process is complete by the time the particles are detected. The incident beam is assumed to be uniform and broad enough to cover all the regions in which the beam particles interact with the target. The incident beam is characterized by its **flux**; this is the rate of flow of particles *per unit time per unit area perpendicular to the beam.*

If we consider a particular scattering experiment (electron-proton scattering, for example), one of the main quantities of interest is the **total cross-section**, σ. This is the total rate at which scattered particles emerge from the target, *per unit time per unit incident flux.* The total cross-section has the dimensions of area. Very loosely, you can think of it as representing the 'effective' area that the target presents to an incident projectile, but you should not give too much weight to this classical interpretation, as most total cross-sections vary markedly with the energy of the incident particles. An acceptable SI unit for the measurement of total cross-sections would be m^2, but measured cross-sections are generally so small that physicists prefer to use a much smaller unit called the **barn**, defined by the relation $1\,\text{barn} = 1 \times 10^{-28}\,\text{m}^2$. The name is intended as a joke, 1 barn being such a large cross-section in particle and nuclear physics that it can be considered to be 'as big as a barn door'. Many cross-sections are measured in millibarn (mb), microbarn (μb) or even nanobarn (nb).

Scattering processes that conserve the total kinetic energy of the colliding particles are said to be examples of **elastic scattering**. They may be contrasted with cases of **inelastic scattering** where the particles may change their internal state of excitation or be absorbed; particles may even be created or destroyed, especially at very high energies. In reality, total cross-sections often contain both elastic and inelastic contributions.

Scattering experiments are often analyzed in great detail. The total cross-section arises as a sum of contributions from particles scattered in different directions. For each direction, we can define a quantity called the **differential cross-section**, which tells us the rate of scattering in a small cone of angles around the given direction. The integral of the differential cross-section, taken over all directions, is equal to the total cross-section. We can also vary the energy of the incident beam. Both the total cross-section and the differential cross-section depend on the energies of the incident particles. There is therefore a wealth of experimental information to collect, interpret and explain.

In exploring the microscopic world of atoms, nuclei and elementary particles, physicists have few options, other than to carry out a scattering experiment. This process has been compared with that of trying to find out how a finely-crafted watch works by the expedient of hurling stones at it and seeing what bits come flying out. It is not a delicate business, but by collecting all the data that a scattering experiment provides, and by comparing these data with the predictions of quantum physics, physicists have learnt an amazing amount about matter on the scale of atoms and below. One early discovery in the scattering of electrons from noble gas atoms (such as xenon) was a sharp dip in the measured cross-section at an energy of about $1\,\text{eV}$. The experimental discovery of this **Ramsauer–Townsend effect** in the early 1920s was an early indication from elastic scattering that some new theoretical insight was needed that would take physics beyond the classical domain. The effect is now recognized as a three-dimensional analogue of the transmission resonance we met earlier.

At the much higher collision energies made available by modern particle accelerators, such as those at the CERN laboratory in Geneva, total cross-sections become dominated by inelastic effects, as new particles are produced. As an example, Figure 7.14 shows some data concerning the scattering of K^- mesons by protons. The upper curve shows the variation of the total cross-section over a very wide range of energies, up to several GeV ($1\,\text{GeV} = 10^9\,\text{eV}$). The lower curve shows the contribution from elastic scattering alone. As the collision energy increases, the contribution from elastic scattering becomes less and less important as inelastic processes become more common.

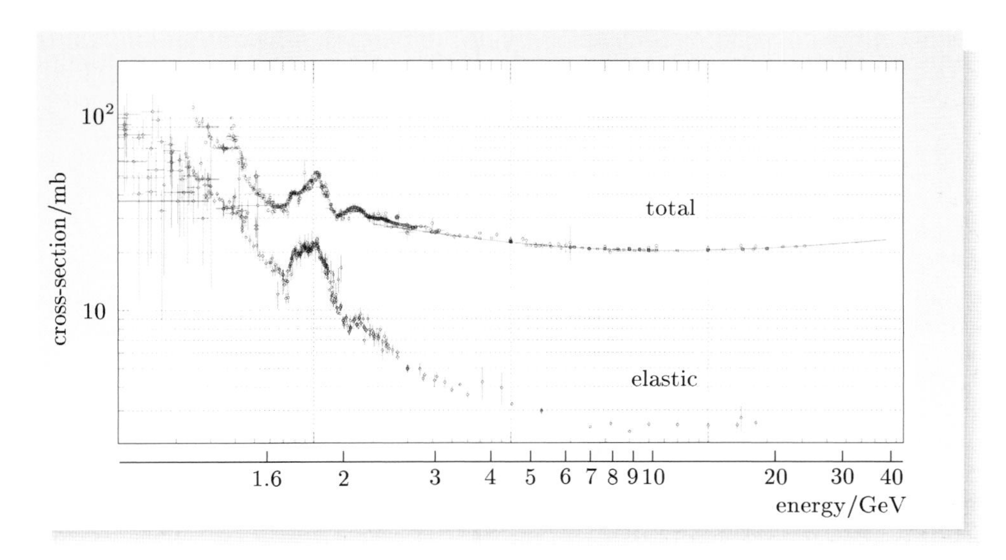

Figure 7.14 The total cross-section (upper curve) and the elastic contribution alone (lower curve), plotted against collision energy, for the scattering of K^- mesons by protons.

7.3 Tunnelling: wave packets and stationary states

One of the most surprising aspects of quantum physics is the ability of particles to pass through regions that they are classically forbidden from entering. This is the phenomenon of quantum-mechanical tunnelling that was mentioned in the Introduction.

In this section we first demonstrate the phenomenon of tunnelling with the aid of wave packets. We then go on to examine some of its quantitative features using stationary-state methods, similar to those used in our earlier discussion of scattering from wells and barriers.

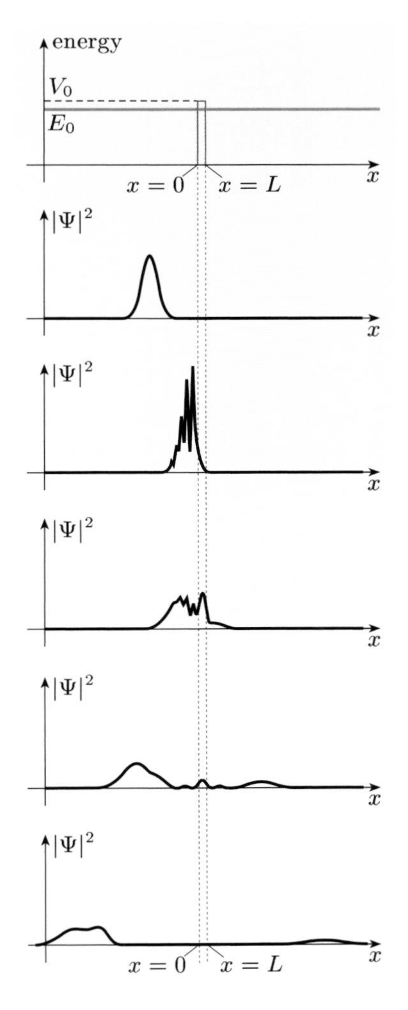

Figure 7.15 The passage of a wave packet with $\langle E \rangle = E_0$ through a finite square barrier of height V_0 when $E_0 < V_0$. The probability density $|\Psi|^2$ is shown in a sequence of snapshots with time increasing from top to bottom.

7.3.1 Wave packets and tunnelling in one dimension

Figure 7.15 shows a sequence of images captured from the wave packet simulation program you used earlier. The sequence involves a Gaussian wave packet, with energy expectation value $\langle E \rangle = E_0$, incident from the left on a finite square barrier of height V_0. The sequence is broadly similar to that shown in Figure 7.6, which involved a similar wave packet and a similar barrier, but with one important difference; in the earlier process E_0 was greater than V_0, so transmission was classically allowed, but in the case of Figure 7.15 E_0 is less than V_0 and transmission is classically forbidden. The bottom image shows that transmission can occur in quantum mechanics.

In the case shown in Figure 7.15, part of the reason for transmission is that the wave packet has a spread of energies, some of which lie above the top of the barrier. However, there is a second reason, which applies even for wave packets with energies wholly below the top of the barrier; there is the possibility that a particle, with insufficient energy to surmount the barrier, may nevertheless *tunnel through* it. For a given wave packet, the probability of tunnelling decreases with the height of the barrier and it also decreases very markedly with its thickness. We shall now use stationary-state methods to investigate this phenomenon.

7.3.2 Stationary states and barrier penetration

The example of tunnelling we have just been examining can be regarded as a special case of scattering; it just happens to have $E_0 < V_0$. As long as we keep this energy range in mind, we can apply the same stationary-state methods to the study of tunnelling that we used earlier when studying scattering.

As before, we shall start by considering the finite square step, whose potential energy function was defined in Equations 7.6 and 7.7. This is shown for the case $E_0 < V_0$ in Figure 7.16. The potential energy function divides the x-axis into two regions: Region 1 ($x \leq 0$) which contains the incident and reflected beams, and Region 2 ($x > 0$) which contains what is effectively an infinitely wide barrier. There is no possibility of tunnelling *through* the barrier in this case since there is no classically allowed region on the far side, but the finite square step nonetheless constitutes a valuable limiting case, as you will see.

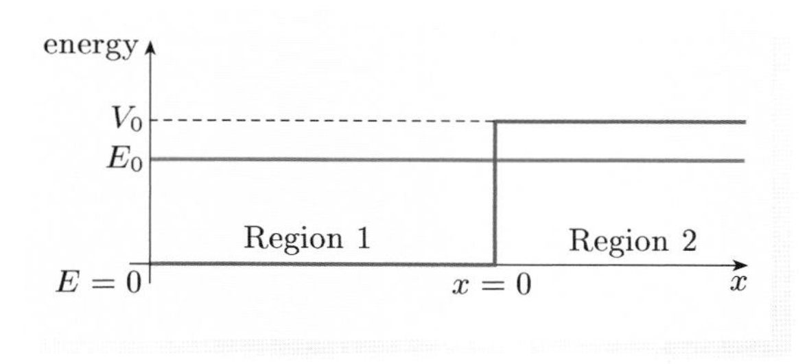

Figure 7.16 A finite square step potential energy function. Also shown is the energy E_0 of an incident particle for the case $E_0 < V_0$, where V_0 is the height of the step.

Proceeding as before, we seek stationary-state solutions of the Schrödinger equation of the form $\Psi(x,t) = \psi(x)e^{-iE_0t/\hbar}$, where $\psi(x)$ is a solution of the corresponding time-independent Schrödinger equation. In this case we might try to use exactly the same solution as in Section 7.2.2 but doing so would complicate the analysis since we would find that in the region $x > 0$ the wave number $k_2 = \sqrt{2m(E_0 - V_0)}/\hbar$ would be imaginary. In view of this, it is better to use the approach taken in a similar situation in Chapter 3; recognize that $E_0 < V_0$ generally implies a combination of exponentially growing and exponentially decaying terms in the region $x > 0$, and write the solutions as

$$\psi(x) = A\mathrm{e}^{\mathrm{i}k_1x} + B\mathrm{e}^{-\mathrm{i}k_1x} \quad \text{for } x \leq 0, \tag{7.46}$$

$$\psi(x) = C\mathrm{e}^{-\alpha x} + D\mathrm{e}^{\alpha x} \quad \text{for } x > 0, \tag{7.47}$$

where, A, B, C and D are arbitrary complex constants, while k_1 and α are *real* quantities given by

$$k_1 = \frac{\sqrt{2mE_0}}{\hbar} \quad \text{and} \quad \alpha = \frac{\sqrt{2m(V_0 - E_0)}}{\hbar}. \tag{7.48}$$

We require D to be zero on physical grounds, to avoid having any part of the solution that grows exponentially as x approaches infinity. To determine the values of B and C relative to that of A we impose the usual requirement (for a finite potential energy function) that both $\psi(x)$, and its derivative $\mathrm{d}\psi/\mathrm{d}x$, must be continuous everywhere. Applying these conditions at $x = 0$ we find:

$$A + B = C$$

$$\mathrm{i}k_1A - \mathrm{i}k_1B = -\alpha C,$$

from which it follows that

$$C = \frac{2\mathrm{i}k_1}{\mathrm{i}k_1 - \alpha}A \quad \text{and} \quad B = \frac{\mathrm{i}k_1 + \alpha}{\mathrm{i}k_1 - \alpha}A.$$

The reflection coefficient is given by

$$R = \left|\frac{B}{A}\right|^2 = \left|\frac{\mathrm{i}k_1 + \alpha}{\mathrm{i}k_1 - \alpha}\right|^2 = (-1)^2\left|\frac{\alpha + \mathrm{i}k_1}{\alpha - \mathrm{i}k_1}\right|^2 = 1.$$

For any complex number, z,

$$\left|\frac{z}{z^*}\right|^2 = \frac{zz^*}{z^*z} = 1.$$

So, if particles of energy $E_0 < V_0$ encounter a finite square step of height V_0, reflection is certain. There is no transmission and no possibility of particles becoming lodged inside the step; everything must eventually be reflected. Note however that $\psi(x)$ is not zero inside the step (see Figure 7.17). Rather, it decreases exponentially over a length scale determined by the quantity $\alpha = \sqrt{2m(V_0 - E_0)}/\hbar$, which is usually called the **attenuation coefficient**. This is another example of the phenomenon of *barrier penetration* that you met in Chapter 3. It is not the same as tunnelling since there is no transmitted beam, but it is what makes tunnelling possible, and the occurrence of exponentially decaying solutions in a classically forbidden region suggests why tunnelling probabilities decline rapidly as barrier width increases.

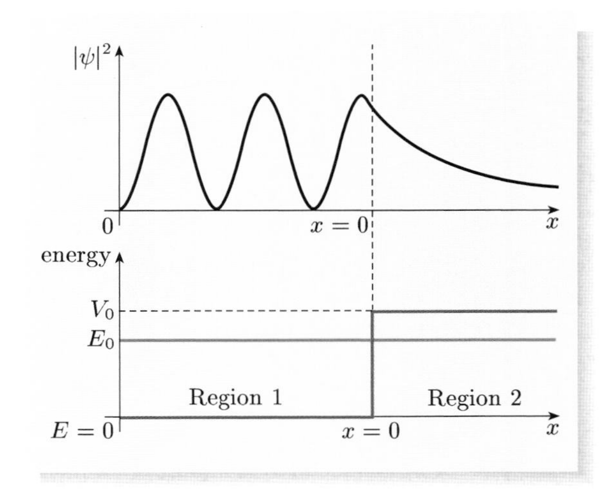

Figure 7.17 The quantity $|\psi(x)|^2$ plotted against x for a finite square step in the case $E_0 < V_0$. There is a finite probability that the particle will penetrate the step, even though there is no possibility of tunnelling through it.

Exercise 7.10 Show that the stationary-state probability density $|\psi(x)|^2$ in Region 1 of Figure 7.17 is a periodic function of x with minima separated by π/k_1.

Exercise 7.11 Show that the probability current in Region 2 of Figure 7.17 is zero. ■

7.3.3 Stationary states and tunnelling in one dimension

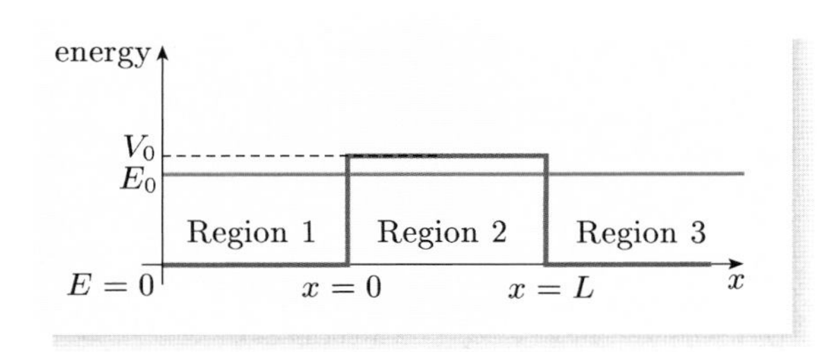

Figure 7.18 A finite square barrier of width L and height V_0, together with the energy E_0 of each tunnelling particle.

We will now use the stationary-state approach to analyze the tunnelling of particles of energy E_0 through a finite square barrier of width L and height V_0 when $E_0 < V_0$ (see Figure 7.18).

Our main aim will be to find an expression for the transmission coefficient. By now, you should be familiar with the general technique for dealing with problems of this kind, including the existence of exponentially growing and decaying solutions in the classically forbidden region, so we shall immediately write down the solution of the relevant time-independent Schrödinger equation:

$$\psi(x) = A\mathrm{e}^{\mathrm{i}k_1 x} + B\mathrm{e}^{-\mathrm{i}k_1 x} \quad \text{for } x < 0 \text{ (Region 1)}, \tag{7.49}$$

$$\psi(x) = C\mathrm{e}^{-\alpha x} + D\mathrm{e}^{\alpha x} \quad \text{for } 0 \le x \le L \text{ (Region 2)}, \tag{7.50}$$

$$\psi(x) = F\mathrm{e}^{\mathrm{i}k_1 x} \quad \text{for } x > L \text{ (Region 3)}, \tag{7.51}$$

where A, B, C, D and F are arbitrary constants and

$$k_1 = \frac{\sqrt{2mE_0}}{\hbar} \quad \text{while} \quad \alpha = \frac{\sqrt{2m(V_0 - E_0)}}{\hbar}.$$

Note that the term that might describe a leftward moving beam in Region 3 has already been omitted, and the wave number in Region 3 has been set equal to that in Region 1.

Requiring the continuity of $\psi(x)$ and $\mathrm{d}\psi/\mathrm{d}x$ at $x = 0$ and at $x = L$ leads to the following four relations:

$$A + B = C + D,$$

$$\mathrm{i}k_1 A - \mathrm{i}k_1 B = -\alpha C + \alpha D,$$

$$C\mathrm{e}^{-\alpha L} + D\mathrm{e}^{\alpha L} = F\mathrm{e}^{\mathrm{i}k_1 L},$$

$$-C\alpha\mathrm{e}^{-\alpha L} + D\alpha\mathrm{e}^{\alpha L} = \mathrm{i}k_1 F\mathrm{e}^{\mathrm{i}k_1 L}.$$

After some lengthy algebra, similar to that in Worked Example 7.1, these four equations can be reduced to a relationship between F and A, from which it is possible to obtain the following expression for the transmission coefficient $T = |F/A|^2$.

$$T = \frac{4E_0(V_0 - E_0)}{4E_0(V_0 - E_0) + V_0^2 \sinh^2(\alpha L)}. \tag{7.52}$$

When $\alpha L \gg 1$, the denominator of this expression can be approximated by $\frac{1}{4}V_0^2\mathrm{e}^{2\alpha L}$, and the transmission coefficient through the barrier is well-described by the useful relationship

Remember: $\sinh x = (\mathrm{e}^x - \mathrm{e}^{-x})/2$.

$$T \simeq 16\left(\frac{E_0}{V_0}\right)\left(1 - \frac{E_0}{V_0}\right)\mathrm{e}^{-2\alpha L}. \tag{7.53}$$

This shows the exponential behaviour that might have been expected on the basis of our earlier results for barrier penetration into a finite square step, but in this case it is a true tunnelling result. It tells us that tunnelling will occur, but indicates that the tunnelling probability will generally be rather small when $\alpha L \gg 1$, and will decrease rapidly as the barrier width L increases.

A graph of $|\psi^2|$ plotted against x for a finite square barrier in the case $E_0 < V_0$ will look something like Figure 7.19. Note that because the incident and reflected beams have different intensities, the minimum value of $|\psi^2|$ in Region 1 is always greater than zero. Also note that for a square barrier of finite width the declining curve in Region 2 is not described by a simple exponential function; there are both exponentially decreasing and exponentially increasing contributions to $\psi(x)$ in that region.

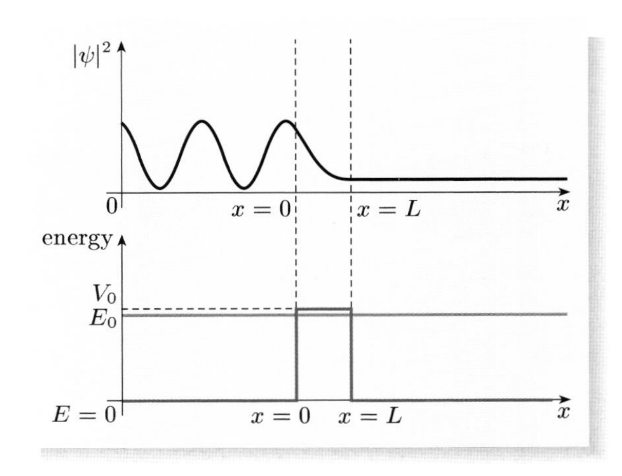

Figure 7.19 The quantity $|\psi^2|$ plotted against x for a finite square barrier in the case $E_0 < V_0$.

Worked Example 7.2

Electrons with a kinetic energy of 5 eV are incident upon a finite square barrier with a height of 10 eV and a width of 0.5 nm. Estimate the value of T and hence the probability that any particular electron will tunnel through the barrier.

Essential skill

Evaluating transmission coefficients for wide barriers

Solution

In this case, $V_0 - E_0 = (10 - 5)\,\text{eV} = 5 \times 1.6 \times 10^{-19}\,\text{J}$. It follows that

$$\begin{aligned}\alpha &= \sqrt{2m(V_0 - E_0)}/\hbar \\ &= \frac{(2 \times 9.11 \times 10^{-31}\,\text{kg} \times 5 \times 1.60 \times 10^{-19}\,\text{J})^{1/2}}{1.06 \times 10^{-34}\,\text{J s}} \\ &= 1.14 \times 10^{10}\,\text{m}^{-1}.\end{aligned}$$

With $L = 5 \times 10^{-10}\,\text{m}$, it follows that

$$\alpha L = 1.14 \times 10^{10}\,\text{m}^{-1} \times 5 \times 10^{-10}\,\text{m} = 5.7.$$

Since this is much larger than 1, we can use Equation 7.53 to estimate

$$T \approx 16\left(\frac{1}{2}\right)\left(1 - \frac{1}{2}\right)\text{e}^{-2\times 5.7} \approx 4 \times 10^{-5}.$$

This is the probability that any particular electron will tunnel through the barrier. (About one electron in 25 000 will pass through.)

Exercise 7.12 Find expressions for the probability current in each of the three regions in the case of tunnelling through a finite square barrier. Comment on the significance of your result for Region 2. ■

7.3.4 Simulations using stationary states

Now would be a good time to study the computer package *Stationary states for tunnelling and scattering in one dimension*. The package solves the time-independent Schrödinger equation for a variety of one-dimensional potential energy functions and for a range of energy eigenvalues E_0. It displays the results graphically by plotting the real and imaginary parts of $\psi(x)$ together with $|\psi(x)^2|$. For a given potential energy function, the program will also plot the corresponding transmission coefficient.

For finite square barriers of specified width L and height V_0, you can choose values of E_0 that are less than V_0 as well as values that are greater than V_0, so you can study the transition from scattering to tunnelling. The package also allows you to study scattering from steps and wells.

This simple but powerful program can provide significant insight into the stationary-state approach to tunnelling and scattering in one dimension. It will reward detailed study, but at this stage you should only spend about half an hour on it. On-screen instructions will guide you through the activity, but you should especially note the following points:

- For $E < V_0$, the strong dependence of the transmission coefficient T on the height and width of the barrier.
- For $E_0 > V_0$, the appearance of transmission resonances and their effect on $|\psi(x)|^2$.
- For a narrow square barrier of fixed height V_0, the effect of gradually reducing the incident particle energy E_0 from a value far above V_0 to one that is far below.
- The differences in the graphs of $|\psi(x)|^2$ (especially in wavelength and amplitude) for scattering from wells and barriers.

7.4 Applications of tunnelling

The discovery that quantum mechanics permits the tunnelling of particles was of great significance. It has deep implications for our understanding of the physical world and many practical applications, particularly in electronics and the developing field of nanotechnology. This section introduces some of these implications and applications. Applications naturally involve the three dimensions of the real world, and realistic potential energy functions are never perfectly square. Despite these added complexities, the principles developed in the last section provide a good basis for the discussion that follows.

7.4.1 Alpha decay

Chapter 1 discussed the law of radioactive decay. You saw that, given a sample of $N(0)$ similar nuclei at time $t = 0$, the number remaining at time t is $N(t) = N(0)\,\mathrm{e}^{-\lambda t}$, where λ, the decay constant for a particular kind of nucleus, determines the rate at which the nuclei decay. The half-life is the time needed for *half* of any sufficiently large sample to decay. It is related to the decay constant by $T_{1/2} = (\ln 2)/\lambda$.

We shall now consider an important type of radioactive decay called **alpha decay** in which an atomic nucleus emits an energetic alpha particle. The emitted alpha particle consists of two protons and two neutrons and is structurally identical to a helium-4 nucleus ($^4_2$He). Alpha decay is the dominant decay mode for a number of heavy nuclei (typically those with atomic numbers greater than 84); a specific example is the decay of uranium to thorium represented by

$$^{238}_{92}\text{U} \rightarrow {}^{234}_{90}\text{Th} + \alpha,$$

where α denotes the alpha particle. Note that the atomic number of the parent nucleus decreases by two and its mass number decreases by four.

Alpha decay was discovered and named by Rutherford in 1898. It was soon established that each type of alpha-decaying nucleus emits an alpha particle with a characteristic energy, E_α. While these alpha emission energies cover a fairly narrow range of values (from about 2 MeV to 8 MeV), the half-lives of the corresponding nuclei cover an enormous range (from 10^{-12} s to 10^{17} s). Experiments showed that, within certain families of alpha-emitting nuclei, the half-lives and alpha emission energies were related to one another. Written in terms of the decay constant, $\lambda = (\ln 2)/T_{1/2}$, this relationship can be expressed in the form

$$\lambda = A\text{e}^{-B/E_\alpha^{1/2}}, \qquad (7.54)$$

where A is a constant that characterizes the particular family of nuclei, and B depends on the charge of the individual nucleus. We shall refer to this empirical law as the **Geiger–Nuttall relation**.

Despite all this information, by the early 1920s alpha decay had become a major puzzle to physicists. The cause of the observed Geiger–Nuttall relation was not understood. Attempts to explain it on the basis of classical physics, with the alpha particle initially confined within the nucleus by an energy barrier that it eventually manages to surmount, did not work. In some cases the observed emission energies were too low to be consistent with surmounting the energy barrier at all. So, how could the alpha particles escape, why did their emission lead to such a staggering range of half-lives, and what was the origin of the Geiger–Nuttall relation?

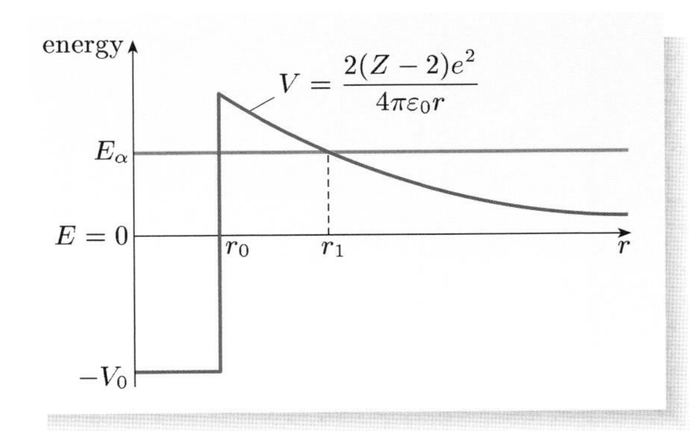

Figure 7.20 The potential energy function $V(r)$ for an alpha particle in the vicinity of an atomic nucleus of atomic number Z.

Answering these questions was one of the early triumphs of wave mechanics. In 1928 the Russian–American physicist George Gamow, and then, independently, Gurney and Condon, proposed a successful theory of alpha decay based on quantum tunnelling. In a simplified version of their approach, the potential energy function responsible for alpha decay has the form shown in Figure 7.20.

Note that $V(r)$ is a function of a radial coordinate r; this is because we are dealing with a three-dimensional problem in which the potential energy function is spherically symmetric, and r represents the distance from an origin at the centre of the nucleus. Initially, an alpha particle of energy E_α is confined within a distance $r = r_0$ of the origin by the well-like part of the potential energy function. This well is due to the powerful but short-range interaction known as the *strong nuclear force*. In addition, a long-range *electrostatic force* acts between the positively-charged alpha particle and the remainder of the positively-charged nucleus, and has the effect of repelling the alpha particle from the nucleus. The electrostatic force corresponds to the potential energy function

$$V(r) = \frac{2(Z-2)e^2}{4\pi\varepsilon_0 r}, \tag{7.55}$$

where Z is the atomic number of the nucleus, $2e$ is the charge of the alpha particle, $(Z-2)e$ is the charge of the nucleus left behind after the decay and ε_0 is a fundamental constant called the **permittivity of free space**. This potential energy function is often called the **Coulomb barrier**. Notice that the Coulomb barrier exceeds the energy of the alpha particle in the region between $r = r_0$ and $r = r_1$ in Figure 7.20 (where $V(r_1) = E_\alpha$). In classical physics, the alpha particle does not have enough energy to enter this region, but in quantum physics it may tunnel through. Once beyond the point $r = r_1$, the alpha particle is electrostatically repelled from the nucleus.

To apply the quantum-mechanical theory of tunnelling to alpha decay, we first note that a classically confined particle would oscillate back and forth inside the well; the combination of its energy (E_α) and the nuclear diameter ($2r_0$) implying that it is incident on the barrier about 10^{21} times per second. Taking this idea over into quantum mechanics, we shall regard each of these encounters as an escape attempt. The small probability of escape at each attempt is represented by the transmission coefficient for tunnelling, T. To estimate T we must take account of the precise shape of the Coulomb barrier. We shall not go through the detailed arguments used to estimate T in this case, but we shall note that they involve the approximation

$$T \approx \exp\left(-2\int_{r_0}^{r_1} \frac{\sqrt{2m(V(r)-E_\alpha)}}{\hbar}\,\mathrm{d}r\right), \tag{7.56}$$

where r_0 and r_1 are the minimum and maximum values of r for which $V(r) \geq E_\alpha$. Equation 7.56 is closely related to the expression for tunnelling through a finite square barrier given in Section 7.3.3. If the potential energy function $V(r)$ happened to be constant over a region of length L, then Equation 7.56 would reproduce the exponential term of Equation 7.53. The other factors in Equation 7.53 are not reproduced, but they vary so slowly compared with the exponential factor that they can be ignored for present purposes.

For given values of E_α, Z and r_0, Equation 7.56 can be evaluated using the Coulomb barrier potential energy function of Equation 7.55. After a lengthy calculation, including some approximations, the final result is of the form

$$T \approx a\mathrm{e}^{-b(Z-2)/E_\alpha^{1/2}}, \tag{7.57}$$

where a and b are constants. Multiplying T by the number of escape attempts per second gives the rate of *successful* escape attempts, and this can be equated to the decay constant, λ. So, according to the quantum tunnelling theory of alpha decay, we have

$$\lambda \propto e^{-b(Z-2)/E_\alpha^{1/2}}. \qquad (7.58)$$

This agrees with the Geiger–Nuttall relation and a detailed comparison with experimental data is shown in Figure 7.21. The exponential-dependence on $E_\alpha^{-1/2}$ implies that a very wide range of decay constants is associated with a small range of emission energies. The sensitivity to energy is far greater than for a square barrier because of the shape of the Coulomb barrier; increasing the energy of the alpha particle decreases the effective width that must be tunnelled through.

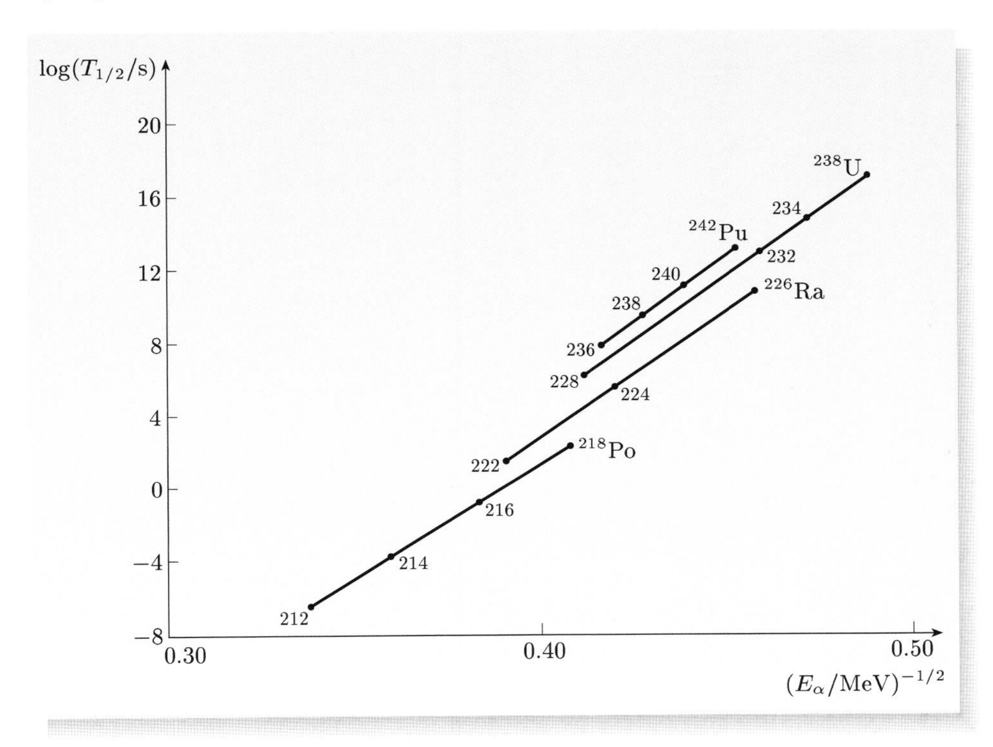

Figure 7.21 A comparison of the Geiger–Nuttall relation with experimental data for different families of nuclei. In this plot, the straight lines confirm the exponential dependence of $T_{1/2}$ (and hence λ) on $E_\alpha^{-1/2}$.

7.4.2 Stellar astrophysics

If tunnelling out of nuclei is possible then so is tunnelling in! As a consequence it is possible to trigger nuclear reactions with protons of much lower energy than would be needed to climb over the full height of the Coulomb barrier. This was the principle used by J. D. Cockcroft and E. T. S. Walton in 1932 when they caused lithium-7 nuclei to split into pairs of alpha particles by bombarding them with high-energy protons. Their achievement won them the 1951 Nobel prize for physics. The same principle is also at work in stars, such as the Sun, where it facilitates the nuclear reactions that allow the stars to shine. Indeed, were it not for the existence of quantum tunnelling, it's probably fair to say that the Sun would not shine and that life on Earth would never have arisen.

The nuclear reactions that allow stars to shine are predominantly **fusion reactions** in which low mass nuclei combine to form a nucleus with a lower mass than the total mass of the nuclei that fused together to form it. It is the difference between the total nuclear masses at the beginning and the end of the fusion process that (via $E = mc^2$) is ultimately responsible for the energy emitted by a star. The energy released by each individual fusion reaction is quite small, but in the hot dense cores of stars there are so many fusing nuclei that they collectively account for the prodigious energy output that is typical of stars (3.8×10^{26} W in the case of the Sun).

In order to fuse, two nuclei have to overcome the repulsive Coulomb barrier that tends to keep them apart. The energy they need to do this is provided by the kinetic energy associated with their thermal motion. This is why the nuclear reactions are mainly confined to a star's hot central core. In the case of the Sun, the core temperature is of the order of 10^7 K. Multiplying this by the Boltzmann constant indicates that the typical thermal kinetic energy of a proton in the solar core is about $1.4 \times 10^{-16}\,\text{J} \approx 1\,\text{keV}$. However, the height of the Coulomb barrier between two protons is more than a thousand times greater than this. Fortunately, as you have just seen, the protons do not have to climb over this barrier because they can tunnel through it. Even so, and despite the hectic conditions of a stellar interior, where collisions are frequent and above-average energies not uncommon, the reliance on tunnelling makes fusion a relatively slow process.

Again taking the Sun as an example, its energy comes mainly from a process called the **proton–proton chain** that converts hydrogen to helium. The first step in this chain involves the fusion of two protons and is extremely slow, taking about 10^9 years for an average proton in the core of the Sun. This is one of the reasons why stars are so long-lived. The Sun is about 4.6×10^9 years old, yet it has only consumed about half of the hydrogen in its core. So, we have quantum tunnelling to thank, not only for the existence of sunlight, but also for its persistence over billions of years.

7.4.3 The scanning tunnelling microscope

The **scanning tunnelling microscope** (STM) is a device of such extraordinary sensitivity that it can reveal the distribution of individual atoms on the surface of a sample. It can also be used to manipulate atoms and even to promote chemical reactions between specific atoms. The first STM was developed in 1981 at the IBM Laboratories in Zurich by Gerd Binnig and Heinrich Rohrer. Their achievement was recognized by the award of the 1986 Nobel prize for physics.

In an STM the sample under investigation is held in a vacuum and a very fine tip, possibly only a single atom wide, is moved across its surface (see Figure 7.22). Things are so arranged that there is always a small gap between the tip and the surface being scanned. An applied voltage between the tip and the sample tends to cause electrons to cross the gap, but the gap itself constitutes a potential energy barrier that, classically, the electrons would not be able to surmount. However, thanks to quantum physics, they can tunnel through the barrier and thereby produce a measurable electric current. Since the current is caused by a tunnelling process, the magnitude of the current is very sensitive to the size of the gap (detailed estimates can again be obtained using Equation 7.56). This sensitivity is the key to finding the positions of tiny irregularities in the surface, including individual atoms.

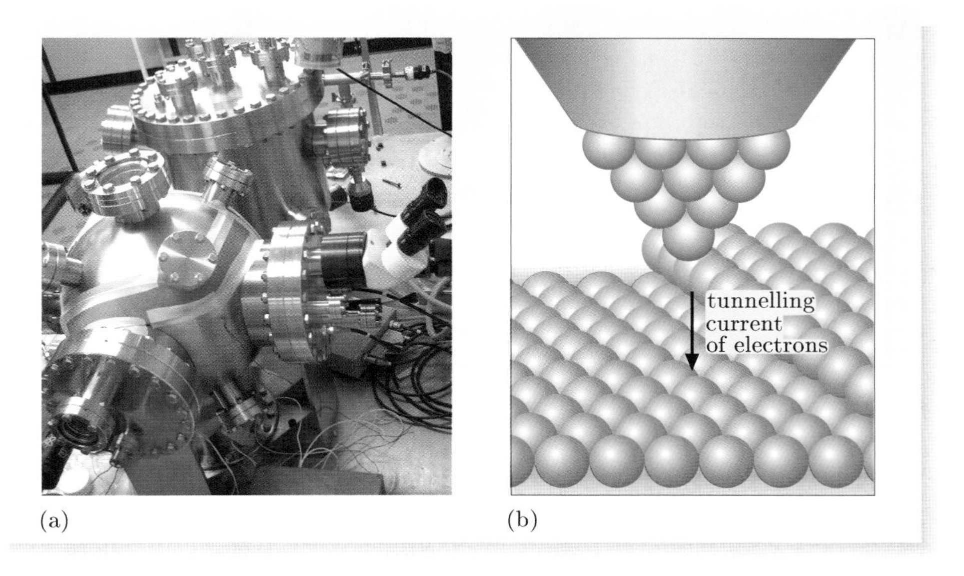

Figure 7.22 (a) A scanning tunnelling microscope (STM) surrounded by vacuum chambers. (b) A schematic diagram showing an STM tip in operation.

In practice, the STM can operate in two different ways. In *constant-height mode*, the tip moves at a constant height and the topography of the surface is revealed by changes in the tunnelling current. In the more common *constant-current mode* the height of the tip is adjusted throughout the scanning process to maintain a constant current and the tiny movements of the tip are recorded. In either mode the structure of the sample's surface can be mapped on an atomic scale, though neither mode involves imaging of the kind that takes place in a conventional optical or transmission electron microscope.

STMs have now become a major tool in the developing field of nanotechnology. This is partly because of the images they supply, but even more because of their ability to manipulate individual atoms, and position them with great accuracy. One of the products of this kind of nano-scale manipulation is shown in Figure 7.23, the famous 'quantum corral' formed by positioning iron atoms on a copper surface.

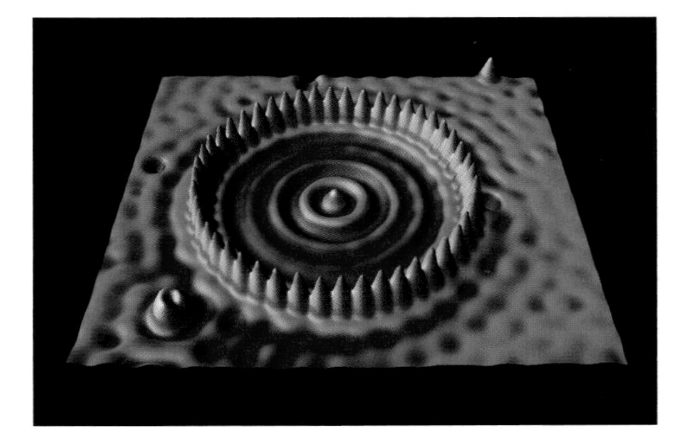

Figure 7.23 Iron atoms on a copper surface forming a ring called a 'quantum corral'. Standing waves of surface electrons trapped inside the corral are also visible.

You can learn more about the science and technology of the STM by studying the multimedia package *The scanning tunnelling microscope*. Now would be the ideal time to do so. Note that the package contains a computer simulation and a video.

Summary of Chapter 7

Introduction Scattering is a process in which incident particles interact with a target and are changed in nature, number, speed or direction of motion as a result. Tunnelling is a quantum phenomenon in which particles that are incident on a classically impenetrable barrier are able to pass through the barrier and emerge on the far side of it.

Section 7.1 In one dimension, wave packets scattered by finite square barriers or wells generally split into transmitted and reflected parts, indicating that there are non-zero probabilities of both reflection and transmission. These probabilities are represented by the reflection and transmission coefficients R and T. The values of R and T generally depend on the nature of the target and the properties of the incident particles. If there is no absorption, creation or destruction of particles, $R + T = 1$.

Section 7.2 Unnormalizable stationary-state solutions of Schrödinger's equation can be interpreted in terms of steady beams of particles. A term such as $Ae^{i(kx-\omega t)}$ can be associated with a beam of linear number density $n = |A|^2$ travelling with speed $v = \hbar k/m$ in the direction of increasing x. Such a beam has intensity $j = nv$. In this approach, $T = j_{\text{trans}}/j_{\text{inc}}$ and $R = j_{\text{ref}}/j_{\text{inc}}$.

For particles of energy $E_0 > V_0$, incident on a finite square step of height V_0, the transmission coefficient is

$$T = \frac{4k_1k_2}{(k_1 + k_2)^2},$$

where

$$k_1 = \frac{\sqrt{2mE_0}}{\hbar} \quad \text{and} \quad k_2 = \frac{\sqrt{2m(E_0 - V_0)}}{\hbar}$$

are the wave numbers of the incident and transmitted beams. For a finite square well or barrier of width L, the transmission coefficient can be expressed as

$$T = \frac{4E_0(E_0 \pm V_0)}{4E_0(E_0 \pm V_0) + V_0^2 \sin^2(k_2L)},$$

where $k_2 = \sqrt{2m(E_0 \pm V_0)}/\hbar$, with the plus signs being used for a well and the minus signs for a barrier. Transmission resonances, at which $T = 1$ and the transmission is certain, occur when $k_2L = N\pi$ where N is an integer.

Travelling wave packets and steady beams of particles can both be thought of as representing flows of probability. In one dimension such a flow is described by the probability current

$$j_x(x,t) = -\frac{i\hbar}{2m}\left(\Psi^*\frac{\partial\Psi}{\partial x} - \Psi\frac{\partial\Psi^*}{\partial x}\right).$$

In three dimensions, scattering is described by the total cross-section, σ, which is the rate at which scattered particles emerge from the target per unit time per unit incident flux. For any chosen direction, the differential cross-section tells us the rate of scattering into a small cone of angles around that direction. At very high energies, total cross-sections are dominated by inelastic effects due to the creation of new particles.

Section 7.3 Wave packets with a narrow range of energies centred on E_0 can tunnel though a finite square barrier of height $V_0 > E_0$. In a stationary-state approach, solutions of the time-independent Schrödinger equation in the classically forbidden region contain exponentially growing and decaying terms of the form $C\mathrm{e}^{-\alpha x}$ and $D\mathrm{e}^{\alpha x}$, where $\alpha = \sqrt{2m(V_0 - E_0)}/\hbar$ is the attenuation coefficient. The transmission coefficient for tunnelling through a finite square barrier of width L and height V_0 is approximately

$$T \simeq 16\left(\frac{E_0}{V_0}\right)\left(1 - \frac{E_0}{V_0}\right)\mathrm{e}^{-2\alpha L} \qquad \text{provided that } \alpha L \gg 1.$$

Such a transmission probability is small and decreases rapidly as the barrier width L increases.

Section 7.4 Square barriers and wells are poor representations of the potential energy functions found in Nature. However, if the potential $V(x)$ varies smoothly as a function of x, the transmission coefficient for tunnelling of energy E_0 can be roughly represented by

$$T \approx \exp\left(-2\int_{r_0}^{r_1} \frac{\sqrt{2m(V(r) - E_0)}}{\hbar}\,\mathrm{d}x\right).$$

This approximation can be used to provide a successful theory of nuclear alpha decay as a tunnelling phenomenon. It can also account for the occurrence of nuclear fusion in stellar cores, despite the relatively low temperatures there. In addition, it explains the operation of the scanning tunnelling microscope which can map surfaces on the atomic scale.

Achievements from Chapter 7

After studying this chapter you should be able to:

7.1 Explain the meanings of the newly defined (emboldened) terms and symbols, and use them appropriately.

7.2 Describe the behaviour of wave packets when they encounter potential energy steps, barriers and wells.

7.3 Describe how stationary state-solutions of the Schrödinger equation can be used to analyze scattering and tunnelling.

7.4 For a range of simple potential energy functions, obtain the solution of the time-independent Schrödinger equation and use continuity boundary conditions to find reflection and transmission coefficients.

7.5 Present information about solutions of the time-independent Schrödinger equation in graphical terms.

7.6 Evaluate probability density currents and explain their significance.

7.7 Describe and comment on applications of scattering and tunnelling in a range of situations including: three-dimensional scattering, alpha decay, nuclear fusion in stars, and the scanning tunnelling microscope.

Chapter 8 Mathematical toolkit

Introduction

This mathematical toolkit provides support for some mathematical topics you will meet in the other chapters of the book. It deals with basic techniques involving complex numbers, ordinary and partial differential equations and probability. This is important material that you will need to know in order to progress freely with the physics of wave mechanics.

8.1 Complex numbers

This section supports all the chapters in this book and is best studied in conjunction with Chapter 1.

In this section, ordinary numbers, stretching continuously from $-\infty$ to $+\infty$, are called *real numbers*. Then a **complex number** is something that can be written in the form

$$z = x + \mathrm{i}y, \tag{8.1}$$

where x and y are real numbers and i is a special quantity with the property that $\mathrm{i}^2 = -1$. We shall refer to i as *the square root of minus 1*. Complex numbers of the form $\mathrm{i}y$ are called **imaginary numbers**. Real numbers, such as 2.7 or 4.8 are also special cases of complex numbers, with $y = 0$.

Each complex number $z = x + \mathrm{i}y$ has a **real part**, $\mathrm{Re}(z) = x$, and an **imaginary part**, $\mathrm{Im}(z) = y$. Note that the imaginary part of a complex number is the real number y, not the imaginary number $\mathrm{i}y$. Two complex numbers z and w are equal if, and only if, $\mathrm{Re}(z) = \mathrm{Re}(w)$ *and* $\mathrm{Im}(z) = \mathrm{Im}(w)$.

We also define the **modulus** of the complex number $z = x + \mathrm{i}y$ to be

$$|z| = \sqrt{x^2 + y^2}. \tag{8.2}$$

This is a real, non-negative quantity.

Exercise 8.1 What are the real part, the imaginary part and the modulus of the complex number $3 - 4\mathrm{i}$? ■

When you first meet complex numbers, you may need to suspend your disbelief; it is clear that no real number can play the role of i, because the square of any real number is non-negative. The traditional terminology of 'imaginary numbers' scarcely helps since it seems to warn us that we are entering a mystical world that has nothing to do with reality. This is unfair and untrue. Admittedly, complex numbers do not play the role of ordinary numbers in counting and measuring things; no one has $3 + 2\mathrm{i}$ eggs for breakfast! However, complex numbers can describe other aspects of the world we live in.

To take a simple example, suppose that we associate $+1$ with looking East and -1 with looking West; then looking North may be represented by the complex number i, and looking South by the complex number $-\mathrm{i}$. The fact that $\mathrm{i}^2 = -1$ expresses the fact that the $90°$ rotation that converts looking East into looking North can be repeated twice in a row, and the end-result will be looking West.

Complex numbers sometimes appear in classical physics, but always in an auxiliary role, to help simplify calculations. The situation is quite different in quantum physics. Basic equations of quantum physics (including Schrödinger's equation) involve complex numbers in an essential way and the quantum-mechanical wave function, $\Psi(x, t)$, has complex values. This wave function cannot itself be measured, but it is used to calculate real-valued quantities that can be measured. So the situation in quantum physics is rather unusual: complex numbers are unavoidable because they appear in fundamental equations, but real numbers are eventually extracted from the formalism so that quantum-mechanical predictions can be compared with real measured values.

8.1.1 The arithmetic of complex numbers

It is easy to carry out arithmetic using complex numbers. All the normal rules of arithmetic apply, with the additional rule that i is treated as an algebraic symbol with the special property that $\mathrm{i}^2 = -1$. For example, we can add or subtract two complex numbers by adding or subtracting their real and imaginary parts separately:

$$(2 + 3\mathrm{i}) + (1 + 4\mathrm{i}) = (2 + 1) + (3\mathrm{i} + 4\mathrm{i}) = 3 + 7\mathrm{i},$$

$$(2 + 3\mathrm{i}) - (1 + 4\mathrm{i}) = (2 - 1) + (3\mathrm{i} - 4\mathrm{i}) = 1 - \mathrm{i}.$$

When multiplying complex numbers we simply need to remember to replace every occurrence of i^2 by -1. For example

$$\begin{aligned}(2 + 3\mathrm{i})(1 + 4\mathrm{i}) &= (2)(1) + (3\mathrm{i})(4\mathrm{i}) + (2)(4\mathrm{i}) + (3\mathrm{i})(1)\\ &= 2 - 12 + 8\mathrm{i} + 3\mathrm{i} = -10 + 11\mathrm{i}.\end{aligned}$$

It is slightly harder to take the reciprocal of a complex number. To evaluate $1/(x + \mathrm{i}y)$, we multiply by the fraction $(x - \mathrm{i}y)/(x - \mathrm{i}y)$ which, of course, is equal to 1:

$$\frac{1}{x + \mathrm{i}y} = \frac{1}{x + \mathrm{i}y} \times \frac{x - \mathrm{i}y}{x - \mathrm{i}y}. \tag{8.3}$$

The advantage of doing this is that the denominator simplifies to a *real* number:

$$(x + \mathrm{i}y)(x - \mathrm{i}y) = x^2 - (\mathrm{i}y)(\mathrm{i}y) = x^2 + y^2,$$

so we have

$$\frac{1}{x + \mathrm{i}y} = \frac{x - \mathrm{i}y}{x^2 + y^2},$$

which is a complex number with real part $x/(x^2 + y^2)$ and imaginary part $-y/(x^2 + y^2)$.

The quotient of any two complex numbers can be found in a similar way:

$$\frac{a + \mathrm{i}b}{c + \mathrm{i}d} = \frac{a + \mathrm{i}b}{c + \mathrm{i}d} \times \frac{c - \mathrm{i}d}{c - \mathrm{i}d} = \frac{ac + bd}{c^2 + d^2} + \mathrm{i}\,\frac{bc - ad}{c^2 + d^2}.$$

Exercise 8.2 Given that $z_1 = 1 + 2\mathrm{i}$ and $z_2 = 4 - 3\mathrm{i}$, find: (a) $z_1 + z_2$, (b) $2z_1 - 3z_2$, (c) $z_1 z_2$ and (d) z_1/z_2. ■

8.1.2 Complex conjugation

The **complex conjugate** of a complex number $z = x + \mathrm{i}y$ is the complex number z^* obtained by reversing the sign of i. Thus,

$$z^* = x - \mathrm{i}y.$$

More generally, the complex conjugate of any expression involving complex numbers is obtained by reversing the sign of i throughout the expression, or equivalently, by replacing each part of the expression by its complex conjugate. This means, for example, that

$$\left(a + \frac{bc}{d}\right)^* = a^* + \frac{b^*c^*}{d^*}. \tag{8.4}$$

Two successive sign-reversals have no net effect, so we always have

$$(z^*)^* = z.$$

Complex conjugates are useful in finding the real and imaginary parts of complex expressions. Since

$$z = \mathrm{Re}(z) + \mathrm{i}\,\mathrm{Im}(z) \quad \text{and} \quad z^* = \mathrm{Re}(z) - \mathrm{i}\,\mathrm{Im}(z),$$

we have

$$\mathrm{Re}(z) = \frac{z + z^*}{2} \quad \text{and} \quad \mathrm{Im}(z) = \frac{z - z^*}{2\mathrm{i}}.$$

These equations provide useful tests for whether complex expressions are real or imaginary. If an expression z is real, it has $\mathrm{Im}(z) = 0$, and is characterized by $z^* = z$. By contrast, an imaginary expression has $\mathrm{Re}(z) = 0$, and is characterized by $z^* = -z$.

From Equation 8.3, we also have

$$zz^* = (x + \mathrm{i}y)(x - \mathrm{i}y) = x^2 + y^2 = |z|^2, \tag{8.5}$$

so the modulus of any complex expression z is given by $|z| = \sqrt{zz^*}$.

Exercise 8.3 Show that $z = \mathrm{i}(ab^* - a^*b)$ is real. ■

8.1.3 The geometry of complex numbers

Complex numbers were invented as early as 1545, when Gerolamo Cardano used them to solve cubic equations, but it took much longer for them to be interpreted in geometric terms. Around 1800, Caspar Wessel and Jean Argand realized that complex numbers can be represented as points in a plane (Figure 8.1).

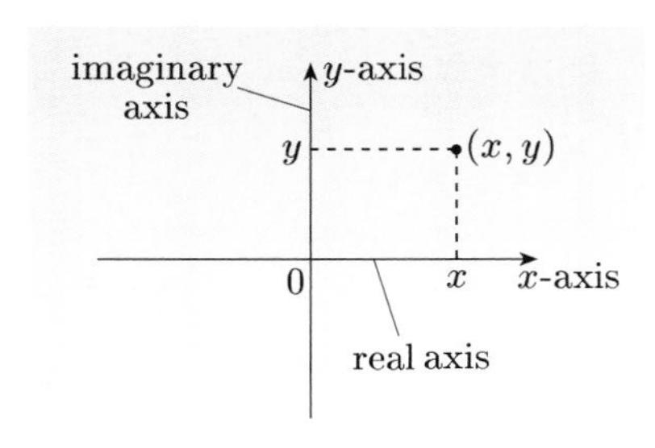

Figure 8.1 The complex plane or Argand diagram.

In geometric terms, the complex number $x + \mathrm{i}y$ is represented by a point with Cartesian coordinates (x, y), and the plane consisting of all such points is called the **complex plane** or the **Argand diagram**. In this plane, the x-axis, which contains only real numbers, is called the **real axis**, and the y-axis, which contains only imaginary numbers, is called the **imaginary axis**.

The geometric relationships linking a complex number $z = x + \mathrm{i}y$, its complex conjugate z^*, and the complex numbers $-z$ and $-z^*$ are shown in Figure 8.2. The operations of adding and subtracting complex numbers can also be visualized geometrically. Two complex numbers z and w are represented by the points shown in Figure 8.3. Their sum is represented by the point marked $z + w$ at the corner of the blue parallelogram, while their difference is equal to the sum of z and $-w$ and so is represented by the point marked $z - w$ at the corner of the red parallelogram.

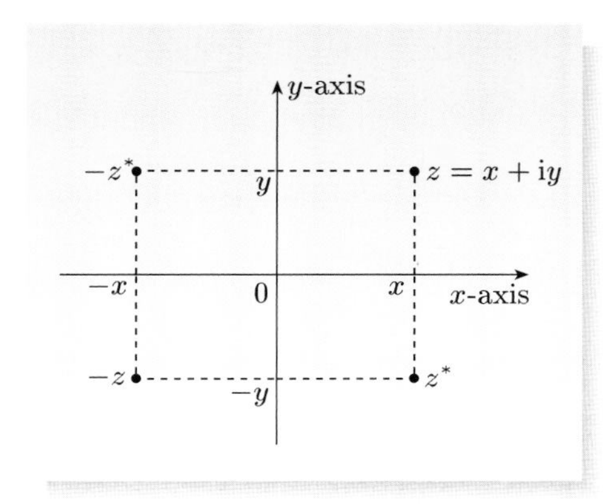

Figure 8.2 The points z, z^*, $-z$ and $-z^*$ in the complex plane.

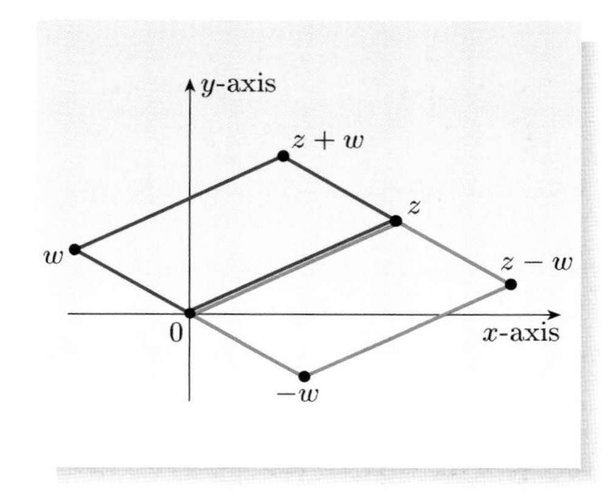

Figure 8.3 Adding and subtracting two complex numbers z and w.

So far, we have specified points in the complex plane by their Cartesian coordinates (x, y), and the corresponding complex numbers written as $x + iy$ are then said to be in **Cartesian form**. However, we can also specify points in the complex plane by the polar coordinates r and θ as shown in Figure 8.4. Anticlockwise rotations increase the value of θ, while clockwise rotations decrease it. Throughout this chapter, we shall assume that angles are expressed in radians, and that θ is a pure number (of radians). Then, values of θ that differ by 2π are equivalent in that they refer to the same point in the complex plane.

We have

$$x = r\cos\theta \quad \text{and} \quad y = r\sin\theta, \tag{8.6}$$

so any complex number can be written in the **polar form**:

$$z = r(\cos\theta + \mathrm{i}\sin\theta). \tag{8.7}$$

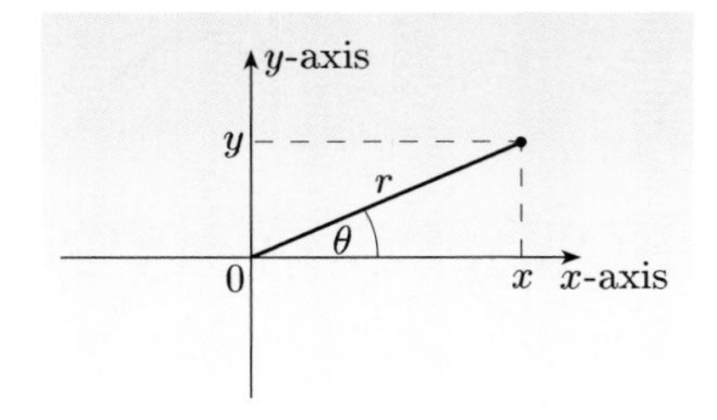

Figure 8.4 Polar coordinates for a point in the complex plane. The angle θ shown has a positive value.

The radial coordinate is given by $r = \sqrt{x^2 + y^2}$, which is the modulus of the complex number. The angular coordinate θ is called the **phase**, or sometimes the **argument**, of the complex number. It can be found by solving either $\tan\theta = y/x$ or $\cos\theta = x/r$ for θ, taking care to choose the angle to be in the appropriate quadrant of the complex plane. For example, the complex number $z = 1 - \mathrm{i}$ lies in the fourth quadrant, so we choose $r = \sqrt{1^2 + (-1)^2} = \sqrt{2}$ and $\theta = -\pi/4$ in this case, giving

$$1 - \mathrm{i} = \sqrt{2}\left[\cos(-\pi/4) + \mathrm{i}\sin(-\pi/4)\right].$$

Exercise 8.4 Express the following complex numbers in polar form: (a) $1 + \mathrm{i}\sqrt{3}$, (b) $-1 + \mathrm{i}\sqrt{3}$.

Exercise 8.5 Use Equation 8.7 to confirm that $|z| = r$. ■

8.1.4 Euler's formula and exponential form

There is yet another way of expressing complex numbers, which is very important in quantum physics. To derive this alternative representation, we shall first consider a complex number of unit modulus (that is, one with $|z| = r = 1$). Such a complex number takes the simple form

$$z = \cos\theta + \mathrm{i}\sin\theta. \tag{8.8}$$

Since all the normal rules of algebra apply to complex numbers, it is reasonable to suppose that the normal rules of differentiation apply too, provided that we treat the symbol i as a constant. We therefore have

$$\begin{aligned}\frac{\mathrm{d}z}{\mathrm{d}\theta} &= \frac{\mathrm{d}}{\mathrm{d}\theta}(\cos\theta + \mathrm{i}\sin\theta)\\ &= (-\sin\theta + \mathrm{i}\cos\theta)\\ &= \mathrm{i}(\cos\theta + \mathrm{i}\sin\theta) = \mathrm{i}z.\end{aligned}$$

The general solution of this differential equation is

$$z = C\mathrm{e}^{\mathrm{i}\theta}, \tag{8.9}$$

where C is a constant. Comparing Equations 8.8 and 8.9, we therefore have

$$\cos\theta + \mathrm{i}\sin\theta = C\mathrm{e}^{\mathrm{i}\theta}.$$

Setting $\theta = 0$ on both sides of this equation, we see that

$$1 = C\mathrm{e}^0 = C,$$

so we can identify $C = 1$ and write

$$\mathrm{e}^{\mathrm{i}\theta} = \cos\theta + \mathrm{i}\sin\theta. \tag{8.10}$$

This result is known as **Euler's formula**. It is a major mathematical discovery since it links two apparently different types of function — the exponential function and the trigonometric functions. For a real variable, θ, the exponential function e^{θ} is monotonically increasing, and does not display the periodic oscillations of $\sin\theta$ or $\cos\theta$. However, the exponential function with an *imaginary* argument $\mathrm{i}\theta$ is directly related to sines and cosines via Euler's formula.

Exercise 8.6 Show that the cosine and sine functions can be represented as

$$\cos\theta = \frac{\mathrm{e}^{\mathrm{i}\theta} + \mathrm{e}^{-\mathrm{i}\theta}}{2} \quad \text{and} \quad \sin\theta = \frac{\mathrm{e}^{\mathrm{i}\theta} - \mathrm{e}^{-\mathrm{i}\theta}}{2\mathrm{i}}.$$

■

Combining Euler's formula with Equation 8.7, we see that any complex number can be written in the form

$$z = r\mathrm{e}^{\mathrm{i}\theta}, \tag{8.11}$$

where r is the radial coordinate (the modulus) of the complex number, and θ is the angular coordinate (the phase). A complex number written in this way is said to be in **exponential form**. In quantum mechanics, a complex number of the form $\mathrm{e}^{\mathrm{i}\theta}$, where θ is real, is sometimes called a **phase factor**. Such complex numbers have unit modulus.

Exercise 8.7 Express the complex number $z = \mathrm{e}^3\,\mathrm{e}^{-\mathrm{i}\pi/6}$ in Cartesian form. ■

8.1.5 Multiplying and dividing complex numbers

Although the Cartesian form, $z = x + iy$ is useful for adding and subtracting complex numbers, other operations, such as multiplication, division and taking powers, are most easily carried out using the exponential form, as we shall now demonstrate.

The exponential function has the property that

$$e^a \times e^b = e^{a+b}.$$

This rule applies even if a and b are complex, so we have

$$r_1 e^{i\theta_1} \times r_2 e^{i\theta_2} = r_1 r_2 \, e^{i(\theta_1+\theta_2)}. \tag{8.12}$$

To form the *product* of two complex numbers, we multiply their moduli and add their phases.

Moduli is the plural of modulus.

As before, the complex conjugate of a complex number is found by changing the sign of i wherever it appears, so we have

$$z^* = \left(re^{i\theta}\right)^* = re^{-i\theta}. \tag{8.13}$$

To form the *complex conjugate* of a complex number we keep the modulus unchanged and change the sign of the phase.

The product of a complex number z and its own complex conjugate z^* is therefore given by

$$z \times z^* = re^{i\theta} \times re^{-i\theta} = r^2 e^{i(\theta-\theta)} = r^2 e^0 = r^2,$$

which in just what we would expect from Equation 8.5.

The *reciprocal* of a complex number $z = re^{i\theta}$ is given by

$$\frac{1}{z} = \frac{1}{re^{i\theta}} = \frac{1}{r}\,e^{-i\theta}, \tag{8.14}$$

and we can check that this does the right job by verifying that

$$z \times \frac{1}{z} = re^{i\theta} \times \frac{1}{r}\,e^{-i\theta} = e^{i(\theta-\theta)} = e^0 = 1.$$

Note that the complex conjugate $z^* = re^{-i\theta}$ and the reciprocal $1/z = e^{-i\theta}/r$ are not the same thing: they are complex numbers with the same phase but different moduli.

Dividing one complex number by another is also easy in exponential form. Using Equations 8.12 and 8.14, we see that

$$\frac{r_1 e^{i\theta_1}}{r_2 e^{i\theta_2}} = \frac{r_1}{r_2}\,e^{i(\theta_1-\theta_2)}. \tag{8.15}$$

To form the *quotient* of two complex numbers, we divide their moduli and subtract their phases.

Figure 8.5a shows the geometric effect of multiplying one complex number by another: multiplying by $z = re^{i\theta}$ causes stretching by a factor of r and rotation through an angle of θ radians. Similarly, Figure 8.5b shows the geometric effect of dividing one complex number by another: dividing by $re^{i\theta}$ causes stretching by a factor of $1/r$ and rotation through an angle of $-\theta$ radians.

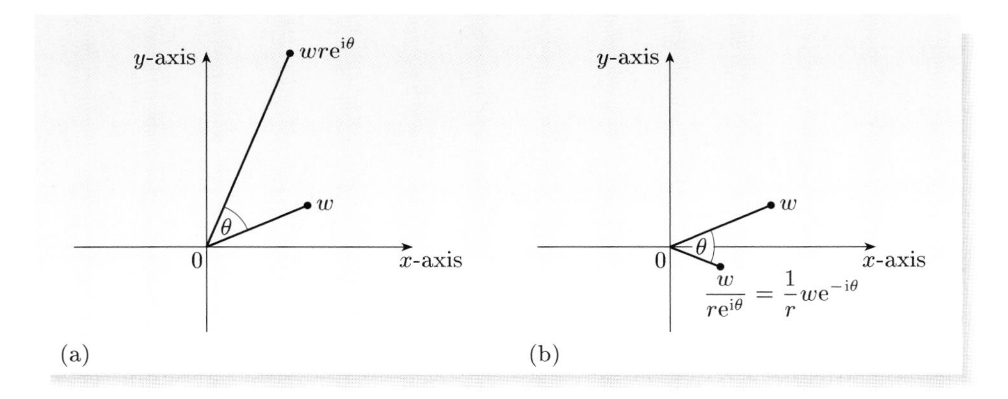

Figure 8.5 (a) The effect of multiplying a complex number w by $re^{i\theta}$. (b) The effect of dividing a complex number w by $re^{i\theta}$. In the cases shown we have taken $r = 2$ and $\theta = \pi/4$.

Exercise 8.8 Describe the geometric effects of (a) multiplying and (b) dividing a complex number by i. ■

8.1.6 Powers of complex numbers

The exponential function has the property that

$$\left(e^a\right)^n = e^{na},$$

where n is any number. This rule applies even if a is complex, so it is easy to take the square of a complex number in exponential form:

$$z^2 = \left(re^{i\theta}\right)^2 = r^2e^{i2\theta},$$

and more generally,

$$z^n = \left(re^{i\theta}\right)^n = r^ne^{in\theta}, \tag{8.16}$$

for any integer n.

Exercise 8.9 Given that $z = (1 + i)/\sqrt{2}$, what is z^{100}? ■

A slight complication arises if n is fractional. For example, suppose we want to find the square root of $z = re^{i\theta}$. Direct use of Equation 8.16 gives

$$z^{1/2} = r^{1/2}e^{i\theta/2}, \tag{8.17}$$

where we take $r^{1/2}$ to be real and positive. This answer is certainly appropriate for a square root because we have

$$\left(r^{1/2}e^{i\theta/2}\right)^2 = re^{i\theta} = z.$$

However, this is not the *only* square root of z. To see why, remember that any integer multiple of 2π can be added to the phase of $re^{i\theta}$ without affecting the complex number being represented. We therefore have

$$z = r\,e^{i\theta} = r\,e^{i(\theta+2\pi)}.$$

Hence, taking the square root also gives

$$z^{1/2} = r^{1/2}\,e^{i(\theta+2\pi)/2} = r^{1/2}\,e^{i\theta/2}e^{i\pi}.$$

Since $e^{i\pi} = \cos\pi + i\sin\pi = -1$, we conclude that $z^{1/2}$ has the alternative value

$$z^{1/2} = -r^{1/2}e^{i\theta/2}, \tag{8.18}$$

and this also makes good sense, since

$$\left(z^{1/2}\right)^2 = \left(-r^{1/2}e^{i\theta/2}\right)^2 = re^{i\theta} = z.$$

You might well have guessed that the square root could appear with a plus sign or minus sign, but the procedure we have just introduced can be generalized to find the nth root of any complex number. For example, the complex number $z = re^{i\theta}$ has three cube roots:

$$r^{1/3}e^{i\theta/3}, \quad r^{1/3}e^{i(\theta+2\pi)/3} \quad \text{and} \quad r^{1/3}e^{i(\theta+4\pi)/3}.$$

Figure 8.6 shows the three cube roots of -1 as points in the complex plane.

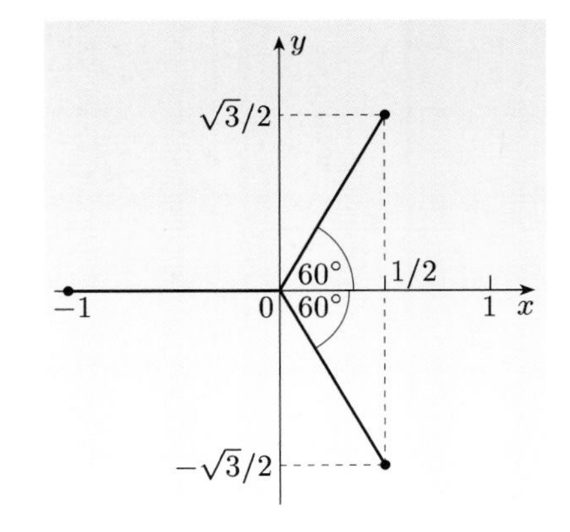

Figure 8.6 The three cube roots of -1 are shown as dots at $(-1, 0)$, at $(1/2, +\sqrt{3}/2)$ and at $(1/2, -\sqrt{3}/2)$.

8.2 Ordinary differential equations

This section supports the study of Chapters 2, 3 and 7.

A **differential equation** is an equation that involves *derivatives* of an unknown function that we would like to find. For example, the aim of solving the differential equation

$$\frac{dy}{dx} = x^2 y(x),$$

is to find an expression for the dependent variable y in terms of the independent variable x. If the equation involves functions of more than one variable and their partial derivatives, it is a partial differential equation. If it involves only ordinary derivatives with respect to a single independent variable, it is an **ordinary differential equation** — although the adjective 'ordinary' is often omitted for brevity. Ordinary differential equations are the subject of this section.

8.2.1 Types of differential equation

The only type of differential equation we need to consider takes the form

$$a(x)\frac{d^2y}{dx^2} + b(x)\frac{dy}{dx} + c(x)\,y(x) = f(x). \tag{8.19}$$

This is a general *second-order linear differential equation.*

- The **order** of a differential equation is the order of the highest derivative that appears in it. Assuming that $a(x)$ is not identically equal to zero, the derivative of highest order in Equation 8.19 is d^2y/dx^2, so this equation is of second order.
- A **linear differential equation** is one that can be written in the form

$$\widehat{L}y(x) = f(x), \tag{8.20}$$

where $\widehat{\mathrm{L}}$ is a **linear differential operator** — that is, one that can be expressed as

$$\widehat{\mathrm{L}} = a_n(x)\frac{\mathrm{d}^n}{\mathrm{d}x^n} + a_{n-1}(x)\frac{\mathrm{d}^{n-1}}{\mathrm{d}x^{n-1}} + \ldots + a_1(x)\frac{\mathrm{d}}{\mathrm{d}x} + a_0(x), \qquad (8.21)$$

where $a_n(x), a_{n-1}(x) \ldots a_0(x)$ are functions of x (some of which may be constant or equal to zero). The crucial feature of a linear differential operator is that it gives

$$\widehat{\mathrm{L}}\big[\alpha\, y_1(x) + \beta\, y_2(x)\big] = \alpha\, \widehat{\mathrm{L}}y_1(x) + \beta\, \widehat{\mathrm{L}}y_2(x) \qquad (8.22)$$

for any smooth functions $y_1(x)$ and $y_2(x)$ and any constants α and β.

We are usually interested in the special case of Equation 8.19 that arises when $f(x) = 0$ for all x, so that

$$a(x)\frac{\mathrm{d}^2y}{\mathrm{d}x^2} + b(x)\frac{\mathrm{d}y}{\mathrm{d}x} + c(x)y(x) = 0. \qquad (8.23)$$

An equation of this type is said to be **homogeneous**. It is not sufficient to say that a homogeneous equation is one with zero on the right-hand side, as this could always be achieved by a rearrangement of terms. The distinguishing feature of a homogeneous equation is that each of its terms is proportional to y or one of its derivatives. The simplest case occurs when the coefficients, a, b and c are constants. We then have a *second-order linear homogeneous equation with constant coefficients*. The description is certainly a mouthful, but such equations are relatively easy to solve, and you will see examples of this shortly. It is usually much harder to solve equations with non-constant coefficients, but special tricks can sometimes be used, and you will see an example of this in Chapter 5.

8.2.2 General properties of solutions

In general it is found that a given differential equation is satisfied by a family of functions that can be described by a single comprehensive formula involving **arbitrary constants**. A solution to a differential equation of order n contains n arbitrary constants. The arbitrary constants do not appear in the differential equation itself but serve to distinguish one solution from another. For example, the differential equation

$$\frac{\mathrm{d}y}{\mathrm{d}t} = -\lambda y(t) \qquad (8.24)$$

is satisfied by a family of functions of the form $y(t) = A\mathrm{e}^{-\lambda t}$, where A is an arbitrary constant; λ is *not* an arbitrary constant but is a parameter of the equation.

An example of Equation 8.24 can be found in Chapter 1, where λ is the *decay constant* of a radioactive nucleus.

The solution that includes the arbitrary constants is called the **general solution** of the differential equation, while a solution that corresponds to a particular choice of arbitrary constants is called a **particular solution**. In a given physical situation, we may have enough additional information to determine the values of the arbitrary constants. For example, in the situation described by Equation 8.24, if we know that $y = 10$ when $t = 0$, we can deduce that $A = 10$, and therefore select the particular solution $y(t) = 10\,\mathrm{e}^{-\lambda t}$.

Linear homogeneous differential equations have an important property that follows from Equation 8.22. If $y_1(x)$ and $y_2(x)$ are two solutions of a linear homogeneous differential equation, we have

$$\widehat{\mathrm{L}}y_1(x) = 0 \quad \text{and} \quad \widehat{\mathrm{L}}y_2(x) = 0,$$

so it follows that

$$\widehat{\mathrm{L}}\big[\alpha y_1(x) + \beta y_2(x)\big] = \alpha\widehat{\mathrm{L}}y_1(x) + \beta\widehat{\mathrm{L}}y_2(x) = 0,$$

for any constants α and β. Thus, any linear combination of solutions is also a solution of the differential equation. This fact is called the **principle of superposition**.

8.2.3 Solutions of some basic differential equations

When considering the quantum-mechanical behaviour of a particle in a region of constant potential energy, you will meet second-order differential equations of the form

$$\frac{\mathrm{d}^2y}{\mathrm{d}x^2} \pm k^2 y(x) = 0, \tag{8.25}$$

where k is a real constant. We will consider first the case where the plus sign appears and Equation 8.25 takes the form

$$\frac{\mathrm{d}^2y}{\mathrm{d}x^2} + k^2 y(x) = 0. \tag{8.26}$$

To solve an equation of this type, we can substitute in a trial function of the form $y(x) = \mathrm{e}^{ax}$, where a is a constant that remains to be determined. Evaluating the second derivative of this trial function, we see that

$$\frac{\mathrm{d}^2y}{\mathrm{d}x^2} = \frac{\mathrm{d}^2}{\mathrm{d}x^2}\,\mathrm{e}^{ax} = a^2\mathrm{e}^{ax}.$$

Substituting into Equation 8.26 then leads to the equation

$$(a^2 + k^2)\mathrm{e}^{ax} = 0.$$

which implies that

$$a^2 + k^2 = 0. \tag{8.27}$$

This algebraic equation is called the **auxiliary equation** corresponding to the original differential equation. Solving the auxiliary equation for a, gives $a = \pm\mathrm{i}k$, so we conclude that solutions of the form $y(x) = \mathrm{e}^{\mathrm{i}kx}$ or $y(x) = \mathrm{e}^{-\mathrm{i}kx}$ will satisfy Equation 8.26. This differential equation is linear and homogeneous, so the principle of superposition tells us that it also is satisfied by any function of the form

$$y(x) = A\mathrm{e}^{\mathrm{i}kx} + B\mathrm{e}^{-\mathrm{i}kx}, \tag{8.28}$$

where A and B are arbitrary constants (which may be real or complex). Equation 8.28 is the *general solution* of Equation 8.26; it contains two arbitrary constants, as expected for the general solution of a second-order differential equation.

Using Euler's formula (Equation 8.10) with $\theta = kx$, allows us to write Equation 8.28 in the alternative form

$$y(x) = C\cos(kx) + D\sin(kx), \tag{8.29}$$

where $C = A + B$ and $D = \mathrm{i}(A - B)$ are also arbitrary constants. Equations 8.28 and 8.29 are alternative descriptions of the general solution of Equation 8.26. You will see in the physics chapters that each of these descriptions has advantages under different circumstances.

A very similar analysis can be carried out for the differential equation

$$\frac{\mathrm{d}^2 y}{\mathrm{d}x^2} - k^2 y(x) = 0. \tag{8.30}$$

This time, substituting in a trial solution of the form $y(x) = \mathrm{e}^{ax}$ leads to the auxiliary equation

$$a^2 - k^2 = 0, \tag{8.31}$$

which has solutions $a = \pm k$. Consequently, Equation 8.30 is satisfied by $y(x) = \mathrm{e}^{kx}$ and $y(x) = \mathrm{e}^{-kx}$, and its general solution can be written as

$$y(x) = A\mathrm{e}^{kx} + B\mathrm{e}^{-kx}, \tag{8.32}$$

where A and B are arbitrary constants.

The exponential functions e^{kx} and e^{-kx} are related to the so-called **hyperbolic functions**:

Equations 8.33 and 8.34 are the hyperbolic analogues of the trigonometric relations derived in Exercise 8.6.

$$\cosh(kx) = \frac{\mathrm{e}^{kx} + \mathrm{e}^{-kx}}{2}, \tag{8.33}$$

$$\sinh(kx) = \frac{\mathrm{e}^{kx} - \mathrm{e}^{-kx}}{2}, \tag{8.34}$$

and these allow us to express the general solution of Equation 8.30 in the alternative form

$$y(x) = C\cosh(kx) + D\sinh(kx), \tag{8.35}$$

where $C = A + B$ and $D = (A - B)$ are further arbitrary constants.

8.2.4 Boundary conditions and eigenvalue equations

It is sometimes possible to refine the general solution of a differential equation to obtain the particular solution that applies under a given set of circumstances. To achieve this we need additional information so that we can determine the arbitrary constants. For a second-order differential equation it is sufficient to know the value of the dependent variable, y, and its derivative, $\mathrm{d}y/\mathrm{d}x$ at some value of the independent variable, x.

For example, the general solution of $\mathrm{d}^2y/\mathrm{d}x^2 + k^2 y(x) = 0$ is

$$y(x) = C\cos(kx) + D\sin(kx), \qquad \text{(Eqn 8.29)}$$

so it follows that

The prime in $y'(x)$ indicates that the function $y(x)$ is differentiated with respect to its argument x.

$$y'(x) \equiv \frac{\mathrm{d}y}{\mathrm{d}x} = -kC\sin(kx) + kD\cos(kx).$$

Now, if we happen to know that $y(0) = 1$ and $y'(0) = 2$, we can substitute these values into the above equations to obtain $C = 1$ and $D = 2/k$. The particular solution consistent with these conditions is therefore

$$y(x) = \cos(kx) + \frac{2}{k}\sin(kx).$$

Conditions that supply extra information about the solution of a differential equation at a *single* value of the independent variable are called **initial conditions**. More generally, any conditions that give extra information about the solutions are called **boundary conditions**. For example, the information that $y(x)$ is a smooth function of x that does not diverge as $x \to \pm\infty$ is typical of the boundary conditions you will meet in quantum mechanics.

In quantum mechanics, boundary conditions also occur in a context that does not lead to a unique particular solution. You will often meet linear homogeneous differential equations of the form

$$\widehat{\mathrm{L}}y(x) = \lambda y(x), \tag{8.36}$$

where $\widehat{\mathrm{L}}$ is a linear operator and the constant λ has an unknown value. In this context, the boundary conditions give us information about the possible values of λ (which are known as **eigenvalues**) and the possible solutions $y(x)$ (which are known as **eigenfunctions**). Equations like this are called **eigenvalue equations**.

An example of an eigenvalue equation is

$$\frac{\mathrm{d}^2y}{\mathrm{d}x^2} = \lambda y(x). \tag{8.37}$$

This is identical to Equation 8.25, except that the constant λ is now regarded as an undetermined parameter that is to be constrained by the boundary conditions. To take a specific case, let us suppose that the boundary conditions are:

$$y(0) = 0 \quad \text{and} \quad y(L) = 0.$$

There are three cases to consider: $\lambda = 0$, $\lambda = k^2$ and $\lambda = -k^2$ (where without any loss in generality we can take k to be positive). First, if $\lambda = 0$, the differential equation reduces to $\mathrm{d}^2y/\mathrm{d}x^2 = 0$, which has the general solution

$$y(x) = A + Bx,$$

where A and B are arbitrary constants. In this case, the boundary conditions give $A = B = 0$, and we are left with the solution $y(x) = 0$ for all x. Trivial solutions such as this are usually of no physical interest and are not counted as acceptable solutions of the eigenvalue equation.

The next case, $\lambda = k^2$, produces the general solution

$$y(x) = A\mathrm{e}^{kx} + B\mathrm{e}^{-kx}. \tag{Eqn 8.32}$$

In this case, the boundary conditions give

$$A + B = 0 \quad \text{and} \quad A\mathrm{e}^{kL} + B\mathrm{e}^{-kL} = 0,$$

but there is no way of satisfying these equations since they imply that $\mathrm{e}^{2kL} = 1$, which cannot be satisfied by any non-zero k.

Finally, we consider the case $\lambda = -k^2$, which produces the general solution

$$y(x) = C\cos(kx) + D\sin(kx). \tag{Eqn 8.29}$$

Now, the boundary conditions give

$$C = 0 \quad \text{and} \quad D\sin kL = 0.$$

Rejecting the trivial solution in which $C = D = 0$, and noting that the case $k = 0$ has already been dealt with, we conclude that $kL = n\pi$, where n is a non-zero integer. Since k has been taken to be positive, we can restrict n to the positive integers $1, 2, 3, \ldots$. It therefore follows that the differential equation has acceptable solutions (consistent with the boundary conditions) if and only if

$$\lambda = -k^2 = -n^2\pi^2/L^2 \quad \text{for } n = 1, 2, 3, \ldots.$$

These special values of λ are the *eigenvalues* of the problem and the corresponding *eigenfunctions* are $y_n(x) = D_n \sin(n\pi x/L)$, where we have used the integer n to label different eigenfunctions and replaced D by D_n to allow for the possibility that different arbitrary constants may be chosen for different values of n.

In an eigenvalue problem such as this, the boundary conditions do not lead to a unique particular solution, free from all arbitrary constants. Instead, they restrict the possible values of the parameter λ that appears in the differential equation, requiring it to be one of the allowed eigenvalues. Different eigenvalues correspond to different eigenfunctions, and each eigenfunction $y_n(x)$ still contains an arbitrary constant, D_n, whose value remains to be decided.

It is also worth noting that the principle of superposition does *not* imply that a linear combination of eigenfunctions *with different eigenvalues* is also an eigenfunction. If two eigenfunctions have different eigenvalues, they satisfy *different* differential equations, so the principle of superposition does not apply in this case.

It is remarkable how the process of solving a differential equation yields a simple set of numbers — the eigenvalues. Schrödinger was well aware of this phenomenon from his knowledge of waves in classical physics, where the frequencies of vibrating strings and organ pipes can be related to the eigenvalues of differential equations. He therefore set himself the task of finding an appropriate eigenvalue equation for atomic physics, where the eigenvalues would give the allowed energies of atoms; the result was the *time-independent Schrödinger equation*, which is a major topic in this book.

Exercise 8.10 Suppose that the eigenvalue equation given in Equation 8.37 is supplemented by the boundary conditions $y(0) = y'(L) = 0$. What are the eigenvalues and eigenfunctions in this case? ■

The above exercise illustrates an important general point: eigenvalues and eigenfunctions depend on the boundary conditions. We need to know the eigenvalue equation *and* its associated boundary conditions to specify an eigenvalue problem.

8.3 Partial differential equations

8.3.1 Partial differentiation

This section supports the study of Chapters 2, 3 and 6.

If $f(x, t)$ is a function of the two independent variables x and t, we can differentiate $f(x, t)$ with respect to x *while treating t as a constant*. The result of this differentiation is called the **partial derivative** of f with respect to x, and is written as $\partial f/\partial x$. Similarly, we can differentiate $f(x, t)$ with respect to t *while treating x as a constant*; this gives the partial derivative of f with respect to t, written as $\partial f/\partial t$.

For example, consider the function

$$f(x, t) = \sin(kx - \omega t), \tag{8.38}$$

where k and ω are constants. In this case, partially differentiating with respect to x gives

$$\frac{\partial f}{\partial x} = k\cos(kx - \omega t), \tag{8.39}$$

while partially differentiating with respect to t gives

$$\frac{\partial f}{\partial t} = -\omega\cos(kx - \omega t). \tag{8.40}$$

This is no harder than ordinary differentiation provided that we remember to treat t as a constant in the first differentiation and x as a constant in the second. In graphical terms, these partial derivatives represent the gradients of $f(x, t)$ in the x- and t-directions respectively, as illustrated in Figure 8.7.Partial derivatives can be evaluated for functions of any number of variables (not just two); each partial

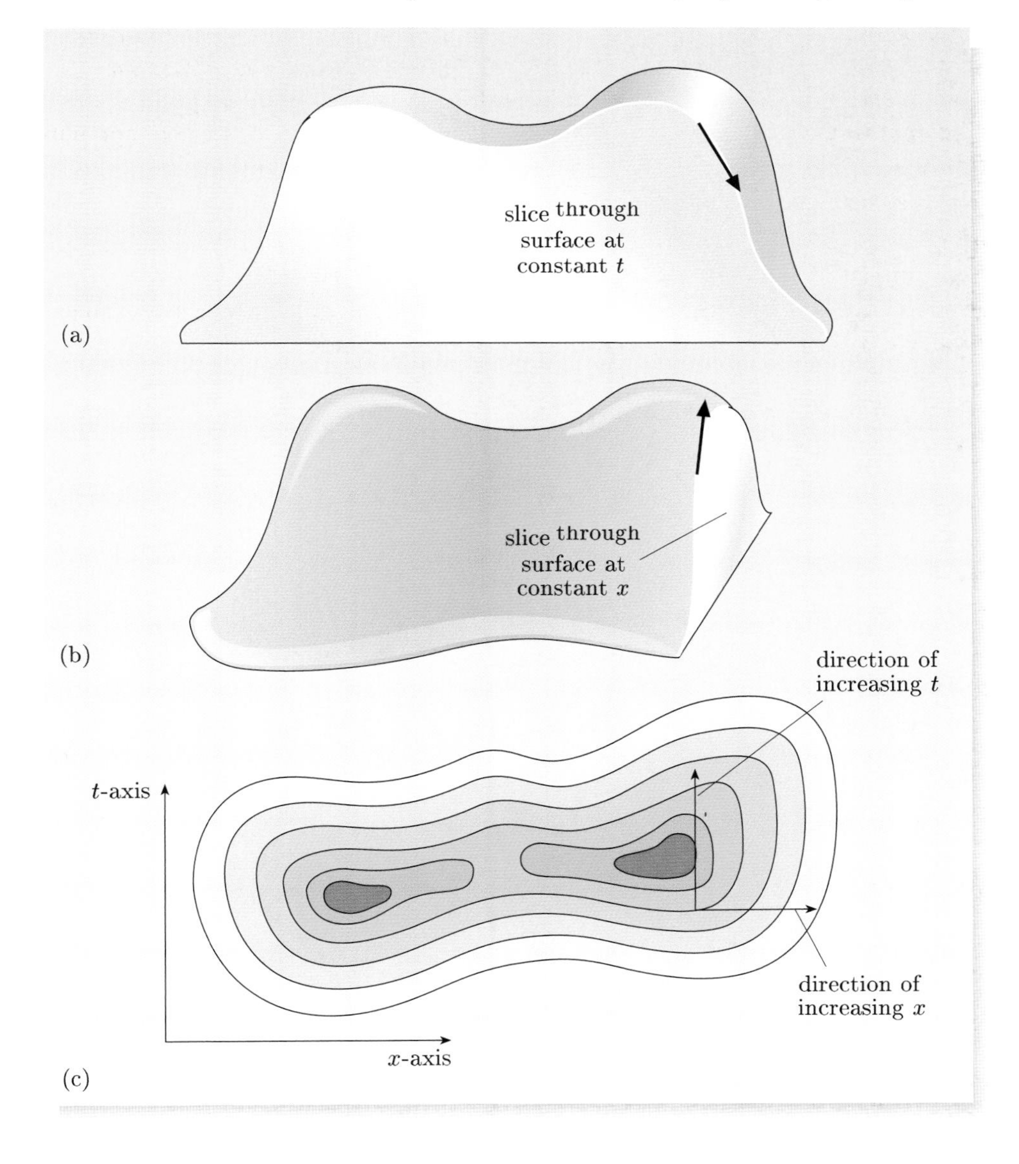

Figure 8.7 Geometrical interpretation of the partial derivatives $\partial f/\partial x$ and $\partial f/\partial t$. Parts (a) and (b) are three-dimensional plots of the function $f(x, t)$ while (c) is the corresponding contour map. The partial derivative $\partial f/\partial x$ is the rate of change of f in the direction of the arrow in (a) while $\partial f/\partial t$ is the rate of change of f in the direction of the arrow in (b).

derivative is obtained by differentiating the function with respect to one of its variables while keeping *all* the other variables constant.

The partial derivatives $\partial f/\partial x$ and $\partial f/\partial t$ are called **first-order partial derivatives** because they involve a single partial differentiation. These partial derivatives are themselves functions of x and t, so they can be partially differentiated again. For a function of two independent variables, $f(x, t)$, there are four **second-order partial derivatives**:

$$\frac{\partial^2 f}{\partial x^2} = \frac{\partial}{\partial x}\left(\frac{\partial f}{\partial x}\right) \qquad \frac{\partial^2 f}{\partial t^2} = \frac{\partial}{\partial t}\left(\frac{\partial f}{\partial t}\right),$$

$$\frac{\partial^2 f}{\partial t \partial x} = \frac{\partial}{\partial t}\left(\frac{\partial f}{\partial x}\right) \qquad \frac{\partial^2 f}{\partial x \partial t} = \frac{\partial}{\partial x}\left(\frac{\partial f}{\partial t}\right). \tag{8.41}$$

However, for well-behaved functions (and this includes any functions you will need to differentiate in this course) it turns out that the sequence of partial differentiation is irrelevant:

$$\frac{\partial^2 f}{\partial t \partial x} = \frac{\partial^2 f}{\partial x \partial t}, \tag{8.42}$$

so, in practice, only three of the second-order partial derivatives are independent.

Exercise 8.11 Find the four second-order partial derivatives of the function in Equation 8.38 and verify that Equation 8.42 is valid in this case. ■

8.3.2 Partial differential equations

A **partial differential equation** is an equation that contains partial derivatives of an unknown function that we would like to find. For example, the equation

$$\frac{\partial^2 f}{\partial x^2} = \frac{1}{c^2}\frac{\partial^2 f}{\partial t^2}, \tag{8.43}$$

where c is a constant, is a partial differential equation for the function $f(x, t)$. This equation governs the propagation of some one-dimensional waves.

Partial differential equations can be classified in much the same way as ordinary differential equations. For example, Equation 8.43 is a second-order linear homogeneous partial differential equation. It can be expressed in the form

$$\widehat{\mathrm{L}} f(x) = 0,$$

where

$$\widehat{\mathrm{L}} = \frac{\partial^2}{\partial x^2} - \frac{1}{c^2}\frac{\partial^2}{\partial t^2}$$

is a linear (partial) differential operator.

Linear homogeneous partial differential equations satisfy the principle of superposition (the switch from ordinary differentiation to partial differentiation in no way affects our previous argument). So, if $f_1(x, t)$ and $f_2(x, t)$ are both solutions of a linear homogeneous partial differential equation, then so is the linear combination $\alpha f_1(x, t) + \beta f_2(x, t)$, where α and β are any constants. This principle is a central feature in quantum physics, and it underlies some important physical effects, including interference phenomena and the behaviour of quantum computers.

The general solution of a partial differential equation can be very general indeed, encompassing wide classes of function that cannot be characterized by a few arbitrary constants. For example, Equation 8.43 is satisfied by *any* well-behaved function of the form $f(x,t) = g(x \pm ct)$. The only restriction on the function g is that we should be able to differentiate it twice. Using the principle of superposition, we see that any function of the form

$$f(x,t) = \alpha\, g(x - ct) + \beta\, g(x + ct), \tag{8.44}$$

where α and β are constants, satisfies Equation 8.43 (and is, in fact, the general solution of this partial differential equation).

The boundary conditions needed to specify a particular solution of a partial differential equation are much more extensive than those needed for an ordinary differential equation. In the case of Equation 8.43, sufficient information would be provided by the initial ($t = 0$) values of the function $f(x,t)$ and its partial derivative $\partial f/\partial t$, but we would need to know these values for all x.

Exercise 8.12 By differentiating twice with respect to x, and twice with respect to t, show that any function $f(x,t) = g(x - ct)$ of the single variable $x - ct$ satisfies Equation 8.43. ■

8.3.3 The method of separation of variables

Finally, we shall describe a method that allows us to find *some* solutions of *some* partial differential equations (we can put it no more strongly than that). The method will be illustrated for Equation 8.43.

The basic idea is to look for solutions that are products of functions of a single variable. In the present case, we shall look for solutions of the form

$$f(x,t) = X(x)T(t), \tag{8.45}$$

and then try to obtain and solve ordinary differential equations for the functions $X(x)$ and $T(t)$. This tactic is called the method of **separation of variables**.

Partially differentiating Equation 8.45 twice with respect to x gives $X''(x)T(t)$, where the primes denote differentiation of a function with respect to its argument. Similarly, partially differentiating twice with respect to t gives $X(x)T''(t)$. When these results are substituted into Equation 8.43, we obtain

$$X''(x)T(t) = \frac{1}{c^2}\, X(x)T''(t).$$

Dividing both sides by $X(x)T(t)$ then gives

$$\frac{X''(x)}{X(x)} = \frac{1}{c^2}\,\frac{T''(t)}{T(t)}. \tag{8.46}$$

Now comes the inspired step. We notice that the left-hand side of Equation 8.46 depends only on x, while the right-hand side depends only on t. How can a function of x be equal to a function of t for all values of x and t? This can only happen if both sides of the equation are equal to the *same* constant, λ, say. We can therefore write

$$\frac{X''(x)}{X(x)} = \frac{1}{c^2}\frac{T''(t)}{T(t)} = \lambda,$$

where the undetermined constant λ is called the **separation constant**. Our partial differential equation therefore splits into two ordinary differential equations:

$$\frac{\mathrm{d}^2X}{\mathrm{d}x^2} = \lambda X(x) \quad \text{and} \quad \frac{\mathrm{d}^2T}{\mathrm{d}t^2} = c^2\lambda T(t). \tag{8.47}$$

These are *eigenvalue equations* of the type met in Section 8.2. The allowed values of λ may be restricted by the boundary conditions. To take a definite case, let us suppose that λ is negative. Then we can write $\lambda = -k^2$ and express the solutions as

$$X(x) = A\cos(kx) + B\sin(kx)$$

$$T(t) = C\cos(kct) + D\sin(kct).$$

Equation 8.43 then has a solution of the form

$$f(x,t) = \left[A\cos(kx) + B\sin(kx)\right]\left[C\cos(kct) + D\sin(kct)\right], \tag{8.48}$$

where A, B, C and D are arbitrary constants.

The method of separation of variables has two significant limitations. First, it cannot be used for all partial differential equations. If the equation involves a term such as e^{-x^2/c^2t^2}, for example, substitution of a product function $X(x)T(t)$, followed by rearrangement, will never produce an equation like Equation 8.46 where the terms in x appear on one side of the equation, and the terms in t appear on the other side. Equations for which such a separation of variables can be achieved are called **separable**. The second limitation is that the method does not give the most general solution of the partial differential equation. It focuses attention on solutions of the form $X(x)T(t)$, but it would be idle to suppose that *all* solutions have this form — clearly they do not, as demonstrated by Equation 8.44.

Nevertheless, the method of separation of variables is of major importance. One reason is that product solutions of the form $X(x)T(t)$ are useful in many physical applications. In classical physics, they represent standing waves with fixed nodes; in quantum physics, they represent states of definite energy. Another reason is based on the principle of superposition. If the partial differential equation is linear and homogeneous, the principle of superposition allows us to use product solutions as building blocks from which more complicated solutions can be constructed. In many cases, the most general solution of a partial differential equation can be obtained by taking an arbitrary linear combination of the product solutions that emerge from the method of separation of variables.

8.4 Probability

8.4.1 The concept of probability

This section supports the study of Chapter 4.

Many everyday *experiments* or *trials* have uncertain results or outcomes; examples include the result of tossing a coin, the hand you are dealt in a card game or tomorrow's weather in London. The first aim of a theory of probability is to assign numerical probabilities to outcomes in a way that quantifies their likelihood.

The scale of probability runs from zero (for outcomes that are impossible) to 1 (for outcomes that are certain). If the probability of an outcome is 0.5, it is as likely to happen as not (Figure 8.8). More generally, an outcome has a **probability** of p if it is expected to happen $p \times 100\%$ of the time. For experiments that can be repeated endlessly, the fraction of times a particular outcome occurs is expected to approach its probability as the number of repetitions of the experiment approaches infinity. In the long run, for example, we expect to get heads in about 50% of the tosses of an unbiased coin, corresponding to $p = 0.5$ for this outcome.

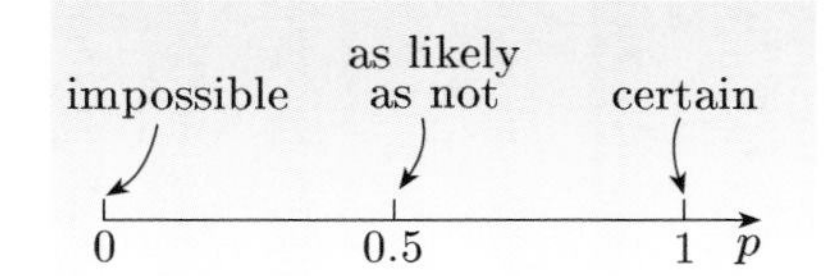

Figure 8.8 The scale of probability $0 \leq p \leq 1$.

It is remarkable that unpredictable events can be described by such a simple concept as probability. A commonly-held (but erroneous) view is that a long run of tails will increase the chances of getting heads on the next coin toss, because (it is said) 'the laws of chance tend to even things up'. This is not how blind chance works: the probability of tossing heads is 0.5 *irrespective* of the past history of coin tosses. It is reasonable to suppose that heads will occur roughly 50% of the time in the very long run, but this happens because any initial run of tails will be *diluted* and eventually swamped by a much longer run of roughly equal numbers of heads and tails that follows.

Some experiments (such as the running of an Olympic race) can never be repeated under exactly the same conditions. In such cases, a more suitable definition of probability is based on the fair odds for a wager. Suppose that you place $0 \leq p \leq 1$ pounds into a kitty, and I place $(1 - p)$ pounds into it, on the understanding that you will collect the kitty if one particular outcome occurs, and I will collect the kitty if it does not. Then, p is equal to the probability of the outcome if the bet is fair to both of us. This way of looking at probability implies that it has a meaning even for one-off, unrepeatable events. When we use the concept of probability in quantum physics, it suggests that we can talk about probability for a single atom or electron.

If a given outcome has a definite probability, and no further information is available to us about whether it will occur or not, the outcome is said to be **random**. In everyday life, tossing heads on a coin may be taken to be random. If we took enough trouble to observe exactly how the coin was tossed we could, in principle, predict whether it would land with heads uppermost, but the necessary information is not usually collected, so we can treat the outcome as random. A very different situation arises in quantum physics. Quantum physics predicts the probabilities of the results of measurements, but these probabilities are regarded as being fundamental; we believe that the basic laws of Nature are expressed in terms of probabilities. In quantum physics, probabilities do not arise through lazy or inadequate observations, or because of a lack of computational power, but are believed to be intrinsic to way the world works.

8.4.2 Adding probabilities

Suppose that we know the probabilities of a set of outcomes. For example, we normally assign a probability of 1/6 to each of the six spot numbers on a die, trusting that the die is not loaded. We still have the problem of assigning probabilities to more complicated (compound) outcomes. What is the probability of rolling a number less than 4, for example?

We say that the outcomes of a trial are **mutually exclusive** if the occurrence of one of the outcomes automatically excludes the possibility of any of the other outcomes in the same trial. For example, the six different spot numbers on a die are mutually exclusive because, if you roll a 5 you cannot, at the same time, roll a 3. The following rule applies to mutually exclusive outcomes:

The addition rule for probability

If $A_1, A_2, \ldots, A_n$ are a set of mutually exclusive outcomes with probabilities $p_1, p_2, \ldots, p_n$, then the probability P of obtaining *either* A_1 *or* A_2 *or* ... A_n in a single trial is the sum of the individual probabilities:

$$P = p_1 + p_2 + \ldots + p_n = \sum_{i=1}^{n} p_i. \tag{8.49}$$

A set of mutually exclusive outcomes is said to be **complete** if it covers every possible outcome of the experiment. In this case, we are certain to get one of these outcomes, so $P = 1$ in Equation 8.49. We therefore have

$$\sum_{i=1}^{n} p_i = 1, \tag{8.50}$$

where the sum is over all the mutually exclusive outcomes. This is called the **normalization rule** for probabilities. This condition requires us to sum over *all* the possible outcomes in a *mutually exclusive* set. Failing to sum over all the outcomes would give a probability less than 1, and if the outcomes were not mutually exclusive, there would be some double-counting. Fortunately, experiments in quantum mechanics generally do produce mutually exclusive outcomes — when the energy of an atom is measured, for example, the result of the measurement is a definite energy value, and this automatically excludes other values.

Exercise 8.13 An experiment has only three outcomes and they are mutually exclusive. The outcomes occur with probabilities in the ratio $1 : 3 : 6$. What are their probabilities? ■

8.4.3 Average values and expectation values

Average values

If a given quantity A, with a discrete set of possible values, A_1, A_2, ..., A_n, is repeatedly measured in a given situation, we can record the number of times each value occurs. If the measurement of A is repeated N times, and the value A_i is obtained on N_i occasions, we define the **average value** or **mean value** of A to be

$$\overline{A} = \frac{N_1A_1 + N_2A_2 + \ldots + N_nA_n}{N} = \frac{1}{N}\sum_{i=1}^{n} N_iA_i. \tag{8.51}$$

We can also define the **relative frequency** f_i of value A_i to be the fraction of times that this value occurs. We then have

$$f_i = \frac{N_i}{N} \quad \text{where} \quad \sum_{i=1}^{n} f_i = 1.$$

In terms of the relative frequencies, the average value is given by

$$\overline{A} = f_1 A_1 + f_2 A_2 + \ldots + f_n A_n = \sum_{i=1}^{n} f_i A_i. \tag{8.52}$$

Average values provide a very convenient way of characterizing the results of repeated measurements.

Expectation values

When predicting the result of measuring of a quantity A, we may know that a number of different values, $A_1, A_2, \ldots, A_n$ are possible, with each value A_i being characterized by a probability p_i. The complete set of probabilities is said to be a **probability distribution** for the measurement. For a set of discrete results, a probability distribution may be represented by a bar chart, as in Figure 8.9.

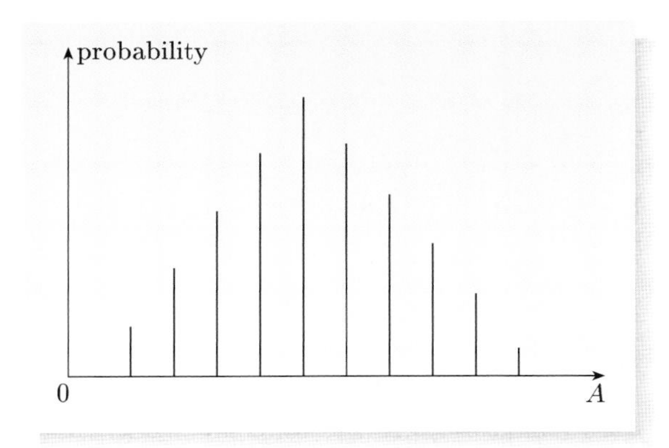

Figure 8.9 A bar chart representing a discrete probability distribution.

Sometimes, we do not need to know the full detail of a probability distribution; it is sufficient to characterize the distribution by a value that can be compared easily with the average value of a set of measurements. We therefore define the **expectation value** of a probability distribution to be

$$\langle A \rangle = p_1 A_1 + p_2 A_2 + \ldots + p_n A_n = \sum_{i=1}^{n} p_i A_i, \tag{8.53}$$

where the sum runs over all the possible values for A.

The expectation value $\langle A \rangle$ is a theoretical prediction for the average value $\overline{A}$. You should not be too surprised if, over a finite sequence of measurements, the average value and the expectation value differ from one another slightly; random fluctuations are in the nature of chance. However, as the number of measurements increases, we expect the relative frequency f_i to approach the probability p_i, so in the long run, the average value is expected to approach the expectation value.

It is worth noting that the expectation value need not be any of the possible values $A_1, A_2, \ldots, A_n$. For example, if A is the score rolled on a die, the expectation value of A is

$$\langle A \rangle = \frac{1}{6} \times 1 + \frac{1}{6} \times 2 + \frac{1}{6} \times 3 + \frac{1}{6} \times 4 + \frac{1}{6} \times 5 + \frac{1}{6} \times 6 = 3.5.$$

This result could never be obtained on a single roll. Under some circumstances, the expectation value need not be close to the most likely value. For the skew distribution shown in Figure 8.10, the expectation value $\langle A \rangle$ is shifted far away from the peak, into the long tail of the distribution. Nevertheless, the expectation value represents our best estimate of the *average* value for a long sequence of measurements of A.

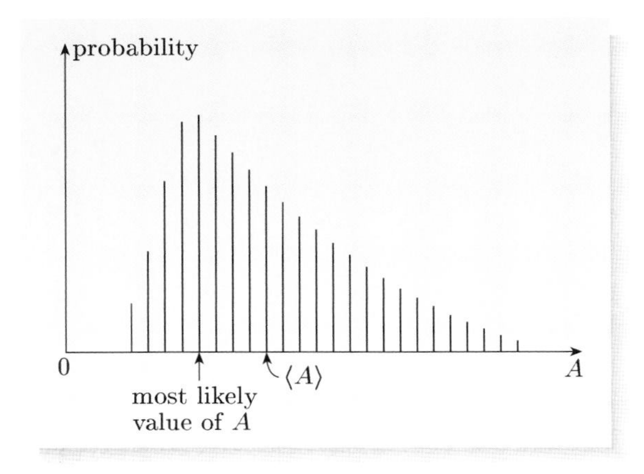

Figure 8.10 The expectation value $\langle A \rangle$ need not coincide with the peak (most likely) value of the distribution.

8.4.4 Standard deviations and uncertainties

Standard deviations

It is often important to characterize the *spread* of values of a quantity around its average value. This is generally done by introducing a quantity called the *standard deviation*. If the quantity A has a discrete set of possible values, A_i, which occur with relative frequencies f_i, we define the **standard deviation** of A to be

$$\sigma(A) = \left[\sum_{i=1}^{n} f_i (A_i - \overline{A})^2\right]^{1/2} = \left[\overline{(A - \overline{A})^2}\right]^{1/2}.$$

In words: we take the deviation $A_i - \overline{A}$ of each value from the average value, square the results, find the average of these squared deviations, and finally take the square root. Taking the square root in the last step ensures that the standard deviation has the same units as A. (Note that it would be useless to consider the average value of the deviations without taking their square first, since this would always give zero, by definition of the average value.)

Uncertainties

The spread of a probability distribution around its expectation value is characterized by a quantity called the *uncertainty*. This is defined in a similar way to the standard deviation, but using the theoretical probabilities rather than the measured relative frequencies. For a quantity A, whose possible values A_i have probabilities p_i, the **uncertainty** in A is defined by

$$\Delta A = \left[\sum_{i=1}^{n} p_i(A_i - \langle A \rangle)^2\right]^{1/2} = \left[\langle (A - \langle A \rangle)^2 \rangle\right]^{1/2}. \tag{8.54}$$

From this definition, it is clear that uncertainty is a theoretical prediction for the standard deviation. The standard deviation is expected to approach the uncertainty in the long run, as the number of measurements increases.

An alternative way of calculating uncertainties is often used. Squaring both sides of Equation 8.54 and expanding out $(A - \langle A \rangle)^2$, we see that

$$\begin{aligned}
(\Delta A)^2 &= \langle (A - \langle A \rangle)^2 \rangle \\
&= \langle A^2 - 2A\langle A \rangle + \langle A \rangle^2 \rangle \\
&= \langle A^2 \rangle - 2\langle A \rangle\langle A \rangle + \langle A \rangle^2 \\
&= \langle A^2 \rangle - \langle A \rangle^2,
\end{aligned}$$

so the uncertainty in A can also be expressed as

$$\Delta A = \left[\langle A^2 \rangle - \langle A \rangle^2\right]^{1/2}. \tag{8.55}$$

This formula generally provides the most efficient way of calculating uncertainties in quantum mechanics.

We stated earlier that the average value $\overline{A}$ need not be *exactly* equal to the expectation value, $\langle A \rangle$, but that the difference between these two quantities is expected to become very small when the number of measurements becomes very large. We can now quantify this statement. If N measurements are taken (with N greater than about 25), it can be shown that the probability of finding an average value $\overline{A}$ that deviates from the expectation value $\langle A \rangle$ by more than a fraction f of the uncertainty ΔA is *less than one in a million* if $N > 25/f^2$. This is true no matter what the shape of the probability distribution.

This result is a consequence of the *central limit theorem* in statistics.

Exercise 8.14 Calculate the standard deviation associated with rolling a die by (a) using Equation 8.54 and (b) using Equation 8.55. ■

8.4.5 Continuous probability distributions

So far, we have considered a discrete set of possible outcomes, which can be labelled $A_1, A_2, \ldots$. However, we can also consider situations in which the set of possible outcomes forms a continuum.

For simplicity, we shall consider the example of measuring the position x of a particle in one dimension, where x can take any value in a continuous range. In this case, it is appropriate to introduce a **probability density function** $\rho(x)$, defined so that the probability of obtaining a value of x in a small range of width δx, centred on x, is $\rho(x)\,\delta x$. Figure 8.11 shows an example of a probability density function. The probability that x lies between a and b is given by the integral

$$P = \int_a^b \rho(x)\,\mathrm{d}x, \tag{8.56}$$

which is the area of the shaded strip in Figure 8.11. The area under the whole curve is then given by

$$\int_{-\infty}^{\infty} \rho(x)\,\mathrm{d}x = 1. \tag{8.57}$$

This is the normalization rule for a continuous set of outcomes.

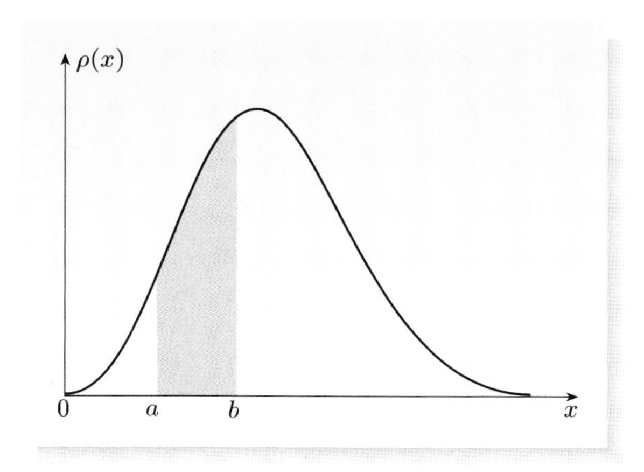

Figure 8.11 A continuous probability distribution $\rho(x)$.

In general, everything is similar to the discrete case, except that sums are replaced by integrals. For example, the expectation value of x is given by

$$\langle x \rangle = \int_{-\infty}^{\infty} \rho(x) x\,\mathrm{d}x, \tag{8.58}$$

and the uncertainty of x is given by

$$\Delta x = \left[\langle x^2 \rangle - \langle x \rangle^2\right]^{1/2}, \tag{8.59}$$

where

$$\langle x^2 \rangle = \int_{-\infty}^{\infty} \rho(x) x^2\,\mathrm{d}x. \tag{8.60}$$

These results can easily be extended to three dimensions, using a probability density function that depends on three coordinates, x, y and z and integrating over all of these coordinates.

Acknowledgements

Grateful acknowledgement is made to the following sources:

Figure 1.2: Mary Evans Picture Library; Figure 1.5: Courtesy of id Quantique.com; Figure 1.6: Mary Evans Picture Library; Figure 1.9: Meggers Gallery/American Institute of Physics/Science Photo Library; Figure 1.12: Reprinted by courtesy of Dr Akira TONOMURA, Hitachi, Ltd., Japan; Figure 1.13: Reprinted by courtesy of Dr Akira TONOMURA, Hitachi, Ltd., Japan; Figure 1.16 & Figure 1.17: Adapted from Grangier, P., Roger, G., and Aspect, A., (1986) 'Experimental evidence for a photon anticorrelation effect on a beam splitter: A new light on single photon interferences', Europhysics Letters 1(4), EDP;

Figure 2.1: Mary Evans Picture Library; Figure 2.4: Mary Evans Picture Library; Figure 2.7: Courtesy of the Astronomical Data Center, and the National Space Science Data Center through the World Data Center A for Rockets and Satellites;

Figure 3.1: Kavli Institute of Nanoscience Delft; Figure 3.2: Courtesy of Dr A. Rogach and Dr D. Talapin, University of Hamburg;

Figure 4.10: AIP Emilio Segre Visual Archives/Gift of Jost Lemmerich; Figure 4.11: Courtesy of Dr Eric Cornell, Jila, University of Colorado;

Figure 5.11: Mary Evans Picture Library;

Figure 6.4: AIP Emilio Segre Visual Archives; Figure 6.6: Science Photo Library;

Figure 7.2: Jon Arnold Images/Alamy; Figure 7.3: Bennett, C. L. et al. First Year Wilkinson Microwave Anisotrophy Probe (WMAP) Observations; Preliminary Maps and Basic Results, Astrophysical Journal (submitted) ©2003 The American Astronomical Society; Figure 7.22: Courtesy of Dr Andrew Flewitt in the Engineering Department University of Cambridge; Figure 7.23: Courtesy of Don Eigler, IBM Research Division;

Every effort has been made to contact copyright holders. If any have been inadvertently overlooked the publishers will be pleased to make the necessary arrangements at the first opportunity.

Solutions to exercises

Ex 1.1 The gamma-ray photon has an energy of 1.60×10^{-13} J. From Equation 1.1, the frequency of the photon is $f = E_{\text{photon}}/h$, so $f = 1.60 \times 10^{-13}\,\text{J}/(6.63 \times 10^{-34}\,\text{J s}) = 2.41 \times 10^{20}$ Hz, which is a frequency roughly 2.4×10^{14} times that of a 1 MHz AM radio photon.

Ex 1.2 The poker would lose energy largely by emitting an enormous number of photons, some visible but mostly infrared. The energy of each photon would be so small compared to the energy of the poker that detecting anything but a continuous cooling down would be hard; yet, in principle, the cooling of the poker is discrete.

Ex 1.3 The power produced is proportional to $\text{d}N/\text{d}t$, which is proportional to N, the number of radioactive nuclei present. Using Equation 1.7, the power produced after 7.30 years would be $\exp(-0.693 \times 7.30/87.7) = 0.944$ times the initial power.

Ex 1.4 (a) The energy and the magnitude of the momentum of a photon are respectively hf and hf/c, so the magnitude of the momentum is the energy divided by c which is $1 \times 10^6 \times 1.60 \times 10^{-19}\,\text{J}/(3.00 \times 10^8\,\text{m s}^{-1}) = 5.33 \times 10^{-22}\,\text{kg m s}^{-1}$.

(b) Force is the rate of change of momentum. Let p be the magnitude of the momentum of each photon. If the rate of absorption of photons is $\text{d}N/\text{d}t$, the rate of transfer of momentum to the lead block is $p\,\text{d}N/\text{d}t$; this is the force exerted on the lead, which is $1\,\text{N} = 1\,\text{kg m s}^{-2}$. Hence,

$$\frac{\text{d}N}{\text{d}t} = \frac{1\,\text{kg m s}^{-2}}{5.33 \times 10^{-22}\,\text{kg m s}^{-1}} = 1.88 \times 10^{21}\,\text{s}^{-1}.$$

(c) Using the results of Exercise 1.1 and noting that the momentum of a photon is proportional to its frequency, the number of AM radio photons required per second is 2.4×10^{14} times greater than the answer to part (b), i.e. $4.5 \times 10^{35}\,\text{s}^{-1}$. An AM radio photon carries much less momentum, as well as much less energy, than a gamma-ray photon.

Ex 1.5 The photon is described by a wave that goes *both ways* through the half-silvered mirror until it manifests itself at one or the other of the two detectors. This is similar to the way that each photon must progress as a wave through the entire apparatus of Taylor until it manifests itself at one location on the photographic plate.

Ex 1.6 The electron would be most likely to be found in region B, least likely in region C, and somewhere in between in region A. Since $|\Psi(x,t)|^2$ has moved as a whole to the left, we conclude that it is describing an electron moving in the direction of decreasing x. But see the next exercise!

Ex 1.7 No! The regions where $|\Psi(x,t)|^2$ is substantially non-zero overlap at the two instants of time. In particular, there is a significant chance of finding the electron in region C at the later time.

Ex 1.8 The probability of being found in a small interval δx, centred on position x, is $|\Psi_{\text{dB}}(x,t)|^2\,\delta x$. Let us use the symbol θ to stand for the phase of the wave, so $\theta = kx - \omega t$. Then $|\Psi_{\text{dB}}|^2 = |A|^2|\text{e}^{\text{i}\theta}|^2 = |A|^2\text{e}^{-\text{i}\theta}\text{e}^{+\text{i}\theta} = |A|^2\text{e}^0 = |A|^2$. So the probability of finding the particle in the small interval δx is independent of x.

Ex 1.9 The function $q(x)$ is never zero, being unity for all x since $\sin^2 x + \cos^2 x = 1$ for all x. The function $p(x)$ is equal to zero when $\sin x + \cos x = 0$. This happens when $\sin x = -\cos x$, that is when $\tan x = -1$, which is satisfied by $x = 3\pi/4 + n\pi$, where n is any integer.

Ex 1.10 Blocking path P1 (or P2, for that matter) would lead to photons being counted at Db. The probability of detecting a photon is $p_{\text{b}} = |b_2|^2$. (This would be 0.25 since a photon has probability 0.5 of reaching H2 and any photon that does so has a probability 0.5 of being detected at Db.)

Ex 1.11 For photons: if only one path is open, only one probability amplitude will contribute, and this is never zero, so an incident photon can appear at either output. But if both paths are open, the probability amplitude for a photon to appear at one of the outputs will be a sum of two probability amplitudes and this sum will be zero if the two probability amplitudes have equal moduli and opposite phases.

For a classical wave: if the path lengths are appropriate, the wave that reaches the detector by one path can exactly cancel the wave that arrives by the other path.

Ex 2.1 (a) In this case:

$$\widehat{\mathrm{O}}_1 g(x,t) = \frac{\partial}{\partial x}(3x^2t^3) = 6xt^3,$$

$$\widehat{\mathrm{O}}_2 g(x,t) = \left(3\frac{\partial}{\partial x} + 3x^2\right)(3x^2t^3) = 18xt^3 + 9x^4t^3,$$

$$\widehat{\mathrm{O}}_3 g(x,t) = \left(\frac{\partial^2}{\partial x^2} + 5\right)(3x^2t^3) = 6t^3 + 15x^2t^3.$$

(b) In this case:

$$\widehat{\mathrm{O}}_1 h(x,t) = \frac{\partial}{\partial x}[\alpha\sin(kx-\omega t)] = k\alpha\cos(kx-\omega t),$$

$$\widehat{\mathrm{O}}_2 h(x,t) = \left(3\frac{\partial}{\partial x} + 3x^2\right)[\alpha\sin(kx-\omega t)] = 3k\alpha\cos(kx-\omega t) + 3x^2\alpha\sin(kx-\omega t),$$

$$\widehat{\mathrm{O}}_3 h(x,t) = \left(\frac{\partial^2}{\partial x^2} + 5\right)[\alpha\sin(kx-\omega t)] = -k^2\alpha\sin(kx-\omega t) + 5\alpha\sin(kx-\omega t) = (5-k^2)\alpha\sin(kx-\omega t).$$

Ex 2.2 (a) $\widehat{\mathrm{Q}}$ is not linear because

$$\widehat{\mathrm{Q}}(\alpha f + \beta g) = \log(\alpha f + \beta g) \neq \alpha\log f + \beta\log g.$$

(b) $\widehat{\mathrm{R}}$ is linear because

$$\widehat{\mathrm{R}}(\alpha f + \beta g) = \frac{\mathrm{d}^2}{\mathrm{d}x^2}(\alpha f + \beta g) = \alpha\frac{\mathrm{d}^2 f}{\mathrm{d}x^2} + \beta\frac{\mathrm{d}^2 g}{\mathrm{d}x^2}.$$

(c) $\widehat{\mathrm{S}}$ is not linear because

$$\widehat{\mathrm{S}}(\alpha f + \beta g) = (\alpha f + \beta g)^2 \neq \alpha f^2 + \beta g^2.$$

Ex 2.3 Letting the given operators act on the given function $\mathrm{e}^{\mathrm{i}\alpha x}$, we obtain

$$\frac{\mathrm{d}}{\mathrm{d}x}(\mathrm{e}^{\mathrm{i}\alpha x}) = \mathrm{i}\alpha\mathrm{e}^{\mathrm{i}\alpha x}$$

and

$$\frac{\mathrm{d}^2}{\mathrm{d}x^2}(\mathrm{e}^{\mathrm{i}\alpha x}) = (\mathrm{i}\alpha)^2\mathrm{e}^{\mathrm{i}\alpha x} = -\alpha^2\mathrm{e}^{\mathrm{i}\alpha x}.$$

So $\mathrm{e}^{\mathrm{i}\alpha x}$ is an eigenfunction of both operators, with eigenvalues $\mathrm{i}\alpha$ and $-\alpha^2$ respectively.

Ex 2.4 Operating with $\widehat{\mathrm{p}}_x$ on the given wave function, we obtain

$$\widehat{\mathrm{p}}_x\Psi(x,t) = -\mathrm{i}\hbar\frac{\partial}{\partial x}\left(A\mathrm{e}^{\mathrm{i}(-kx-\omega t)}\right) = -\hbar k\Psi(x,t).$$

So, $\Psi(x,t)$ is an eigenfunction of the momentum operator, with eigenvalue $-\hbar k$. We interpret this by saying that the momentum of the particle described by the wave is $p_x = -\hbar k$. The negative sign makes sense because the wave $A\mathrm{e}^{\mathrm{i}(-kx-\omega t)}$ propagates in the negative x-direction.

Ex 2.5 A free particle is one that is not subject to forces, so we can put $V(x) = 0$: The Hamiltonian function is then $H = p_x^2/2m$.

Ex 2.6 With $V(x) = \frac{1}{2}Cx^2$, we get $F_x = -\partial V/\partial x = -Cx$, as required. Adding this potential energy to the kinetic energy expressed in terms of momentum, the Hamiltonian function is

$$H = \frac{p_x^2}{2m} + \tfrac{1}{2}Cx^2.$$

Ex 2.7 The Hamiltonian function for this system was obtained in Exercise 2.6. Carrying out the standard replacement of Equation 2.24, the corresponding Hamiltonian operator is

$$\widehat{\mathrm{H}} = -\frac{\hbar^2}{2m}\frac{\partial^2}{\partial x^2} + \tfrac{1}{2}Cx^2.$$

Ex 2.8 (a) With no potential energy, we have

$$\mathrm{i}\hbar\frac{\partial\Psi(x,t)}{\partial t} = -\frac{\hbar^2}{2m}\frac{\partial^2\Psi(x,t)}{\partial x^2}.$$

This is exactly Equation 2.7 as you would expect.

(b) Since the particle is subject to Hooke's law, we must substitute the one-dimensional harmonic oscillator potential energy function $V(x) = \frac{1}{2}Cx^2$ into Equation 2.19. Doing this gives

$$i\hbar\frac{\partial\Psi(x,t)}{\partial t} = -\frac{\hbar^2}{2m}\frac{\partial^2\Psi(x,t)}{\partial x^2} + \tfrac{1}{2}Cx^2\Psi(x,t).$$

Ex 2.9 Using the properties of partial differentiation, we have

$$\frac{\partial}{\partial t}\left[a\Psi_1 + b\Psi_2\right] = a\,\frac{\partial\Psi_1}{\partial t} + b\,\frac{\partial\Psi_2}{\partial t}$$

and

$$\frac{\partial^2}{\partial x^2}\left[a\Psi_1 + b\Psi_2\right] = a\,\frac{\partial^2\Psi_1}{\partial x^2} + b\,\frac{\partial^2\Psi_2}{\partial x^2}.$$

In addition, it is clear that

$$V(x)\left[a\Psi_1 + b\Psi_2\right] = aV(x)\Psi_1 + bV(x)\Psi_2.$$

Combining all these results, we have

$$\left(-\frac{\hbar^2}{2m}\frac{\partial^2}{\partial x^2} + V(x) - i\hbar\frac{\partial}{\partial t}\right)\left[a\Psi_1 + b\Psi_2\right]$$

$$= a\left(-\frac{\hbar^2}{2m}\frac{\partial^2}{\partial x^2} + V(x) - i\hbar\frac{\partial}{\partial t}\right)\Psi_1 + b\left(-\frac{\hbar^2}{2m}\frac{\partial^2}{\partial x^2} + V(x) - i\hbar\frac{\partial}{\partial t}\right)\Psi_2,$$

which is equal to zero because both Ψ_1 and Ψ_2 satisfy Schrödinger's equation. This shows that $a\Psi_1 + b\Psi_2$ also satisfies Schrödinger's equation, as required.

Ex 2.10 Not quite. Any linear combination of normalized wave functions satisfies Schrödinger's equation, but to represent a state, the linear combination must itself be normalized. This can always be achieved by multiplying the whole linear combination by a suitable normalization constant.

Ex 2.11 Substituting $T(t) = e^{-iEt/\hbar}$ into the left-hand side of Equation 2.33 and performing the differentiation gives $i\hbar(-iE/\hbar)e^{-iEt/\hbar} = Ee^{-iEt/\hbar}$, which is equal to $ET(t)$ as required.

Ex 3.1 Equation 3.1 would be replaced by $V(x) = 0$ for $-L/2 \leq x \leq L/2$ and Equation 3.2 would be replaced by $V(x) = \infty$ for $x < -L/2$ and $x > L/2$.

Ex 3.2 An energy eigenfunction $\psi(x)$ does not depend on time. It satisfies the time-independent Schrödinger equation, which is the eigenvalue equation for energy. A wave function $\Psi(x,t)$ depends on time as well as position. It satisfies Schrödinger's equation and provides the most complete possible description of the state of a system at all times.

Comment: A *stationary-state* wave function, $\Psi_n(x,t) = \psi_n(x)e^{-iE_nt/\hbar}$ is an energy eigenfunction *at any fixed time*. For example, at $t = 0$, we have $\Psi_n(x,0) = \psi_n(x)$. However an energy eigenfunction should never be called a wave function because it does not contain the necessary information about time-dependence.

The distinction between wave functions and energy eigenfunctions is maintained in our notation: upper case Ψ is used for wave functions and lower case ψ is used for energy eigenfunctions.

Ex 3.3 Substituting the suggested form for the eigenfunction $\psi(x)$ into both sides of Equation 3.7 and carrying out the differentiations gives

$$\frac{\hbar^2}{2m}k^2\Big[A\sin(kx) + B\cos(kx)\Big] = E\Big[A\sin(kx) + B\cos(kx)\Big]$$

for $0 \leq x \leq L$, which will be true if $E = \hbar^2k^2/2m$. This last condition is met provided that Equation 3.11 is satisfied.

Ex 3.4 If we replace k by $-k$ in the general solution of Equation 3.10, and use the relations $\sin(-kx) = -\sin(kx)$ and $\cos(-kx) = \cos(kx)$ we see that

$$\begin{aligned}\psi(x) &= A\sin(-kx) + B\cos(-kx)\\ &= -A\sin(kx) + B\cos(kx)\end{aligned}$$

for $0 \leq x \leq L$.

So, from a mathematical point of view, using $k = -\sqrt{2mE}/\hbar$ instead of $k = +\sqrt{2mE}/\hbar$ is equivalent to reversing the sign of A. So far, we have placed no restrictions on the arbitrary constants A and B, which could be any complex numbers. Hence there is nothing to be learned by considering negative values of k that will not emerge from studying positive values alone.

Ex 3.5 From Equation 3.16, we have

$$E_n = \frac{n^2\pi^2\hbar^2}{2mL^2} \quad \text{and} \quad E_{n+1} = \frac{(n+1)^2\pi^2\hbar^2}{2mL^2}.$$

So the energy difference between two neighbouring energy levels with quantum numbers $n+1$ and n is

$$E_{n+1} - E_n = \frac{[(n+1)^2 - n^2]\pi^2\hbar^2}{2mL^2} = \frac{[2n+1]\pi^2\hbar^2}{2mL^2}.$$

This energy difference tends to zero as L tends to infinity, as claimed.

Ex 3.6 For the region inside the box, Equations 3.26 and 3.27 give

$$\left[\mathrm{Re}(\Psi_n(x,t))\right]^2 = \frac{2}{L}\sin^2\left(\frac{n\pi x}{L}\right)\cos^2\left(\frac{E_n t}{\hbar}\right)$$

$$\left[\mathrm{Im}(\Psi_n(x,t))\right]^2 = \frac{2}{L}\sin^2\left(\frac{n\pi x}{L}\right)\sin^2\left(\frac{E_n t}{\hbar}\right).$$

Adding the corresponding sides of these equations and remembering that $\cos^2(E_n t/\hbar) + \sin^2(E_n t/\hbar) = 1$ gives

$$|\Psi_n(x,t)|^2 = \frac{2}{L}\sin^2\left(\frac{n\pi x}{L}\right),$$

which is independent of time, as required.

Ex 3.7 The coordinate x that appears in the argument of $\Psi_n(x,t)$ is used to label positions at which the particle might be found. If the position of the particle is measured, then $|\Psi_n(x,t)|^2\,\delta x$ is the probability of finding the particle in a small interval δx, centred on the position x, at time t.

Ex 3.8 Since $\mathrm{e}^{-\mathrm{i}(E_n - V_0)t/\hbar} = \mathrm{e}^{-\mathrm{i}E_n t/\hbar}\mathrm{e}^{\mathrm{i}V_0 t/\hbar}$, the new wave functions are the same as the original ones, apart from a time-dependent phase factor $\mathrm{e}^{\mathrm{i}V_0 t/\hbar}$.

It is always possible to multiply a wave function $\Psi(x,t)$ by an arbitrary phase factor without making any change to the predictions made using the wave function. For example, the probability of finding the particle in a small interval δx, centred on x, is given by $|\Psi(x,t)|^2\,\delta x$, and this is unchanged by multiplying $\Psi(x,t)$ by a phase factor. The modified stationary-state wave functions therefore describe the same behaviour as the original ones.

Ex 3.9 Dropping the indices n_x and n_y to simplify the notation, a stationary-state wave function takes the form $\Psi(x,y,t) = \psi(x,y)\mathrm{e}^{-\mathrm{i}Et/\hbar}$. The corresponding probability density is

$$\begin{aligned}\left|\Psi(x,y,t)\right|^2 &= \left|\psi(x,y)\mathrm{e}^{-\mathrm{i}Et/\hbar}\right|^2\\ &= \left|\psi(x,y)\right|^2\left|\mathrm{e}^{-\mathrm{i}Et/\hbar}\right|^2\\ &= \left|\psi(x,y)\right|^2,\end{aligned}$$

which is independent of time.

Ex 3.10 The energy of the $n_x = 7$, $n_y = 4$ state is given by

$$\begin{aligned}E_{n_x,n_y} &= \left(n_x^2 + n_y^2\right)\frac{\pi^2\hbar^2}{2mL^2}\\ &= 65\frac{\pi^2\hbar^2}{2mL^2}.\end{aligned}$$

To identify other states that are degenerate with this, we must find all the other pairs of positive integers (n_x, n_y) such that $n_x^2 + n_y^2 = 65$. The only possibilities are (4,7), (8,1) and (1,8). The energy level that corresponds to the $(7,4)$ state is therefore four-fold degenerate.

Ex 3.11 (a) From Equation 3.47, the lowest energy level and next-to-lowest energy level have energies: $E_{1,1,1} = (1+1+1)\pi^2\hbar^2/2mL^2$ and $E_{1,1,2} = (1+1+4)\pi^2\hbar^2/2mL^2$. It follows that the difference in energy between these two levels is given by $3\pi^2\hbar^2/2mL^2$.

(b) The lowest energy level is not degenerate, it corresponds to the unique state with the quantum numbers $(1,1,1)$. The next to lowest energy level is triply degenerate since it corresponds to three different states $(1,1,2)$, $(1,2,1)$ and $(2,1,1)$.

Ex 3.12 According to Equations 3.24, 3.46, and 3.49, stationary-state wave functions have acceptable SI units of $\mathrm{m}^{-1/2}$, m^{-1} and $\mathrm{m}^{-3/2}$ in one, two and three dimensions. Born's rule then shows that the corresponding probability densities, $|\Psi|^2$, have SI units of m^{-1}, m^{-2} and m^{-3}. This is appropriate because a probability density describes a probability *per unit length* in one dimension, a probability *per unit area* in two dimensions and a probability *per unit volume* in three dimensions. It means that integrals of probability density over lines, areas and volumes in one, two and three dimensions are equal to (unitless) probabilities.

Ex 3.13 The relevant time-independent Schrödinger equation is given in Equation 3.55. Substituting the given form of $\psi(x)$ into this equation and performing the necessary differentiations gives

$$\begin{aligned}&-\frac{\hbar^2\alpha^2}{2m}\left[A\mathrm{e}^{\alpha x} + B\mathrm{e}^{-\alpha x}\right]\\ &\quad = E\left[A\mathrm{e}^{\alpha x} + B\mathrm{e}^{-\alpha x}\right].\end{aligned}$$

This equation is satisfied provided that $\alpha^2 = -2mE/\hbar^2$. Since we are considering cases in which E is negative, the condition on α^2 is satisfied if $\alpha = \sqrt{2m(-E)}/\hbar$. Thus, $\psi(x)$ is a solution to the time-independent Schrödinger equation; moreover, it is the *general* solution because it contains two arbitrary constants, A and B, which is appropriate for a second-order differential equation.

Ex 4.1 The Hamiltonian operator involves differentiation with respect to x, but does not involve any differentiation with respect to time. When this operator is applied to a stationary-state wave function $\Psi_n(x,t) = \psi_n(x)\,e^{-iE_nt/\hbar}$, it does not affect the exponential phase factor, so we have

$$\begin{aligned}\widehat{H}\,\Psi_n(x,t) &= (\widehat{H}\,\psi_n(x))\,e^{-iE_nt/\hbar}\\ &= (E_n\,\psi_n(x))\,e^{-iE_nt/\hbar}\\ &= E_n\Psi_n(x,t).\end{aligned}$$

The stationary-state wave function satisfies the time-independent Schrödinger equation, and is therefore an energy eigenfunction with eigenvalue E_n. This is true at any time t, so if the energy of the system is measured, the value E_n will be obtained no matter what the time of the measurement.

Ex 4.2 There is no possibility of obtaining this energy because it is not one of the energy eigenvalues of a particle in a one-dimensional infinite square well.

Ex 4.3 The proof is essentially the same as that given in Exercise 2.9 of Chapter 2. We have

$$\begin{aligned}\widehat{H}\Psi - i\hbar\frac{\partial\Psi}{\partial t} &= \left(\widehat{H} - i\hbar\frac{\partial}{\partial t}\right)\Psi\\ &= \left(\widehat{H} - i\hbar\frac{\partial}{\partial t}\right)(a_1\Psi_1 + a_2\Psi_2 + \cdots)\\ &= a_1\left(\widehat{H}\Psi_1 - i\hbar\frac{\partial\Psi_1}{\partial t}\right)\\ &\quad + a_2\left(\widehat{H}\Psi_2 - i\hbar\frac{\partial\Psi_2}{\partial t}\right) + \cdots,\end{aligned}$$

where the last step follows because the Hamiltonian $\widehat{H}$ and $i\hbar\,\partial/\partial t$ are both linear operators. In the final expression, each of the terms in brackets vanishes because the stationary-state wave functions Ψ_i separately satisfy Schrödinger's equation. Hence Ψ also satisfies Schrödinger's equation.

Ex 4.4 We need to consider the integral

$$\begin{aligned}I_{nm} &= \int_{-\infty}^{\infty}\psi_n^*(x)\,\psi_m(x)\,dx\\ &= \frac{2}{L}\int_0^L \sin\left(\frac{n\pi x}{L}\right)\sin\left(\frac{m\pi x}{L}\right)dx.\end{aligned}$$

The integral can be evaluated by changing the variable of integration to $u = \pi x/L$. We have $dx = (L/\pi)\,du$, and the limits of integration become $u = 0$ and $u = \pi$, so

$$I_{nm} = \frac{2}{L}\times\frac{L}{\pi}\int_0^{\pi}\sin(nu)\sin(mu)\,du.$$

Finally, using a standard integral given inside the back cover of the book, we conclude that

$$I_{nm} = \frac{2}{L}\times\frac{L}{\pi}\times\frac{\pi}{2}\,\delta_{nm} = \delta_{nm} \quad \text{for all positive integers } n \text{ and } m,$$

which confirms that the eigenfunctions are orthonormal.

Ex 4.5 Neither the normalization condition nor the probabilities depend on time. We can therefore save some effort by performing the calculation at $t = 0$, when the time-dependent phase factors are both equal to 1. The normalization condition then gives $1 = |2A|^2 + |-3A|^2 = 13|A|^2$, so we can choose $A = 1/\sqrt{13}$. The probabilities of getting energies E_1 and E_2 are then $p_1 = |2/\sqrt{13}|^2 = 4/13$ and $p_2 = |-3/\sqrt{13}|^2 = 9/13$. As a final check, note that the sum of these two probabilities is equal to 1, as expected.

Ex 4.6 (a) Let

$$I = \int_{-a}^{a} f(x)\,dx,$$

where $f(x)$ is an odd function. If we change the variable to $y = -x$, the limits of integration reverse sign and $dx = -dy$, so the integral becomes

$$I = \int_{a}^{-a} f(-y)\,(-dy).$$

Swapping the limits of integration, and remembering that $f(-y) = -f(y)$, we then obtain

$$I = (-1)^3\int_{-a}^{a} f(y)\,dy = -I,$$

which is possible only if $I = 0$.

Comment: Figure S4.1 is a graphical interpretation of this result. It is clear that each contribution to the integral from a small interval with $x > 0$ is cancelled by a contribution of equal magnitude and opposite sign from a similar interval with $x < 0$.

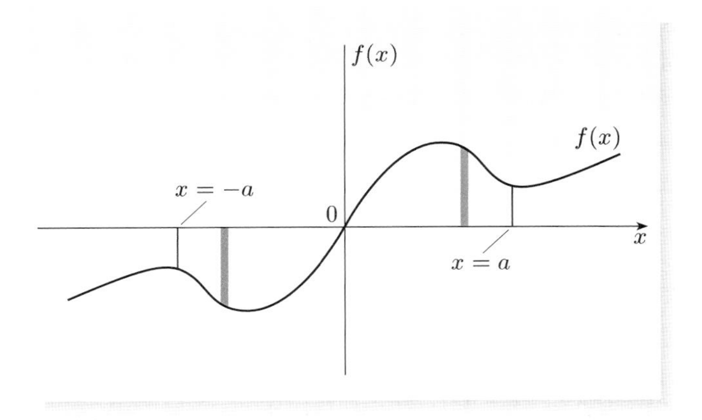

Figure S4.1 The definite integral of an odd function $f(x)$, taken over a range that is centred on $x = 0$, is equal to zero.

(b) The probability amplitude for getting energy E_2 in the given state is

$$c_2(0) = \int_{-\infty}^{\infty} \psi_2^*(x)\,\Psi(x,0)\,\mathrm{d}x$$
$$= \sqrt{\frac{2}{L}}\sqrt{\frac{30}{L^5}} \int_{-L/2}^{L/2} \sin\left(\frac{2\pi x}{L}\right)\left(x^2 - \frac{L^2}{4}\right)\mathrm{d}x.$$

This is equal to zero because the integrand is an odd function and the range of integration is centred on the origin, $x = 0$. The probability of getting energy E_2 is therefore equal to zero.

Ex 4.7 The probability amplitude for energy E_3 is

$$c_3(0) = \int_{-\infty}^{\infty} \psi_3^*(x)\,\Psi(x,0)\,\mathrm{d}x$$
$$= \sqrt{\frac{2}{L}}\sqrt{\frac{30}{L^5}} \int_{-L/2}^{L/2} \cos\left(\frac{3\pi x}{L}\right)\left(x^2 - \frac{L^2}{4}\right)\mathrm{d}x.$$

Following the method given in Worked Example 4.1, we change the variable of integration to $y = 3\pi x/L$. This gives

$$c_3(0) = \frac{\sqrt{60}}{L^3}\left[\left(\frac{L}{3\pi}\right)^3 \int_{-3\pi/2}^{3\pi/2} y^2 \cos y\,\mathrm{d}y - \frac{L^2}{4} \times \frac{L}{3\pi}\int_{-3\pi/2}^{3\pi/2} \cos y\,\mathrm{d}y\right].$$

Then, using a standard integral given inside the back cover of the book,

$$c_3(0) = \frac{\sqrt{60}}{L^3}\left[-\left(\frac{L}{3\pi}\right)^3 \frac{3^2\pi^2 - 8}{2} + \frac{L^3}{12\pi} \times 2\right]$$
$$= 4\frac{\sqrt{60}}{(3\pi)^3},$$

so the probability of energy E_3 is

$$p_3 = |c_3(0)|^2 = \frac{960}{(3\pi)^6} = 1.37 \times 10^{-3}.$$

Comment: Just over one measurement in a thousand gives the energy E_3. Energies like E_5 and E_7 are also possible, but even less likely.

Ex 4.8 According to the solution to Exercise 4.5, the probabilities of energies E_1 and E_2 are $p_1 = 4/13$ and $p_2 = 9/13$, and the probabilities of all other energies are equal to zero. Hence the expectation value of the energy is $\langle E \rangle = (4E_1 + 9E_2)/13$.

Ex 4.9 The wave function is the same as that in Worked Example 4.2. The time-dependent phase factors cancel out as before, so for p_x we are left with

$$\langle p_x \rangle = \frac{2}{L}\int_{-L/2}^{L/2} \cos\left(\frac{\pi x}{L}\right)\left(-\mathrm{i}\hbar\frac{\partial}{\partial x}\right)\cos\left(\frac{\pi x}{L}\right)\mathrm{d}x$$
$$= \frac{\mathrm{i}\hbar\pi}{L} \times \frac{2}{L}\int_{-L/2}^{L/2} \cos\left(\frac{\pi x}{L}\right)\sin\left(\frac{\pi x}{L}\right)\mathrm{d}x.$$

This integral vanishes because the integrand is an odd function of x and the range of integration is centred on the origin. Hence

$$\langle p_x \rangle = 0.$$

The expectation value of p_x^2 is given by

$$\langle p_x^2 \rangle = \frac{2}{L}\int_{-L/2}^{L/2} \cos\left(\frac{\pi x}{L}\right)\left(-\mathrm{i}\hbar\frac{\partial}{\partial x}\right)^2\cos\left(\frac{\pi x}{L}\right)\mathrm{d}x$$
$$= \frac{\pi^2\hbar^2}{L^2} \times \frac{2}{L}\int_{-L/2}^{L/2} \cos^2\left(\frac{\pi x}{L}\right)\mathrm{d}x.$$

The remaining integral is easily evaluated, but direct calculation can be avoided by noting that $(2/L)\int_{-L/2}^{L/2}\cos^2(\pi x/L)\,\mathrm{d}x$ is the normalization integral for the wave function, and so must be equal to 1. Hence

$$\langle p_x^2 \rangle = \frac{\pi^2\hbar^2}{L^2}.$$

Ex 4.10 The expectation value of the energy is given by

$$\langle E \rangle = \int_{-\infty}^{\infty} \Psi^*(x,0)\,\widehat{\mathrm{H}}\,\Psi(x,0)\,\mathrm{d}x$$
$$= \frac{30}{L^5}\int_{-L/2}^{L/2}\left(x^2 - \frac{L^2}{4}\right) \times \left(-\frac{\hbar^2}{2m}\frac{\partial^2}{\partial x^2}\right)\left(x^2 - \frac{L^2}{4}\right)\mathrm{d}x,$$

where we have used the fact that the wave function vanishes outside the well. Evaluating the second derivative, we obtain

$$\begin{aligned}\langle E\rangle &= \frac{30}{L^5}\int_{-L/2}^{L/2}\left(x^2-\frac{L^2}{4}\right)\left(-\frac{\hbar^2}{m}\right)\mathrm{d}x\\ &= -\frac{30\hbar^2}{mL^5}\left[\frac{x^3}{3}-\frac{L^2x}{4}\right]_{-L/2}^{L/2}\\ &= -\frac{30\hbar^2}{mL^5}\times 2\left(\frac{L^3}{3\times 8}-\frac{L^3}{8}\right)\\ &= \frac{5\hbar^2}{mL^2}.\end{aligned}$$

Comment: The expectation value is slightly greater than the ground-state energy, $\pi^2\hbar^2/2mL^2$. This agrees with the results of Worked Example 4.1 and Exercises 4.6 and 4.7, which showed that there is a very high probability of obtaining the ground-state energy in this state, and much smaller probabilities of obtaining higher energies. No matter what the state of the system, the expectation value of the energy is always greater than or equal to the ground-state energy.

Ex 4.11 Any measurement of energy in the ith stationary state is certain to yield the corresponding energy eigenvalue E_i. Consequently, $\langle E\rangle = E_i$ and $\langle E^2\rangle = E_i^2$. The uncertainty in energy is therefore

$$\Delta E = \sqrt{\langle E^2\rangle - \langle E\rangle^2} = \sqrt{E_i^2 - E_i^2} = 0.$$

Ex 4.12 The square of the uncertainty in energy is

$$(\Delta E)^2 = \langle E^2\rangle - \langle E\rangle^2.$$

The probabilities of getting energies E_1 and E_2 in the given state are each equal to $1/2$, so

$$\langle E\rangle = \tfrac{1}{2}E_1 + \tfrac{1}{2}E_2 \quad\text{and}\quad \langle E^2\rangle = \tfrac{1}{2}E_1^2 + \tfrac{1}{2}E_2^2.$$

Hence

$$\begin{aligned}(\Delta E)^2 &= \tfrac{1}{2}(E_1^2+E_2^2) - \tfrac{1}{4}(E_1+E_2)^2\\ &= \tfrac{1}{4}(E_1^2+E_2^2-2E_1E_2)\\ &= \tfrac{1}{4}(E_2-E_1)^2,\end{aligned}$$

so

$$\Delta E = \tfrac{1}{2}(E_2-E_1).$$

Ex 4.13 In the absence of thermal energy, the kinetic energy is entirely due to quantum effects (arising from the uncertainty principle). Using Equation 4.37, the average kinetic energy of a helium atom has a minimum value of order

$$\begin{aligned}\frac{23\hbar^2}{2mL^2} &= \frac{23\times(1.06\times 10^{-34}\,\mathrm{J\,s})^2}{2\times 6.7\times 10^{-27}\,\mathrm{kg}\times(3.5\times 10^{-10}\,\mathrm{m})^2}\\ &= 1.6\times 10^{-22}\,\mathrm{J}.\end{aligned}$$

Comment: This is roughly equal to the average thermal energy of an atom at a temperature of 8 K. At temperatures well below this, liquid helium is dominated by quantum effects, and it is called a *quantum liquid*.

Ex 4.14 As suggested in the question, we take $\Delta x = s$, where s is the typical electron–proton separation. The uncertainty principle then shows that a typical momentum component in the x-direction obeys $p_x \simeq \Delta p_x \geq \hbar/2s$, with similar results applying in the other two directions. Consequently, in quantum mechanics, the typical energy obeys the inequality

$$E \geq \frac{3\hbar^2}{8ms^2} - \frac{e^2}{4\pi\varepsilon_0 s}.$$

Differentiating with respect to s to find the minimum of the right-hand side gives

$$-\frac{3\hbar^2}{4ms^3} + \frac{e^2}{4\pi\varepsilon_0 s^2} = 0,$$

so

$$\begin{aligned}s &= \frac{3\pi\varepsilon_0\hbar^2}{me^2}\\ &= \frac{3\pi\times 8.85\times 10^{-12}\times(1.06\times 10^{-34})^2}{9.11\times 10^{-31}\times(1.60\times 10^{-19})^2}\,\mathrm{m}\\ &= 4.0\times 10^{-11}\,\mathrm{m}.\end{aligned}$$

Substituting into the expression for the energy, we obtain

$$\begin{aligned}E_{\text{min}} &= \frac{3\hbar^2}{8m}\frac{m^2e^4}{(3\pi\varepsilon_0\hbar^2)^2} - \frac{e^2}{4\pi\varepsilon_0}\frac{me^2}{3\pi\varepsilon_0\hbar^2}\\ &= -\frac{2me^4}{3(4\pi\varepsilon_0\hbar)^2}\\ &= -\frac{2\times 9.11\times 10^{-31}\times(1.60\times 10^{-19})^4}{3(4\pi\times 8.85\times 10^{-12}\times 1.06\times 10^{-34})^2}\,\mathrm{J}\\ &= -2.9\times 10^{-18}\,\mathrm{J}.\end{aligned}$$

These values of s and E_{min} are our estimates for the radius and ground-state energy of a hydrogen atom. The ground-state energy is negative because the potential energy is zero when the particles are infinitely far apart. A bound hydrogen atom therefore has negative potential energy.

Comment: Our estimates turn out to be almost correct. Full solutions of Schrödinger's equation show that s is too small by a factor of $3/4$, while $|E_{\text{min}}|$ is too large by a factor of $4/3$. Such errors are only to be expected, given the roughness of our approximations, but we have still achieved something remarkable. Using only a classical expression for the total energy of an electron interacting with a proton, together with the uncertainty principle, we have obtained good order-of-magnitude estimates of the ground-state energy and radius of a hydrogen atom!

Ex 5.1 The particle comes instantaneously to rest at its points of maximum displacement, where $x = \pm A$. At these points, the kinetic energy is zero, and the potential energy is $\frac{1}{2}CA^2$, so the total energy is

$$E = 0 + \tfrac{1}{2}CA^2 = \tfrac{1}{2}CA^2.$$

This remains constant throughout the oscillation, by the conservation of energy.

Comment: Figure S5.1 is a graphical interpretation of this result. This figure compares the potential energy function $\frac{1}{2}Cx^2$ with the total energy E. The kinetic energy is non-negative, so the potential energy is always less than, or equal to, the total energy E. The particle is therefore confined to the region where $\frac{1}{2}Cx^2 \leq E$. This inequality becomes an equality at the points of maximum displacement, $x = \pm A$, where $E = \frac{1}{2}CA^2$.

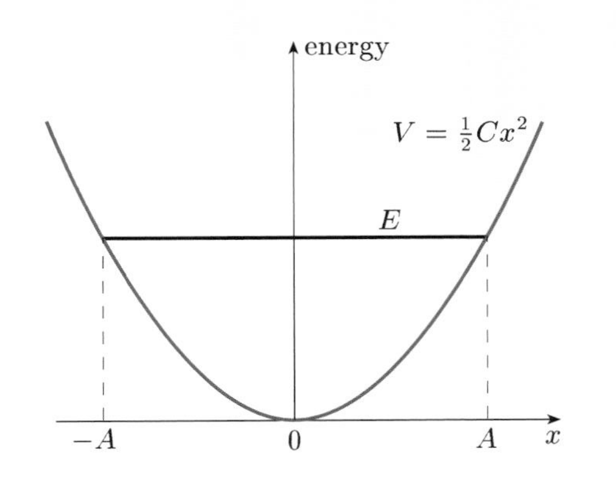

Figure S5.1 An oscillating particle is confined to the region $-A \leq x \leq A$, where the potential energy is less than or equal to the total energy E.

Ex 5.2 Differentiating Equation 5.4 with respect to time,

$$v_x = \frac{\mathrm{d}x}{\mathrm{d}t} = -\omega_0 A \sin(\omega_0 t + \phi),$$

so

$$p_x = -m\omega_0 A \sin(\omega_0 t + \phi).$$

Substituting into Equation 5.6, and using the relationship $\omega_0 = \sqrt{C/m}$, the total energy is

$$\begin{aligned} E &= \tfrac{1}{2}m\omega_0^2 A^2 \sin^2(\omega_0 t + \phi) + \tfrac{1}{2}CA^2 \cos^2(\omega_0 t + \phi) \\ &= \tfrac{1}{2}CA^2\left(\sin^2(\omega_0 t + \phi) + \cos^2(\omega_0 t + \phi)\right) \\ &= \tfrac{1}{2}CA^2, \end{aligned}$$

which is independent of time, as required (and agrees with the result of the preceding exercise).

Ex 5.3 Using Equations 5.3 and 5.5, and recalling that the appropriate mass is the reduced mass, gives

$$\omega_0 = 2\pi f = \sqrt{\frac{C}{\mu}},$$

so

$$C = 4\pi^2 f^2 \mu.$$

Let the mass of a hydrogen atom be m. Then the reduced mass of a hydrogen molecule is

$$\mu = \frac{m^2}{2m} = \frac{m}{2},$$

and we have

$$\begin{aligned} C &= 4\pi^2 f^2 \frac{m}{2} \\ &= 2\pi^2 (1.25 \times 10^{14}\,\text{Hz})^2 \times 1.67 \times 10^{-27}\,\text{kg} \\ &= 520\,\text{N}\,\text{m}^{-1}. \end{aligned}$$

This is comparable to the force constant of a stiff rubber band.

Ex 5.4 Substituting $\Psi(x,t) = \psi(x)\,\mathrm{e}^{-\mathrm{i}Et/\hbar}$ into the two sides of Schrödinger's equation gives

$$\begin{aligned} \mathrm{i}\hbar\frac{\partial\Psi}{\partial t} &= \mathrm{i}\hbar\left(-\frac{\mathrm{i}E}{\hbar}\right)\psi(x)\,\mathrm{e}^{-\mathrm{i}Et/\hbar} \\ &= E\,\psi(x)\,\mathrm{e}^{-\mathrm{i}Et/\hbar} \end{aligned}$$

and

$$\widehat{\mathrm{H}}\Psi = \left[-\frac{\hbar^2}{2m}\frac{\mathrm{d}^2\psi}{\mathrm{d}x^2} + \tfrac{1}{2}m\omega_0^2 x^2\,\psi(x)\right]\mathrm{e}^{-\mathrm{i}Et/\hbar}.$$

Equating these expressions and cancelling the exponential factor on both sides, we see that $\psi(x)$ satisfies the time-independent Schrödinger equation for a harmonic oscillator. (Equation 5.15).

Ex 5.5 The normalization condition gives

$$1 = \int_{-\infty}^{\infty} |\psi_0(x)|^2 \, dx$$
$$= \int_{-\infty}^{\infty} \left| C_0 e^{-x^2/2a^2} \right|^2 dx$$
$$= |C_0|^2 \int_{-\infty}^{\infty} e^{-x^2/a^2} \, dx.$$

Changing the variable of integration from x to $y = x/a$, and noting that $dx = a\,dy$, we obtain

$$1 = |C_0|^2 a \int_{-\infty}^{\infty} e^{-y^2} \, dy.$$

According to the list given inside the back cover, the remaining integral is equal to $\sqrt{\pi}$, so we have

$$|C_0|^2 = \frac{1}{\sqrt{\pi}a}.$$

The result quoted in the question is obtained by taking the square root of both sides of this equation, choosing C_0 to be real and positive as usual.

Ex 5.6 (a) The dimensions of $\hbar$ are those of energy $\times$ time $= [M][L]^2[T]^{-1}$, so the dimensions of $a = \sqrt{\hbar/m\omega_0}$ are

$$\sqrt{\frac{[M][L]^2[T]^{-1}}{[M][T]^{-1}}} = [L].$$

(b) Exercise 5.1 showed that a *classical* harmonic oscillator with amplitude A has energy $E = \frac{1}{2}CA^2$. So, at energy $E_0 = \hbar\omega_0/2$, the classical amplitude is

$$A = \sqrt{\frac{2E_0}{C}} = \sqrt{\frac{\hbar\omega_0}{C}}.$$

Using the relationship $\omega_0 = \sqrt{C/m}$ to eliminate C, we obtain

$$A = \sqrt{\frac{\hbar\omega_0}{m\omega_0^2}} = \sqrt{\frac{\hbar}{m\omega_0}} = a,$$

which is the length parameter of the oscillator.

Ex 5.7 Probabilities do not depend on time for stationary states, so we can carry out the calculation at time $t = 0$, when the stationary-state wave function is equal to the energy eigenfunction $\psi_0(x)$. The probability P_{out} of finding the particle in the classically forbidden region is the total area under the graph of $|\psi_0(x)|^2$ in the regions $-\infty < x < -a$ and $a < x < \infty$. Thus

$$P_{\text{out}} = \int_{-\infty}^{-a} |\psi_0(x)|^2 \, dx + \int_{a}^{\infty} |\psi_0(x)|^2 \, dx.$$

The eigenfunction $\psi_0(x)$ is an even function, so we have

$$P_{\text{out}} = 2\int_{a}^{\infty} |\psi_0(x)|^2 \, dx = \frac{2}{\sqrt{\pi}a} \int_{a}^{\infty} e^{-x^2/a^2} \, dx.$$

Changing the variable of integration to $y = x/a$, as in Exercise 5.5, and using the definite integral given in the question, we conclude that

$$P_{\text{out}} = \frac{2}{\sqrt{\pi}} \int_{1}^{\infty} e^{-y^2} \, dy = \frac{2}{\sqrt{\pi}} \times 0.139 = 0.157.$$

So there is a reasonable chance (nearly 1 in 6) of finding the particle in the classically forbidden region $|x| > a$.

Ex 5.8 Acting on $\widehat{A}\,\psi_n(x)$ with the Hamiltonian operator gives

$$\widehat{H}\big[\widehat{A}\,\psi_n(x)\big] = \big(\widehat{A}^\dagger \widehat{A} + \tfrac{1}{2}\big)\hbar\omega_0 \big[\widehat{A}\,\psi_n(x)\big].$$

We make use of the commutation relation $\widehat{A}\,\widehat{A}^\dagger - \widehat{A}^\dagger \widehat{A} = 1$ (Equation 5.32) to write

$$\widehat{A}^\dagger \widehat{A} = \widehat{A}\,\widehat{A}^\dagger - 1,$$

and hence

$$\widehat{H}\big[\widehat{A}\,\psi_n(x)\big] = \big(\widehat{A}\,\widehat{A}^\dagger - 1 + \tfrac{1}{2}\big)\hbar\omega_0 \big[\widehat{A}\,\psi_n(x)\big].$$

Then, pulling out a factor $\widehat{A}$ from terms on the right-hand side, we obtain

$$\begin{aligned}\widehat{H}\big[\widehat{A}\,\psi_n(x)\big] &= \widehat{A}\big(\widehat{A}^\dagger \widehat{A} - 1 + \tfrac{1}{2}\big)\hbar\omega_0\, \psi_n(x) \\ &= \widehat{A}\big(\widehat{H} - \hbar\omega_0\big)\, \psi_n(x) \\ &= \widehat{A}\big(E_n - \hbar\omega_0\big)\, \psi_n(x) \\ &= \big(E_n - \hbar\omega_0\big)\big[\widehat{A}\,\psi_n(x)\big],\end{aligned}$$

the required result.

Ex 5.9 We use Equation 5.45 to express the required expectation value in terms of raising and lowering operators:

$$\langle p_x^2 \rangle = -\frac{\hbar^2}{2a^2} \int_{-\infty}^{\infty} \psi_n^*(x)\, \big(\widehat{A} - \widehat{A}^\dagger\big)^2\, \psi_n(x) \, dx.$$

Following the same approach as in Worked Example 5.2, we obtain

$$\langle p_x^2 \rangle = \frac{\hbar^2}{2a^2} \int_{-\infty}^{\infty} \psi_n^*(x)\, \big(\widehat{A}^\dagger \widehat{A} + \widehat{A}\,\widehat{A}^\dagger\big)\, \psi_n(x) \, dx.$$

The integral is exactly the same as that evaluated in the worked example, and has value $(2n+1)$, so we conclude that

$$\langle p_x^2 \rangle = \left(n + \tfrac{1}{2}\right) \frac{\hbar^2}{a^2},$$

as required.

Ex 5.10 Let I be the integral on the right-hand side of Equation 5.53. Then, in terms of raising and lowering operators, we have

$$\begin{aligned} I &= \frac{a}{\sqrt{2}} \int_{-\infty}^{\infty} \psi_n^*(x) \left(\widehat{\mathrm{A}} + \widehat{\mathrm{A}}^\dagger\right) \psi_m(x)\,\mathrm{d}x \\ &= \frac{a}{\sqrt{2}} \left[\alpha \int_{-\infty}^{\infty} \psi_n^*(x)\, \psi_{m-1}(x)\,\mathrm{d}x \right. \\ &\qquad \left. + \beta \int_{-\infty}^{\infty} \psi_n^*(x)\, \psi_{m+1}(x)\,\mathrm{d}x \right], \end{aligned}$$

where α and β are constants (whose values are not needed). Using the fact that eigenfunctions with different eigenvalues are orthogonal, we see that this integral is equal to zero unless $n = m-1$ or $n = m+1$, corresponding to transitions between neighbouring energy levels.

Ex 6.1 Integrating both sides of Equation 6.4, we obtain

$$\begin{aligned} \int_{-\infty}^{\infty} \Psi^*(x,t)\Psi(x,t)\,\mathrm{d}x &= |a_1|^2 \int_{-\infty}^{\infty} \psi_1^*(x)\psi_1(x)\,\mathrm{d}x \\ &+ |a_2|^2 \int_{-\infty}^{\infty} \psi_2^*(x)\psi_2(x)\,\mathrm{d}x \\ &+ a_1^* a_2 \mathrm{e}^{-\mathrm{i}(E_2-E_1)t/\hbar} \int_{-\infty}^{\infty} \psi_1^*(x)\psi_2(x)\,\mathrm{d}x \\ &+ a_2^* a_1 \mathrm{e}^{\mathrm{i}(E_2-E_1)t/\hbar} \int_{-\infty}^{\infty} \psi_2^*(x)\psi_1(x)\,\mathrm{d}x. \end{aligned}$$

The integrals in the first two terms are both equal to 1 because the energy eigenfunctions are normalized. The integrals in the last two terms are both equal to zero because different energy eigenfunctions are orthogonal. Using Equation 6.3, we therefore conclude that

$$\int_{-\infty}^{\infty} \Psi^*(x,t)\Psi(x,t)\,\mathrm{d}x = |a_1|^2 + |a_2|^2 = 1,$$

as required.

Comment: The fact that the linear combination of stationary-state wave functions is normalized entitles us to call it a *wave packet*. Note that the wave packet remains normalized at all times because the coefficients a_1 and a_2 are constants.

Ex 6.2 At time $t = T = 2\pi/\omega_0$, the phase factors in Equation 6.7 are $\mathrm{e}^{-\mathrm{i}\omega_0 t/2} = \mathrm{e}^{-\mathrm{i}\pi} = -1$ and $\mathrm{e}^{-3\mathrm{i}\omega_0 t} = \mathrm{e}^{-3\mathrm{i}\pi} = -1$. Hence

$$\Psi_\mathrm{A}(x,T) = -\frac{1}{\sqrt{2}}\left[\psi_0(x) + \psi_1(x)\right] = -\Psi_\mathrm{A}(x,0).$$

(This reversal in sign has no physical consequences; it is an example of multiplying a whole wave function by a phase factor, a process which never has any physical consequences.)

For $t = 2T$, the phase factors are $\mathrm{e}^{-2\mathrm{i}\pi} = 1$ and $\mathrm{e}^{-6\mathrm{i}\pi} = 1$, so $\Psi_\mathrm{A}(x,2T) = \Psi_\mathrm{A}(x,0)$ and the wave function returns to its initial value.

Ex 6.3 Using Equations 6.8 and 6.9, we find that

$$\Psi_\mathrm{A}(x,0) = \frac{1}{\sqrt{2}}\left(\frac{1}{\sqrt{\pi}a}\right)^{1/2}\left[1 + \sqrt{2}\,\frac{x}{a}\right] e^{-x^2/2a^2},$$

and

$$\Psi_\mathrm{A}(x,T/2) = \frac{\mathrm{i}}{\sqrt{2}}\left(\frac{1}{\sqrt{\pi}a}\right)^{1/2}\left[-1 + \sqrt{2}\,\frac{x}{a}\right] e^{-x^2/2a^2}.$$

In $\Psi_\mathrm{A}(x,0)$, the two terms in square brackets have the same sign for $x > 0$ (constructive interference) and opposite signs for $x < 0$ (destructive interference), so the probability density $|\Psi_\mathrm{A}|^2$ is concentrated on the right-hand side of the well. In $\Psi_\mathrm{A}(x,T/2)$, the two terms in square brackets have opposite signs for $x > 0$ (destructive interference) and the same sign for $x < 0$ (constructive interference), so the probability density $|\Psi_\mathrm{A}|^2$ is concentrated on the left-hand side of the well.

Also note that our explicit expressions show that $\Psi(x,T/2) = -\mathrm{i}\Psi(-x,0)$, so the probability density at $T/2$ is the reflection of the probability density at $t = 0$, in agreement with Figure 6.1.

Ex 6.4 We have

$$\langle x \rangle = \frac{a}{\sqrt{2}}\cos(\omega_0 t) \quad \text{and} \quad \langle p_x \rangle = -\frac{\hbar}{\sqrt{2}a}\sin(\omega_0 t)$$

so

$$m\frac{\mathrm{d}\langle x \rangle}{\mathrm{d}t} = -\frac{ma\omega_0}{\sqrt{2}}\sin(\omega_0 t) = \frac{ma^2\omega_0}{\hbar}\langle p_x \rangle.$$

Since $a = \sqrt{\hbar/m\omega_0}$, we conclude that $m\,\mathrm{d}\langle x \rangle/\mathrm{d}t = \langle p_x \rangle$.

Ex 6.5 The energy eigenfunctions $\psi_1(x)$ and $\psi_3(x)$ are both odd functions of x, so $\Psi(x,t)$ and $\Psi^*(x,t)$ are also odd functions, with $\Psi(-x,t) = -\Psi(x,t)$ and $\Psi^*(-x,t) = -\Psi^*(x,t)$. Hence

$$\begin{aligned} |\Psi(-x,t)|^2 &= \Psi^*(-x,t)\Psi(-x,t) \\ &= (-1)^2\Psi^*(x,t)\Psi(x,t) \\ &= |\Psi(x,t)|^2. \end{aligned}$$

So the probability density is an even function of x, and is therefore symmetrical about the centre of the well at all times.

The expectation value of position is given by the sandwich integral

$$\langle x\rangle = \int_{-\infty}^{\infty} \Psi^*(x,t)x\Psi(x,t)\,\mathrm{d}x,$$

which is equal to zero because the integrand is an odd function (being the product of three odd functions) and the range of integration is symmetrical about the origin. (Alternatively, we could note that the operator $\widehat{\mathrm{A}} + \widehat{\mathrm{A}}^\dagger$, acting on a linear combination of $\psi_1(x)$ and $\psi_3(x)$, produces a linear combination of $\psi_0(x)$, $\psi_2(x)$ and $\psi_4(x)$, which is orthogonal to the original function. This again implies that $\langle x\rangle = 0$.)

The wave packet does not oscillate to and fro across the well, but breathes in and out with an uncertainty in position that increases and decreases cyclically.

Ex 6.6 We have $\partial V/\partial x = \partial(mgx)/\partial x = mg$. Combining **E1** and **E2** then gives

$$\frac{\mathrm{d}^2\langle x\rangle}{\mathrm{d}t^2} = \frac{1}{m}\frac{\mathrm{d}\langle p_x\rangle}{\mathrm{d}t} = -\frac{1}{m}\left\langle\frac{\partial V}{\partial x}\right\rangle = -g.$$

The general solution of this differential equation is

$$\langle x\rangle = A + Bt - \tfrac{1}{2}gt^2,$$

where A and B are constants that depend on the initial conditions, i.e. the values of $\langle x\rangle$ and $\mathrm{d}\langle x\rangle/\mathrm{d}t$ at time $t = 0$. This solution remains valid so long as the potential energy function is $V(x) = mgx$. In practice, this ceases to be so when the wave packet describing the state of the particle extends as far as the ground, since forces other than gravity then come into play.

Ex 6.7 The initial wave function $\Psi(x,0)$ is symmetric, so it is an even function of x. The energy eigenfunctions in the harmonic well are either even or odd functions of x: $\psi_0(x), \psi_2(x), \ldots$ are even while $\psi_1(x), \psi_3(x), \ldots$ are odd. If we now consider the overlap integral of Equation 6.30,

$$a_i = \int_{-\infty}^{\infty} \psi_i^*(x)\Psi(x,0)\,\mathrm{d}x$$

we see that when $i = 1, 3, 5, \ldots$, the integrand is the product of an odd function $\psi_i^*(x)$ and an even function $\Psi(x,0)$, and so is an odd function. The range of integration is symmetrical about the origin, so the integral vanishes in this case, and we conclude that $a_i = 0$ for $i = 1, 3, 5, \ldots$. From Equation 6.26, the wave function at future times is therefore given by

$$\Psi(x,t) = a_0\psi_0(x)\mathrm{e}^{-\mathrm{i}E_0t/\hbar} + a_2\psi_2(x)\mathrm{e}^{-\mathrm{i}E_2t/\hbar} + \ldots$$

This only involves the even functions $\psi_0(x), \psi_2(x), \ldots$, so $\Psi(x,t)$ is itself an even function, with $\Psi(-x,t) = \Psi(x,t)$ for all t. Hence, the wave function remains symmetric at all times.

Ex 6.8 The expectation value of momentum is given by the sandwich integral

$$\langle p_x\rangle = \int_{-\infty}^{\infty} \Psi^*(x,0)\widehat{\mathrm{p}}_x\Psi(x,0)\,\mathrm{d}x,$$

where $\Psi(x,0) = a_2\psi_2(x) + a_5\psi_5(x)$. The momentum operator $\widehat{\mathrm{p}}_x$ can be expressed as a linear combination of lowering and raising operators. When these operators act on the harmonic oscillator eigenfunctions $\psi_2(x)$ and $\psi_5(x)$, they produce terms proportional to $\psi_1(x)$ and $\psi_3(x)$ in the first case, and $\psi_4(x)$ and $\psi_6(x)$ in the second case. Each of these is orthogonal to the original eigenfunctions $\psi_2(x)$ and $\psi_5(x)$. Using these results, the above sandwich integral produces a sum of terms each of which is equal to zero (by orthogonality). Hence $\langle p_x\rangle = 0$.

Comment: This answer is complete. If you would prefer a more explicit derivation, you could supply one using the above sandwich integral and Equations 6.12, 6.13, 6.15 and 6.29. Symmetry arguments are ineffective in this case because the wave function $\Psi(x,0)$ is neither even nor odd.

Ex 6.9 Since $E_k = \hbar^2k^2/2m$, the right-hand sides of the two equations given in the question are equal. Hence,

$$-\frac{\hbar^2}{2m}\frac{\partial^2\Psi}{\partial x^2} = \mathrm{i}\hbar\frac{\partial\Psi}{\partial t},$$

so $\Psi(x,t)$ is a solution of the free-particle Schrödinger equation.

Ex 6.10 Born's rule for momentum tells us that the probability of finding the momentum in a small interval of momentum $\hbar\,\delta k$, centred on $\hbar k$, is $|A(k)|^2\,\delta k$.

Consequently, the expectation value of the momentum is

$$\langle p_x \rangle = \int_{-\infty}^{\infty} \hbar k |A(k)|^2 \, dk,$$

and the expectation value of the kinetic energy is

$$\langle E_{\text{kin}} \rangle = \int_{-\infty}^{\infty} \frac{\hbar^2 k^2}{2m} |A(k)|^2 \, dk.$$

In a free-particle wave packet, the momentum amplitude function $A(k)$ is a time-independent coefficient used in a linear superposition of de Broglie wave functions. In more physical terms, the momentum distribution remains fixed because there are no forces acting to change it. It follows that $\langle p_x \rangle$ and $\langle E_{\text{kin}} \rangle$ are independent of time for any free-particle wave packet.

Ex 6.11 Including the factor $1/\sqrt{2\pi}$ in the momentum eigenfunction simplifies several other equations. It allows Born's rule for momentum to be expressed in terms of $|A(k)|^2 \, \delta k$ and it allows the momentum amplitude $A(k)$ to be expressed as a simple overlap integral $\int_{-\infty}^{\infty} \psi_k^*(x)\Psi(x,0) \, dx$.

Ex 6.12 Since $\Delta x = a/\sqrt{2}$, we find that $a = \sqrt{2} \times 10^{-3}$ m at $t = 0$. We wish to know Δx at $t = 5$ hours $= 1.8 \times 10^4$ s. Using Equation 6.42, and taking values of the required constants from inside the back cover, we get:

$$\Delta x \simeq \frac{1.06 \times 10^{-34}\,\text{J s} \times 1.8 \times 10^4\,\text{s}}{\sqrt{2} \times \sqrt{2} \times 10^{-3}\,\text{m} \times 9.11 \times 10^{-31}\,\text{kg}}$$
$$= 1.05 \times 10^3\,\text{m}.$$

Comment: This wave packet would take about 10 minutes to stroll across! This is remarkable, though it does assume that the electron is isolated; even in space, a real electron would encounter atoms that would make the wave packet collapse long before it reached that size. This 'collapse' is what happens when a detecting screen makes an electron 'choose' the place where it appears. Remember that the size of the wave packet is not the size of an electron, only a measure of the size of the region in which it might be found in a position measurement.

Ex 7.1 The two bumps in the probability density have roughly equal areas. This indicates that the probability of transmission is roughly equal to the probability of reflection. Since the sum of these two probabilities must be equal to 1, the probability of each outcome is about 0.5.

Ex 7.2 In view of the quantum behaviour of individual particles (as represented by wave packets) when they meet a finite square barrier, it is reasonable to expect that there is some chance that the particles encountering a finite square step will be reflected. In the case of quantum scattering we should therefore expect the outcome to include a reflected beam as well as a transmitted beam, even though $E_0 > V_0$.

Ex 7.3 (a) From Equations 7.21 and 7.22,

$$R + T = \frac{(k_1 - k_2)^2 + 4k_1k_2}{(k_1 + k_2)^2}$$
$$= \frac{k_1^2 + k_2^2 + 2k_1k_2}{(k_1 + k_2)^2}$$
$$= 1.$$

(b) From Equation 7.13, we have

$$k_1 = \frac{\sqrt{4mV_0}}{\hbar} \quad \text{and} \quad k_2 = \frac{\sqrt{2mV_0}}{\hbar}.$$

So, in this case, $k_1 = \sqrt{2}k_2$. Therefore

$$R = \frac{(k_1 - k_2)^2}{(k_1 + k_2)^2} = \frac{(\sqrt{2} - 1)^2}{(\sqrt{2} + 1)^2} = 0.03,$$

$$T = \frac{4k_1k_2}{(k_1 + k_2)^2} = \frac{4\sqrt{2}}{(\sqrt{2} + 1)^2} = 0.97.$$

So we can check that $R + T = 1$ in this case.

Ex 7.4 (a) When $k_2 = k_1/2$, we have

$$B = \frac{k_1 - k_2}{k_1 + k_2} A = \frac{A}{3} \quad \text{and} \quad C = \frac{2k_1}{k_1 + k_2} A = \frac{4A}{3}.$$

(b) From Equation 7.15 with $D = 0$,

$$|\Psi|^2 = |C e^{i(k_2 x - \omega t)}|^2 = |C|^2 = \frac{16}{9}|A|^2.$$

Similarly, from Equation 7.14,

$$|\Psi|^2 = \left| A e^{i(k_1 x - \omega t)} + B e^{-i(k_1 x + \omega t)} \right|^2$$
$$= \left| e^{-i\omega t} \left(A e^{ik_1 x} + B e^{-ik_1 x} \right) \right|^2$$
$$= \left| e^{-i\omega t} \right|^2 \times \left| A \left(e^{ik_1 x} + \tfrac{1}{3} e^{-ik_1 x} \right) \right|^2.$$

Since $|e^{-i\omega t}| = 1$, we have

$$|\Psi|^2 = |A|^2 \left(e^{ik_1 x} + \tfrac{1}{3} e^{-ik_1 x} \right)^* \left(e^{ik_1 x} + \tfrac{1}{3} e^{-ik_1 x} \right)$$
$$= |A|^2 (e^{-ik_1 x} + \tfrac{1}{3} e^{ik_1 x})(e^{ik_1 x} + \tfrac{1}{3} e^{-ik_1 x}).$$

Multiplying out the brackets, we find

$$|\Psi|^2 = |A|^2\left[1 + \tfrac{1}{9} + \tfrac{1}{3}(e^{-2ik_1x} + e^{2ik_1x})\right]$$
$$= |A|^2\left[\tfrac{10}{9} + \tfrac{2}{3}\cos(2k_1x)\right].$$

(c) The variation indicated by the cosine-dependence to the left of the step is a result of *interference* between the incident and reflected beams. The presence of interference effects was noted earlier when we were discussing the scattering of wave packets but there the effect was transitory. In the stationary-state approach interference is a permanent feature.

(d) The linear number densities in the incident, reflected and transmitted beams are given by $|A|^2$, $|B|^2$ and $|C|^2$. The question tells us that $|A|^2 = 1.00 \times 10^{24}\,\text{m}^{-1}$, so the linear number density in the reflected and transmitted beams are

$$|B|^2 = \frac{|A|^2}{9} = 1.11 \times 10^{23}\,\text{m}^{-1},$$

$$|C|^2 = \frac{16|A|^2}{9} = 1.78 \times 10^{24}\,\text{m}^{-1}.$$

Note that the transmitted beam is denser than the incident beam: $|C|^2 > |A|^2$. However, since $k_2 = k_1/2$, we have $j_{\text{trans}} < j_{\text{inc}}$. The transmitted beam is less intense than the incident beam because it travels much more slowly.

Ex 7.5 A suitable graph is shown in Figure S7.1.

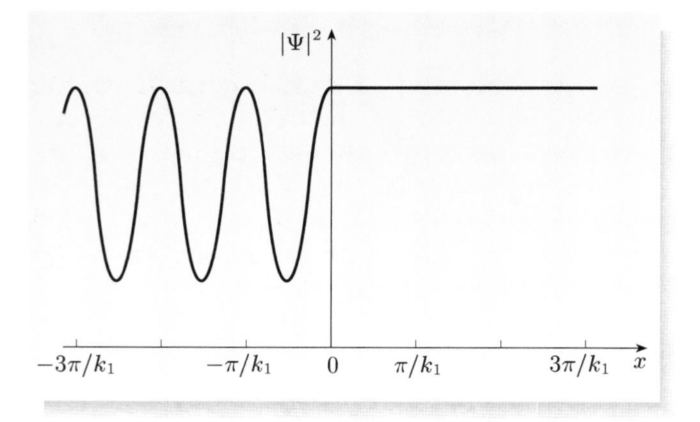

Figure S7.1 $|\Psi|^2$ plotted against x for a finite square step at $x = 0$ when $E_0 > V_0$.

Ex 7.6 One way of showing that a quantity is real is to show that it is equal to its own complex conjugate. Taking the complex conjugate of each factor in j_x, we obtain

$$j_x^*(x,t) = +\frac{i\hbar}{2m}\left(\Psi\frac{\partial\Psi^*}{\partial x} - \Psi^*\frac{\partial\Psi}{\partial x}\right)$$
$$= -\frac{i\hbar}{2m}\left(\Psi^*\frac{\partial\Psi}{\partial x} - \Psi\frac{\partial\Psi^*}{\partial x}\right) = j_x(x,t).$$

Since $j_x^* = j_x$, we conclude that j_x is real.

Ex 7.7 In the region $x > 0$,

$$\psi(x) = Ce^{ik_2x} \quad \text{and} \quad \psi^*(x) = C^*e^{-ik_2x},$$

so Equation 7.29 gives the probability current

$$j_x = -\frac{i\hbar}{2m}\left[C^*e^{-ik_2x}\left(ik_2Ce^{ik_2x}\right) - Ce^{ik_2x}\left(-ik_2C^*e^{-ik_2x}\right)\right]$$
$$= -\frac{i\hbar}{2m}\left[C^*C(ik_2 + ik_2)\right]$$
$$= \frac{\hbar k_2}{m}|C|^2.$$

In the region $x \le 0$,

$$\psi(x) = Ae^{ik_1x} + Be^{-ik_1x}$$

and

$$\psi^*(x) = A^*e^{-ik_1x} + B^*e^{ik_1x},$$

so Equation 7.29 gives the probability current

$$j_x = -\frac{i\hbar}{2m}\Big[\left(A^*e^{-ik_1x} + B^*e^{ik_1x}\right) \times \left(ik_1\right)\left(Ae^{ik_1x} - Be^{-ik_1x}\right) - \left(Ae^{ik_1x} + Be^{-ik_1x}\right) \times \left(-ik_1\right)\left(A^*e^{-ik_1x} - B^*e^{ik_1x}\right)\Big].$$

Simplifying this expression, we obtain

$$j_x = \frac{\hbar k_1}{m}\left[A^*A - B^*B\right] = \frac{\hbar k_1}{m}\left[|A|^2 - |B|^2\right].$$

This can be interpreted as the sum of an incident probability current, $\hbar k_1|A|^2/m$, and a reflected probability current, $-\hbar k_1|B|^2/m$. These two contributions have opposite signs because they flow in opposite directions. Note that, in each region, the probability currents are consistent with the incident, reflected and transmitted beam intensities assumed earlier (and now justified).

Ex 7.8 The main difference is that in the case of a finite square barrier V_0 must be replaced by $-V_0$

throughout the analysis. The wave number in Region 2 is

$$k_2 = \frac{\sqrt{2m(E_0 - V_0)}}{\hbar}.$$

Adapting Equation 7.44, the transmission coefficient is then given by

$$T = \frac{4E_0(E_0 - V_0)}{4E_0(E_0 - V_0) + V_0^2 \sin^2(k_2 L)}.$$

A typical graph of the real part of $\psi(x)$ for scattering by a finite square barrier is shown in Figure S7.2. For a barrier, the wavelength in Region 2 is longer, corresponding to a smaller wave number and a smaller momentum. The amplitude of the wave is generally greater in Region 2 than outside it (unless $T = 1$).

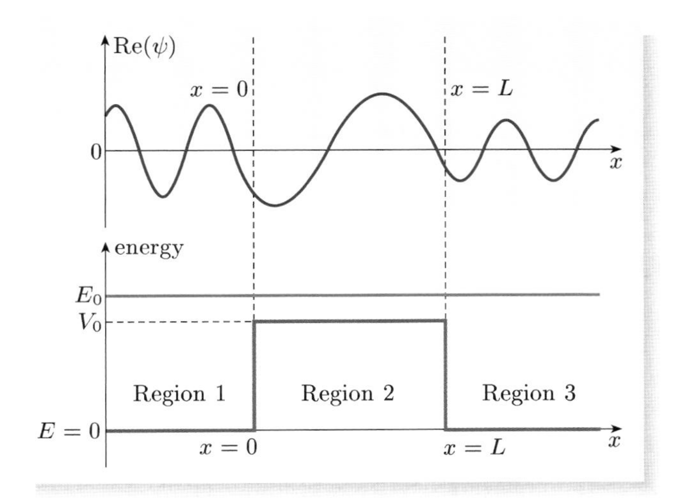

Figure S7.2 A typical graph of the real part $\text{Re}(\psi)$ of $\psi(x)$ for a finite square barrier.

Ex 7.9 The graphs of $|\Psi(x,t)|^2$ plotted against x for each case are shown in Figure S7.3.

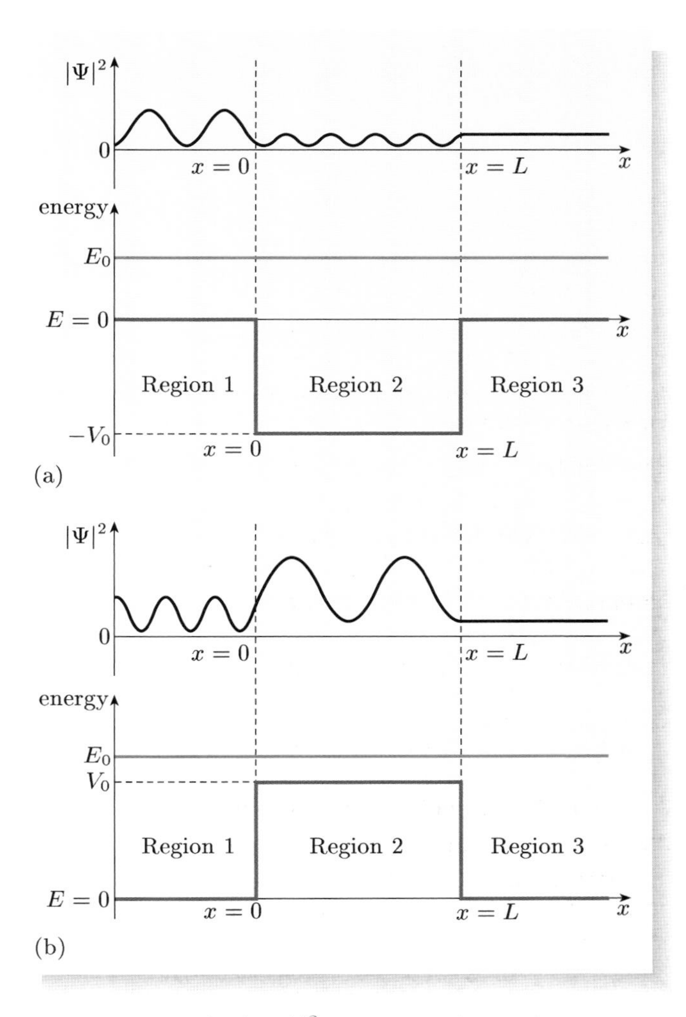

Figure S7.3 $|\Psi(x,t)|^2$ plotted against x for (a) a finite square well and (b) a finite square barrier.

In Regions 1 and 2 interference effects lead to periodic spatial variations with a period related to the relevant wave number. In Region 3 there is only one beam, so there are no interference effects. Note that there is no simple relationship between the plots of $|\Psi(x,t)|^2$ and the corresponding plots of the real part of $\Psi(x,t)$. This is because $|\Psi(x,t)|^2$ is partly determined by the imaginary part of $\Psi(x,t)$. In both cases $|\Psi(x,t)|^2$ is independent of time. This must be the case, despite the time-dependence of $\Psi(x,t)$, because we are dealing with *stationary states*.

Ex 7.10 From Equation 7.46

$$\psi(x) = A\text{e}^{\text{i}k_1 x} + B\text{e}^{-\text{i}k_1 x}$$

and

$$\psi^*(x) = A^*\text{e}^{-\text{i}k_1 x} + B^*\text{e}^{\text{i}k_1 x},$$

so

$$|\psi(x)|^2 = \left(A^*e^{-ik_1x} + B^*e^{ik_1x}\right)\left(Ae^{ik_1x} + Be^{-ik_1x}\right)$$
$$= A^*A + B^*B + A^*Be^{-2ik_1x} + B^*Ae^{2ik_1x}.$$

This is a positive periodic function that runs through the same range of values each time x increases by $2\pi/2k_1$, so its minima are separated by π/k_1.

Ex 7.11 Using Equation 7.47, with $D = 0$ and α real, and recalling the definition of probability current given in Equation 7.28, we have

$$j_x(x,t) =$$
$$-\frac{i\hbar}{2m}\left[C^*e^{-\alpha x}\frac{\partial}{\partial x}\left(Ce^{-\alpha x}\right) - Ce^{-\alpha x}\frac{\partial}{\partial x}\left(C^*e^{-\alpha x}\right)\right]$$
$$= -\frac{i\hbar}{2m}\left[C^*Ce^{-\alpha x}(-\alpha)e^{-\alpha x} - C^*Ce^{-\alpha x}(-\alpha)e^{-\alpha x}\right]$$
$$= 0.$$

Ex 7.12 From Equation 7.29,

$$j_x(x) = -\frac{i\hbar}{2m}\left[\psi^*\frac{\partial\psi}{\partial x} - \psi\frac{\partial\psi^*}{\partial x}\right].$$

In Region 1, where $\psi(x) = Ae^{ik_1x} + Be^{-ik_1x}$ and $\psi^*(x) = A^*e^{-ik_1x} + B^*e^{ik_1x}$, we have

$$j_x(x) =$$
$$-\frac{i\hbar}{2m}\Big[\left(A^*e^{-ik_1x} + B^*e^{ik_1x}\right)\left(ik_1Ae^{ik_1x} - ik_1Be^{-ik_1x}\right)$$
$$-\left(Ae^{ik_1x} + Be^{-ik_1x}\right)\left(-ik_1A^*e^{-ik_1x} + ik_1B^*e^{ik_1x}\right)\Big].$$

Simplifying gives

$$j_x = \frac{\hbar k_1}{m}\left[A^*A - B^*B\right] \quad \text{for } x < 0,$$

which can be interpreted as the sum of the probability density currents associated with the incident and reflected beams.

Similarly, in Region 2, where $\psi(x) = Ce^{-\alpha x} + De^{\alpha x}$ and $\psi^*(x) = C^*e^{-\alpha x} + D^*e^{\alpha x}$, we have

$$j_x = -\frac{i\hbar}{2m}\Big[\left(C^*e^{-\alpha x} + D^*e^{\alpha x}\right)\left(-\alpha Ce^{-\alpha x} + \alpha De^{\alpha x}\right)$$
$$-\left(Ce^{-\alpha x} + De^{\alpha x}\right)\left(-\alpha C^*e^{-\alpha x} + \alpha D^*e^{\alpha x}\right)\Big],$$

which simplifies to

$$j_x = -\frac{i\hbar\alpha}{m}\left[C^*D - CD^*\right] \quad \text{for } 0 \le x \le L.$$

Finally, in Region 3, where $\psi(x) = Fe^{ik_1x}$,

$$j_x = \frac{\hbar k_1}{m}\left[F^*F\right] \quad \text{for } x > L.$$

The result for Region 2 is a real non-zero quantity, so there is a probability current inside the barrier. This is not really surprising since particles must pass through the barrier to produce a transmitted beam in Region 3. In fact, conservation of probability *requires* that the probability current should be the same in all three regions.

Ex 8.1 For $z = 3 - 4i$, the real part is 3, the imaginary part is -4 and the modulus is $|z| = \sqrt{3^2 + (-4)^2} = 5$.

Ex 8.2 Given $z_1 = 1 + 2i$ and $z_2 = 4 - 3i$, we have:

(a) $z_1 + z_2 = (1 + 2i) + (4 - 3i) = 5 - i$,

(b) $2z_1 - 3z_2 = 2(1 + 2i) - 3(4 - 3i)$
$= -10 + 13i$,

(c) $z_1z_2 = (1 + 2i)(4 - 3i)$
$= 4 + (2i)(-3i) + 8i - 3i = 10 + 5i$,

(d)
$$\frac{z_1}{z_2} = \frac{1+2i}{4-3i} = \frac{1+2i}{4-3i} \times \frac{4+3i}{4+3i}$$
$$= \frac{4 + (2i)(3i) + 8i + 3i}{25} = \frac{-2+11i}{25}.$$

Ex 8.3 Taking the complex conjugate of z gives

$$z^* = \left[i(ab^* - a^*b)\right]^*$$
$$= -i(a^*b - ab^*)$$
$$= i(ab^* - a^*b)$$
$$= z.$$

We conclude that z is real because $z^* = z$.

Ex 8.4 (a) For $z = 1 + i\sqrt{3}$, the radial coordinate is $r = |z| = \sqrt{1+3} = 2$. The phase θ satisfies the equation $\cos\theta = x/r = 1/2$. We are looking for a solution in the first quadrant ($0 \leq \theta \leq \pi/2$), so the appropriate choice is $\theta = \pi/3$, giving

$$z = 2\big[\cos(\pi/3) + \mathrm{i}\sin(\pi/3)\big].$$

(b) For $z = -1 + \mathrm{i}\sqrt{3}$, the radial coordinate is again $r = |z| = \sqrt{1+3} = 2$. The phase θ satisfies the equation $\cos\theta = x/r = -1/2$. We are looking for a solution in the second quadrant ($\pi/2 \leq \theta \leq \pi$). Since $\cos(\pi - \phi) = -\cos\phi$, the appropriate choice is $\theta = \pi - \pi/3 = 2\pi/3$, giving

$$z = 2\big[\cos(2\pi/3) + \mathrm{i}\sin(2\pi/3)\big].$$

Ex 8.5 Using Equation 8.5, we have

$$\begin{aligned}|z|^2 = zz^* &= r(\cos\theta + \mathrm{i}\sin\theta) \times r(\cos\theta - \mathrm{i}\sin\theta)\\ &= r^2(\cos^2\theta + \sin^2\theta) = r^2.\end{aligned}$$

Since both $|z|$ and r are real and positive, we conclude that $|z| = r$.

Ex 8.6 Euler's formula, and its complex conjugate, tell us that

$$\mathrm{e}^{\mathrm{i}\theta} = \cos\theta + \mathrm{i}\sin\theta$$

$$\mathrm{e}^{-\mathrm{i}\theta} = \cos\theta - \mathrm{i}\sin\theta.$$

Adding these two equations and dividing by 2 gives the required expression for $\cos\theta$, while subtracting the equations and dividing by 2i gives the required expression for $\sin\theta$.

Ex 8.7 We have

$$\begin{aligned}\mathrm{e}^3\,\mathrm{e}^{-\mathrm{i}\pi/6} &= \mathrm{e}^3\big[\cos(\pi/6) - \mathrm{i}\sin(\pi/6)\big]\\ &= \frac{\mathrm{e}^3\sqrt{3}}{2} - \mathrm{i}\,\frac{\mathrm{e}^3}{2}.\end{aligned}$$

Ex 8.8 The complex number $\mathrm{i} = \mathrm{e}^{\mathrm{i}\pi/2}$, so if $z = \mathrm{e}^{\mathrm{i}\theta}$, then

$$\mathrm{i}z = \mathrm{e}^{\mathrm{i}\pi/2}\mathrm{e}^{\mathrm{i}\theta} = \mathrm{e}^{\mathrm{i}(\theta+\pi/2)}$$

$$\frac{z}{\mathrm{i}} = \frac{\mathrm{e}^{\mathrm{i}\theta}}{\mathrm{e}^{\mathrm{i}\pi/2}} = \mathrm{e}^{\mathrm{i}(\theta-\pi/2)}.$$

Thus, multiplying by i produces an anticlockwise rotation of $\pi/2\,\text{rad} = 90°$, while dividing by i produces an anticlockwise rotation of $-\pi/2\,\text{rad} = -90°$ (that is, a clockwise rotation of $\pi/2\,\text{rad} = 90°$).

Ex 8.9 A lot of unnecessary effort can be avoided by converting z to exponential form. It is easy to see that $z = (1+\mathrm{i})/\sqrt{2} = \mathrm{e}^{\mathrm{i}\pi/4}$. Hence

$$z^{100} = \left(\mathrm{e}^{\mathrm{i}\pi/4}\right)^{100} = \mathrm{e}^{\mathrm{i}25\pi} = \cos(25\pi) + \mathrm{i}\sin(25\pi) = -1.$$

Ex 8.10 The solution given in the main text can be followed unchanged up to the point where the boundary conditions are used. As in the main text, we can reject the possibility that $\lambda = 0$, since this only gives the trivial solution $y(x) = 0$ for all x.

For positive λ, we can set $\lambda = k^2$. Then the general solution of the differential equation is

$$y(x) = A\mathrm{e}^{kx} + B\mathrm{e}^{-kx} \qquad \text{(Eqn 8.32)}$$

so

$$y'(x) = kA\mathrm{e}^{kx} - kB\mathrm{e}^{-kx}.$$

The boundary conditions $y(0) = y'(L) = 0$ give

$$A + B = 0 \quad \text{and} \quad kA\mathrm{e}^{kL} - kB\mathrm{e}^{-kL} = 0.$$

Rejecting the trivial solution $A = B = 0$, there is no way of satisfying these equations since they imply that $\mathrm{e}^{2kL} = -1$, which is impossible. Hence λ cannot be positive.

For negative λ, we can set $\lambda = -k^2$, where k can be taken to be positive. Then the general solution of the differential equation is

$$y(x) = C\cos(kx) + D\sin(kx), \qquad \text{(Eqn 8.29)}$$

so

$$y'(x) = -kC\sin(kx) + kD\cos(kx).$$

The boundary conditions $y(0) = y'(L) = 0$ give

$$C = 0 \quad \text{and} \quad kD\cos(kL) = 0.$$

Rejecting the trivial solution in which $C = D = 0$, we see that k must satisfy the condition $kL = (2n+1)\pi/2$, where $n = 0, 1, 2, \ldots$. It follows that the eigenvalues are

$$\lambda = -k^2 = -(2n+1)^2\pi^2/4L^2 \quad \text{for } n = 0, 1, 2, 3, \ldots,$$

with corresponding eigenfunctions

$$y_n(x) = D_n \sin\left[\frac{(2n+1)\pi x}{2L}\right].$$

Ex 8.11 Using Equation 8.39, we obtain

$$\frac{\partial^2 f}{\partial x^2} = \frac{\partial}{\partial x}\big[k\cos(kx - \omega t)\big] = -k^2\sin(kx - \omega t),$$

$$\frac{\partial^2 f}{\partial t\,\partial x} = \frac{\partial}{\partial t}\big[k\cos(kx - \omega t)\big] = \omega k\sin(kx - \omega t).$$

Similarly, Equation 8.40 gives

$$\frac{\partial^2 f}{\partial t^2} = \frac{\partial}{\partial t}\big[-\omega\cos(kx-\omega t)\big] = -\omega^2\sin(kx-\omega t),$$

$$\frac{\partial^2 f}{\partial x\partial t} = \frac{\partial}{\partial x}\big[-\omega\cos(kx-\omega t)\big] = k\omega\sin(kx-\omega t).$$

The two mixed second-order partial derivatives are equal, as expected.

Ex 8.12 The function $f(x,t)$ is a function $g(z)$ of another function $z = x - ct$. Using the normal rules of calculus for differentiating such a composite function, we obtain

$$\frac{\partial f}{\partial x} = \frac{\mathrm{d}g}{\mathrm{d}z} \times \frac{\partial z}{\partial x} = g'(x-ct) \times 1$$

$$\frac{\partial f}{\partial t} = \frac{\mathrm{d}g}{\mathrm{d}z} \times \frac{\partial z}{\partial t} = g'(x-ct) \times (-c),$$

where $g'(x-ct)$ is the derivative of the function $g(z)$, evaluated at the point $z = x - ct$. Differentiating again,

$$\begin{aligned}\frac{\partial^2 f}{\partial x^2} &= \frac{\partial}{\partial x}\left(\frac{\partial f}{\partial x}\right)\\ &= \frac{\partial}{\partial x}\left[g'(x-ct)\right]\\ &= g''(x-ct)\end{aligned}$$

$$\begin{aligned}\frac{\partial^2 f}{\partial t^2} &= \frac{\partial}{\partial t}\left(\frac{\partial f}{\partial t}\right)\\ &= \frac{\partial}{\partial t}\left[-cg'(x-ct)\right]\\ &= c^2 g''(x-ct),\end{aligned}$$

where $g''(x-ct)$ is the second derivative of the function $g(z)$, evaluated at the point $z = x - ct$. Hence

$$\frac{\partial^2 f}{\partial x^2} - \frac{1}{c^2}\frac{\partial^2 f}{\partial t^2} = g''(x-ct) - \frac{1}{c^2}c^2 g''(x-ct) = 0.$$

So any function of the form $g(x-ct)$ satisfies Equation 8.43 (assuming that the function can be differentiated twice).

Ex 8.13 Let the probability of outcome 1 be p_1. Then the probability of outcome 2 is $p_2 = 3p_1$, and the probability of outcome 3 is $p_3 = 6p_1$. So the normalization rule gives

$$1 = p_1 + p_2 + p_3 = p_1 + 3p_1 + 6p_1 = 10p_1,$$

which gives $p_1 = 1/10$, $p_2 = 3/10$ and $p_3 = 6/10$.

Ex 8.14 (a) The text showed that the expectation value of the score on a die is 3.5. Using Equation 8.54, the *square* of the uncertainty of the score S is

$$\begin{aligned}(\Delta S)^2 &= \frac{1}{6}\Big[(1-3.5)^2 + (2-3.5)^2 + (3-3.5)^2\\ &\qquad + (4-3.5)^2 + (5-3.5)^2 + (6-3.5)^2\Big]\\ &= 35/12,\end{aligned}$$

so the uncertainty of the score is $\Delta S = \sqrt{35/12} = 1.7$.

(b) Using Equation 8.55 we get

$$(\Delta S)^2 = \frac{1}{6}\left(1^2+2^2+3^2+4^2+5^2+6^2\right) - 3.5^2 = \frac{35}{12},$$

which gives the same uncertainty as before:

$$\Delta S = \sqrt{35/12} = 1.7.$$

Index

Items that appear in the Glossary have page numbers in **bold type**. Ordinary index items have page numbers in Roman type.

addition rule for probability **228**
alpha decay **203**–5
alpha particle **12**, 31, 203
amplitude **17**, **125**
angular frequency **10**, **17**, **126**
arbitrary constants **218**
Argand diagram **212**
argument of complex number **213**
artificial atoms 63
atomic number **12**, 203–4
attenuation coefficient **199**
auxiliary equation **219**
average value **228**

barn **196**
barrier penetration **88**, 90, **133**, 179, 199
beam intensity **183**
beam splitter **15**, 30–2
Binnig, Gerd 206
biprism 22–3
Bohr, Niels 10, 135
Bohr's model 10, 38, 118
Boltzmann constant 206
Born, Max 7, 24, 49
Born's rule **24**–5, 30, 49, 50–1, 56, 57, 149
  and indeterminism 97
  and the overlap rule 109–10
Born's rule for momentum **170**, 175
Bose–Einstein condensation 119
bound states **91**, 92
boundary conditions 68, 86, 88, 187, 199, **221**
box
  in one dimension 64
  in two dimensions 64
  in three dimensions 63
bucky balls 21

carbon dioxode 145
carbon monoxide 131
Cartesian form of complex number **213**
centre-of-mass frame **127**
CERN 196
classical limit **160**–1, 188
classical physics **7**, 51, 160
classically forbidden region **66**, 88, 133, 135, 199

Cockcroft, J. D. 205
coefficient rule **105**, 109, 150
collapse of the wave function **27**, 51, **97**, 182
commutation relation **138**, 144
commutator **138**
complete set of functions **164**, 171
complete set of mutually exclusive outcomes **228**
complex conjugate **212**, 215
complex numbers **210**–7
  addition of 211
  argument **213**
  Cartesian form **213**
  complex conjugate of **212**, 215
  cube root of 217
  difference of 211
  exponential form **214**
  imaginary part **210**
  phase **213**
  polar form **213**
  powers of 216–7
  product of 211, 215
  quotient of 211, 215
  real part **210**, 212
  reciprocal of 211, 215
  square root of 216–7
complex plane **212**
Compton effect 18
Compton, Arthur 18
conservation of energy **105**, 126
constructive interference **28**
continuity boundary conditions **68**–9, 86, 88, 187, 199
continuum **12**, 58, 148, 166
Cooper pair 120
Cornell, Eric 119
correspondence principle **135**
Coulomb barrier **204**

Davisson, Clinton 21
de Broglie relationship **21**, 38
de Broglie wave function **27**, 38, 44, 168
de Broglie wavelength **21**
de Broglie waves 20, 37, 180
de Broglie, Louis 20, 37
Debye, Peter 37

decay constant **13**, 203
degeneracy **81**
degenerate energy level **81**
degenerate quantum states **81**
Democritus 12
destructive interference **28**
deterministic **13**, 96
deuteron **91**
diatomic molecule 127–8
differential cross-section **196**
differential equation **217**
  general solution **218**
  homogeneous **218**
  linear **217**
  order of **217**
  particular solution **218**
diffraction **16**–7
  of electrons 21–2, 176
  of light 18
  pattern **18**
Dirac, Paul 138

Ehrenfest's theorem **158**–61, 172
Ehrenfest, Paul 158
eigenfunction 42, **43**, **221**
eigenvalue 42, **43**, **221**
eigenvalue equation 42, **43**, **221**, 226
eigenvector 42
Einstein, Albert 18
elastic scattering **196**
electron diffraction 22, 176
electron interference 22–4, 28–31
electron microscope 21
electronvolt **11**
electrostatic force 204
energy eigenfunctions 56–7
  harmonic oscillator 131–5, 142
  infinite square well 69, 71, 77
  symmetric infinite square well 77
  three-dimensional infinite square well 82
  two-dimensional infinite square well 80
energy eigenvalues 56–7
  harmonic oscillator 130, 141
  infinite square well 69
  symmetric infinite square well 77
  three-dimensional infinite square well 82
  two-dimensional infinite square well 80
energy levels **10**–1, 75, 82, 90, 130, 141
energy operator 48
energy quantization 10, 57, **70**
equation of continuity **189**
Euler's formula **214**
even function **78**, 155
expectation value **111**–5, 142–4, **229**–30
  of a continuous variable 112, 231–2
  of a power of an observable 112
  of momentum 114
  of position 112
  in a stationary state 149
exponential form of complex number **214**
exponential law of radioactive decay **13**–4, 202–3

F-centre **83**
finite square barrier **181**, 197, 200–1
finite square step **184**, 198–9
finite square well **85**, 182, 192–5
finite well **56**
first-order partial derivative **224**
fluorescent marker 63
flux **196**
force constant 47, **125**, 128
Fourier analysis 171
Fourier transform **171**, 175
Fourier, Jean Baptiste 171
free particle 27, 38–40, **166**–74
frequency **10**, **17**, 126
fusion reactions **205**

gamma ray 11
Gamow, George 203
Gaussian function **131**, 172–4
Gaussian wave packet **182**
Geiger counter 19, 21, 97
Geiger–Nuttall relation **203**–5
general solution of differential equation **218**
Germer, Lester 21
ground state **10**, 12, 73, 78, 92, 131–3

half-life **14**, 203
half-silvered mirror **15**, 32
Hamilton's equations 47
Hamilton, William Rowan 47
Hamiltonian function **47**, 66, 129
Hamiltonian operator **48**, 66, 113
  for a harmonic oscillator 129, 136–8
harmonic oscillator 121, **124**–47
  energy eigenfunctions 131–5, 142
  energy eigenvalues 130, 141
  expectation value of kinetic energy 144

expectation value of potential energy 144
potential energy function 47, 126
selection rule 131, 145–6
total energy 126
uncertainty in momentum 145
uncertainty in position 145
wave packets of 151–8, 161–4
Heisenberg uncertainty principle 22, 95, **117**–22, 145, 172
Heisenberg, Werner 7, 35, 38, 117, 135
Hermite polynomials **134**
Hertz, Heinrich 17
homogeneous differential equation **218**
Hooke's law 47, 48, **125**
hydrogen atom 11
ground-state energy 122
radius 122
hyperbolic functions **220**

identical particles 13–4
identity operator **40**
imaginary axis **212**
imaginary number **210**
imaginary part **210**
indeterministic **13**–6, 95, 96–7
inelastic scattering **196**
inertial frame **127**
infinite square well in one dimension **65**–78
energy eigenfunctions 69, 71, 77
energy eigenvalues 69, 75, 79
quantum number 70
initial conditions **221**
interference **16**
maxima **18**
minima **18**
interference rule **32**
interference term 30
inverse Fourier transform **171**
ion **12**, 83
ionic crystal **83**
ionization **11**–2
isotope **13**

kinetic energy operator **44**
Kronecker delta symbol **103**, 113, 163

ladder operators **140**
Laplace, Pierre Simon 96
lattice spacing 84
length parameter
of a harmonic oscillator **133**, 151
light quanta 18
linear differential equation **217**
linear differential operator **218**, 224
linear number density **183**, 190
linear operator 41, **42**, 52, 60
liquid helium 119–20
lithium-11 nucleus 120
lowering operator **137**, 140, 141, 153

Mach–Zehnder interferometer **32**–4
matrix 42
Maxwell's equations **17**
mean value **111**, **228**
measurement 51, 97
modulus 25, **210**
momentum amplitude **170**–5
momentum eigenfunction **168**
momentum operator **44**–5
momentum wave function **170**–5
momentum, of photon 18–9
mutually exclusive outcomes **228**

Newton's laws 46, 51, 127, 159
Newton's second law 125, 159
nodal line 81
nodes **73**, 92
non-commuting operators 138
normal ordering of ladder operators **138**, 140
normalization condition **50**, 70–1, 80, 89, 103, 132
normalization constant **50**, 71, 132
normalization rule for probability **228**
normalized wave function **50**
nuclear energy 11, 15
number operator **141**

observables **44**, 60
odd function **78**, 114, 155
old quantum theory **7**
one-dimensional box 64
one-dimensional infinite square well **65**–78
energy eigenfunctions 69, 71, 77
energy eigenvalues 69, 75, 77
quantum number 70
operator **40**
linear **41**–2, 52, 60, 218
order of a differential equation **217**
ordinary differential equation **217**
orthogonal set of functions **102**–3
orthonormal set of functions **103**

overlap integral **101**, 163
overlap rule **101**, 163
  and orthonormal energy eigenfunctions 102–3
  for any observable 108

partial derivative **222**–4
partial differential equation 217, **224**–6
particle 26
particular solution of differential equation **218**
period **17**, 126, 151–3
permittivity of free space **204**
phase **18**, **213**
phase constant **18**, **126**
phase factor **52**, 71, 76, 149, 152, 155, **214**
phase of complex number **213**
photon **10**, 17, 18, 33, 145
Plancherel's theorem **169**
Planck's constant **10**
polar form of complex number **213**
position operator **45**
potential energy function **46**
  finite square step 184
  finite square well 85
  harmonic oscillator 126
  1D infinite square well 65, 75
  2D infinite square well 78
  zero of 75, 85, 126
potential energy operator **45**
principle of superposition **52**, 58, 150, 162, 169, **219**
probability 13, 16, 25, 30, 31, 51, 100, 101, 105, 182, **227**–32
probability amplitude **31**–4, **102**, 105, 150
probability current **190**–1
probability density **50**, 74, 91, 135, 190
probability density function **231**
probability distribution 100–10, **229**
proton–proton chain **206**

quantization **10**, 57, 70
quantum dot **63**, 84
quantum field theory 179
quantum jump 146
quantum mechanics **7**
  preliminary principles of 60
quantum non-locality 26–7
quantum numbers **57**
  harmonic oscillator 130, 141
  1D infinite square well **70**
  2D infinite square well 80–1
  3D infinite square well 82
quantum physics **7**
quantum random number generators **15**
quantum wafer **64**
quantum wire **63**–4

radiative transition **145**
radioactive decay, exponential law of 13–4, 202–3
raising operator **137**, 140, 141, 153
Ramsauer–Townsend effect **196**
random outcomes **227**
real axis **212**
real part **210**, 212
reduced mass **128**
reflection coefficient **182**
  finite square step 186–7
relative frequency 111, **229**
restoring force **125**
Rohrer, Heinrich 206
rotational energy 11
rotational motion
  of a diatomic molecule 127
Rutherford, Ernest 13, 178, 203

sandwich integral **112**
sandwich integral rule **112**–5
scanning tunnelling microscope 180, **206**
scattering **178**
Schrödinger, Erwin 37, 53
Schrödinger's equation **24**, **46**–9
  finite square step 185
  finite square well 85
  free particle 39
  harmonic oscillator 129
  indeterminism and 96–7
  infinite square well 66
  three-step recipe 46–9
  2D infinite square well 79
second-order partial derivative **224**
selection rule **131**, 145–6
semiconducting materials **63**
separable partial differential equation **226**
separation constant **54**, 55, **226**
separation of variables **53**–5, 66, 79, **225**–6
simple harmonic motion **124**–6
simple harmonic oscillator
  see *harmonic oscillator*
spectral lines **9**–10, 57, 131
spectroscope 9

spectrum **9**, 131
standard deviation **116**, **230**
standing wave 37, **73**, 226
stars 9–10, 205–6
state
  in classical physics **51**
  in quantum physics **51**
stationary state 55, **57**–8, 124, 148, 149–50
  approach to scattering 183–4, 191–5
  approach to tunnelling 197–202
  energy measurements 98–9
stationary-state wave functions 67, 73
  harmonic oscillator 129
  infinite square well 72–4, 77
  2D infinite square well 80
  3D infinite square well 82
strong nuclear force 204
superposition of states 34
superposition principle see *principle of superposition*
surface of last scattering 178
symmetric well **76**–8
  energy eigenfunctions 77
  energy eigenvalues 77
  stationary-state wave functions 77

Taylor, G.I. 20
Thomson, G.P. 21
three-dimensional box 63
three-dimensional infinite square well **82**
  energy eigenfunctions 82
  energy eigenvalues 82
  stationary-state wave functions 82
time-independent Schrödinger equation **55**
  finite square step 185
  finite square well 86, 192–3
  for a harmonic oscillator 129, 138
  infinite square well 66, 75
  2D infinite square well 79
Tonomura, Akira 23
top-hat function 109
total cross-section **196**–7
total energy 45
  of a harmonic oscillator 126
translational motion
  of a diatomic molecule 127
transmission coefficient **182**
  alpha-particle scattering 204
  finite square barrier 200
  finite square well 192–4
  finite step 186
transmission resonances **194**
tunnelling **179**–80, 197–207
  alpha decay 202–5
  fusion reactions in stars 205–6
  scanning tunnelling microscope 206–7
two-dimensional box 64
two-dimensional infinite square well **78**–81
  energy eigenfunctions 80
  energy eigenvalues 80
  stationary-state wave functions 80
two-slit interference pattern
  for light **18**–9
  for electrons 22–4, 29–31

uncertainty **116**, 145, **231**
uncertainty principle 22, 95, 117–22, 145, 172

vibrational motion
  of a diatomic molecule 127

Walton, E. T. S. 205
wave function **24**–6, 37, **49**–53
  normalized 50
  specifying state of system 51, 96
wave mechanics **7**
wave number **17**, 27, 168, 170
wave packet 58, 148, **150**
  free particle 166–74, 176
  in harmonic well 151–8, 161–6
  and scattering 180–3
  and tunnelling 197–8
wave theory of light 16
wave-packet spreading **172**–4
wave–particle duality **17**
  for electrons 22–4, 29–31
  for photons 20
wavelength **17**, 21
well 64
  harmonic 128
  one-dimensional infinite square 65, 75
  symmetric infinite square 75
  three-dimensional infinite square 82
  two-dimensional infinite square 78
which-way information 30
Wieman, Carl 119

Young, Thomas 16

zero of potential energy 75, 85, 126
zero-point energy **130**

Complex numbers

$z = x + \mathrm{i}y = r\mathrm{e}^{\mathrm{i}\theta}$

$z^* = x - \mathrm{i}y = r\mathrm{e}^{-\mathrm{i}\theta}$

$|z|^2 = zz^* = x^2 + y^2 = r^2$

$\mathrm{Re}(z) = \dfrac{z + z^*}{2}$

$\mathrm{Im}(z) = \dfrac{z - z^*}{2\mathrm{i}}$

$z^n = r^n \mathrm{e}^{\mathrm{i}n\theta}$

$\mathrm{e}^{\mathrm{i}\theta} = \cos\theta + \mathrm{i}\sin\theta$

$\cos\theta = \dfrac{\mathrm{e}^{\mathrm{i}\theta} + \mathrm{e}^{-\mathrm{i}\theta}}{2}$

$\sin\theta = \dfrac{\mathrm{e}^{\mathrm{i}\theta} - \mathrm{e}^{-\mathrm{i}\theta}}{2\mathrm{i}}$

$\mathrm{e}^{\pm\mathrm{i}\pi} = -1$

$\mathrm{e}^{\mathrm{i}\pi/2} = \mathrm{i}$

$\mathrm{e}^{-\mathrm{i}\pi/2} = -\mathrm{i}$

Elementary functions $(a > 0, b > 0)$

$\mathrm{e}^x \mathrm{e}^y = \mathrm{e}^{x+y}$

$\ln a + \ln b = \ln(ab)$

$\mathrm{e}^{\ln a} = \ln(\mathrm{e}^a) = a$

$\mathrm{e}^x = \cosh x + \sinh x$

$\cosh x = \dfrac{\mathrm{e}^x + \mathrm{e}^{-x}}{2}$

$\sinh x = \dfrac{\mathrm{e}^x - \mathrm{e}^{-x}}{2}$

$\cos(\theta \pm \pi) = -\cos\theta$

$\sin(\theta \pm \pi) = -\sin\theta$

$\tan(\theta \pm \pi) = \tan\theta$

$\cos(\theta + \pi/2) = -\sin\theta$

$\sin(\theta + \pi/2) = \cos\theta$

$\tan(\theta + \pi/2) = -\cot\theta$

$\cos(\theta - \pi/2) = \sin\theta$

$\sin(\theta - \pi/2) = -\cos\theta$

$\tan(\theta - \pi/2) = -\cot\theta$

$\cos(2\theta) = \cos^2\theta - \sin^2\theta$

$\sin(2\theta) = 2\sin\theta\cos\theta$

$\tan(2\theta) = 2\tan\theta/(1 - \tan^2\theta)$

$\sin(A \pm B) = \sin A\cos B \pm \cos A\sin B$

$\cos(A \pm B) = \cos A\cos B \mp \sin A\sin B$

$\sin A\sin B = \frac{1}{2}\big(\cos(A - B) - \cos(A + B)\big)$

$\cos A\cos B = \frac{1}{2}\big(\cos(A - B) + \cos(A + B)\big)$

$\sin A\cos B = \frac{1}{2}\big(\sin(A - B) + \sin(A + B)\big)$

$\cos^2 A + \sin^2 A = 1$

Physical constants

Planck's constant	h	$6.63 \times 10^{-34}\,\mathrm{J\,s}$	Planck's constant/2π	$\hbar$	$1.06 \times 10^{-34}\,\mathrm{J\,s}$
vacuum speed of light	c	$3.00 \times 10^{8}\,\mathrm{m\,s^{-1}}$	Coulomb law constant	$\frac{1}{4\pi\varepsilon_0}$	$8.99 \times 10^{9}\,\mathrm{m\,F^{-1}}$
permittivity of free space	ε_0	$8.85 \times 10^{-12}\,\mathrm{F\,m^{-1}}$	permeability of free space	μ_0	$4\pi \times 10^{-7}\,\mathrm{H\,m^{-1}}$
Boltzmann's constant	k	$1.38 \times 10^{-23}\,\mathrm{J\,K^{-1}}$	Avogadro's constant	N_m	$6.02 \times 10^{23}\,\mathrm{mol^{-1}}$
electron charge	$-e$	$-1.60 \times 10^{-19}\,\mathrm{C}$	proton charge	e	$1.60 \times 10^{-19}\,\mathrm{C}$
electron mass	m_e	$9.11 \times 10^{-31}\,\mathrm{kg}$	proton mass	m_p	$1.67 \times 10^{-27}\,\mathrm{kg}$
Bohr radius	a_0	$5.29 \times 10^{-11}\,\mathrm{m}$	atomic mass unit	u	$1.66 \times 10^{-27}\,\mathrm{kg}$